Handgefertigtes Vorabexemplar
Bindequalität nicht verbindlich

Polar Oxides

Properties, Characterization, and Imaging

Edited by
R. Waser, U. Böttger, and S. Tiedke

WILEY-VCH Verlag GmbH & Co. KGaA

Editors

Rainer Waser
Department IFF &
CNI – Center of Nanoelectronic Systems
for Information Technology
Research Center Jülich, Germany
and
Institut für Werkstoffe der Elektrotechnik
RWTH Aachen, Germany
e-mail: waser@iwe.rwth-aachen.de

Ulrich Böttger
Institut für Werkstoffe der Elektrotechnik
RWTH Aachen, Germany
e-mail: boettger@ iwe.rwth-aachen.de

Stephan Tiedke
aixACCT Systems GmbH, Aachen, Germany
e-mail: tiedke@aixacct.com

Cover Picture
The picture shows an X-ray diffraction reciprocal space mapping (XRD-RSM) measurement at the 204 diffraction of a PZT 52/48 thin film on a SRO-STO substrate. A tetragonal phase with 90° domains as well as a pseudocubic phase induced by the strain of the substrate are observed.

All books published by Wiley-VCH are carefully produced. Nevertheless, authors, editors, and publisher do not warrant the information contained in these books, including this book, to be free of errors. Readers are advised to keep in mind that statements, data, illustrations, procedural details or other items may inadvertently be inaccurate.

Library of Congress Card No.: applied for
British Library Cataloging-in-Publication Data:
A catalogue record for this book is available from the British Library

Bibliographic information published by
Die Deutsche Bibliothek
Die Deutsche Bibliothek lists this publication in the Deutsche Nationalbibliografie; detailed bibliographic data is available in the Internet at <http://dnb.ddb.de>.

© 2005 WILEY-VCH Verlag GmbH & Co. KGaA, Weinheim

All rights reserved (including those of translation into other languages). No part of this book may be reproduced in any form – nor transmitted or translated into machine language without written permission from the publishers. Registered names, trademarks, etc. used in this book, even when not specifically marked as such, are not to be considered unprotected by law.

Printed in the Federal Republic of Germany
Printed on acid-free paper

Composition Thomas Pössinger, Aachen
Printing betz-druck GmbH, Darmstadt
Bookbinding Litges & Dopf Buchbinderei GmbH, Heppenheim

ISBN 3-527-40532-1

Preface

Polar oxides haved received increasing attention in recent years - both from a fundamental perspective and from a novel applications viewpoint. There are three major trends which led to this emphasis. One trend concerns the scaling of the structure size (particles size, device feature sizes, etc.) into the sub 100-nm regime. Although size effects on the nanometer scale have long been investigated, there has been a trend in recent years to strengthen the utilization of these effects and to synthesize oxide ceramics in which the properties are dominated by size effects. This is true for bulk ceramics and thick films as well as for thin films. A second trend concerns the fact that advanced functional components are made of material systems rather than of discrete materials. Material integration issues play an increasingly important role driven by the interest in integrating functions of polar oxides into conventional semiconductor chips as well as for the evolution of multifunctional components and systems. A third trend concerns the role of theory and modeling. The materials and device design is more and more accompanied and guided by modeling, e. g. by thermodynamics, finite-element methods, and ab-initio calculations.

This book is specifically addressed to the properties of polar oxides as well as to their chacterization and imaging techniques. The dielectric, optic, piezoelectric, pyroelectric behavior of this class of materials is discussed. Emphasis is placed on novel methods in the field of electrical and optical investigations, scanning probe microscopy (SPM) techniques and advanced X-ray analysis. The book starts with tutorial reviews, and arrives at up-to-date results about polar oxides. Therefore, it not only stimulates and further motivates young scientists but is of considerable interest for the members of our community.

The publication of this book follows a Workshop on "Polar Oxides – Properties, Characterizing and Imaging" which was held in Capri, Italy, in June 2003. The chapters published here are up-dated and revised manuscripts of the contributions presented during the workshop. The workshop was organized in the framework of "POLECER – European Thematic Network on Polar Electroceramics" which is part of the "Growth European Community Program".

The realization of the book is based on the work of many people. First of all, we would like to thank the authors, who make the book possible by their excellent contributions and their engaged cooperation. In addition, we would like to thank all members of our group at the RWTH Aachen, especially Thomas Pössinger for the preparing of the entire layout, Dagmar Leisten for editing and preparing the figures, and Maria Garcia for the spelling and format check.

We hope that the book will stimulate interest among a large number of its readers and give some motivation for their future work.

Rainer Waser, Ulrich Böttger, and Stephan Tiedke
Aachen, October 2004

Polar Oxides: Properties, Characterization, and Imaging
Edited by R. Waser, U. Böttger, and S. Tiedke
Copyright © 2005 WILEY-VCH Verlag GmbH & Co. KGaA, Weinheim
ISBN: 3-527-40532-1

Contents

1 Dielectric Properties of Polar Oxides   (*U. Böttger*)   **11**
   1.1 Introduction 11
   1.2 Dielectric polarization 13
         1.2.1 Macroscopic and microscopic view 13
         1.2.2 Mechanisms of polarization 15
   1.3 Ferroelectric polarization 17
   1.4 Theory of Ferroelectric Phase Transition 18
         1.4.1 Ginzburg-Landau Theory 18
         1.4.2 Soft Mode Concept 22
   1.5 Ferroelectric Materials 24
         1.5.1 Basic Compositions 24
         1.5.2 Grain Size effects 26
         1.5.3 Influence of Substitutes and Dopants 27
   1.6 Ferroelectric Domains 30
         1.6.1 Reversible and Irreversible Polarization Contributions 32
         1.6.2 Ferroelectric Switching 35
   Bibliography 37

2 Piezoelectric Characterization   (*S. Trolier-McKinstry*)   **39**
   2.1 Important piezoelectric constants 39
   2.2 Measurements in bulk materials 43
   2.3 Measurements in thin films 46
   2.4 Conclusions 50
   Bibliography 52

3 Electrical Characterization of Ferroelectrics   (*K. Prume, T. Schmitz, S. Tiedke*)   **53**
   3.1 Introduction 53
   3.2 Measurement methods 53
         3.2.1 Sawyer Tower method 56
         3.2.2 Shunt method 57
         3.2.3 Virtual ground method 57
         3.2.4 Current step method 58
   3.3 Measurement types 58
         3.3.1 Hysteresis loop and characteristic values 58
         3.3.2 Dynamic hysteresis measurement 59

		3.3.3	Pulse measurement	61
		3.3.4	Static hysteresis measurement	63
		3.3.5	Leakage measurement	65
		3.3.6	Fatigue measurement	66
		3.3.7	Imprint measurement	68
		3.3.8	Retention measurement	71
		3.3.9	Small signal measurements	73
	Bibliography			74

4 Optical Characterization of Ferroelectric Materials   (*C. Buchal*)   77
   4.1 Introduction: Light propagation within anisotropic crystals 77
       4.1.1 Huyghens's construction for uniaxial crystals 78
       4.1.2 The uniaxial indicatrix 80
       4.1.3 The biaxial indicatrix 83
   4.2 The electro-optic effect 83
       4.2.1 Ferroelectrics have anisotropic electronic bonds: Birefringence 84
       4.2.2 Applied fields change the optical pathlength: Phase modulators . . . 87
       4.2.3 Optical waveguides improve the device efficiency 89
   4.3 Non-linear optics 93
       4.3.1 Nonlinear optical media 94
       4.3.2 The nonlinear wave equation 95
       4.3.3 Second order nonlinear optics 96
   Bibliography 98

5 Microwave Properties and Measurement Techniques   (*N. Klein*)   99
   5.1 Introduction 99
   5.2 Basic relations defining microwave properties of dielectrics
       and normal/superconducting metals 100
   5.3 Surface impedance of normal metals 101
   5.4 Surface impedance of high-temperature superconductor films 101
   5.5 Microwave properties of dielectric single crystals, ceramics and thin films . . 103
   5.6 General remarks about microwave material measurements 108
   5.7 Non resonant microwave measurement techniques 109
   5.8 Resonator measurement techniques 110
   5.9 Conclusions 117
   Bibliography 117

6 Advanced X-ray Analysis of Ferroelectrics
   (*K. Saito, T. Kurosawa, T. Akai, S. Yokoyama, H. Morioka, H. Funakubo*)   **119**
   6.1 Introduction 119
   6.2 Experimental 122
       6.2.1 X-ray diffractometer 122
       6.2.2 Method of X-ray diffraction 122
       6.2.3 Sample preparation 127
   6.3 Results and discussion 127

		6.3.1 Structural characterization of PZT 52/48 thin film	127

        6.3.1   Structural characterization of PZT 52/48 thin film 127
        6.3.2   Distinguishing SBTN phase from fluorite–SBTN phase 131
        6.3.3   Grazing incidence X-ray diffraction study on PZT 52/48 thin films . . 132
   6.4   Conclusions 134
   Bibliography 135

7 Characterization of PZT-Ceramics by High-Resolution X-Ray Analysis
(M. J. Hoffmann, H. Kungl, J.-Th. Reszat, S. Wagner)   **137**

   7.1   Introduction 137
   7.2   Experimental 138
   7.3   Results and discussion 139
        7.3.1   Quantitative analysis of the F_T and F_R phase content 139
        7.3.2   Temperature induced $F_R \leftrightarrow F_T$ phase transition 142
        7.3.3   Analysis of intrinsic and extrinsic contributions to the macroscopic strain 145
   7.4   Summary 149
   Bibliography 150

8 In-Situ Synchrotron X-ray Studies of Processing and Physics of Ferroelectric Thin Films
(G. B. Stephenson, S. K. Streiffer, D. D. Fong, M. V. Ramana Murty, O. Auciello, P. H. Fuoss, J. A. Eastman, A. Munkholm, C. Thompson)   **151**

   8.1   Introduction 151
   8.2   Growth of ultrathin ferroelectric films 152
   8.3   Observation of nanoscale 180° stripe domains 154
   Bibliography 160

9 Characterization of Polar Oxides by Photo-Induced Light Scattering
(M. Imlau, M. Goulkov, M. Fally, Th. Woike)   **163**

   9.1   Introduction 163
   9.2   Fundamentals 165
        9.2.1   The relaxor-ferroelectric $Sr_xBa_{1-x}Nb_2O_6$ 165
        9.2.2   Observation of photo-induced light scattering in SBN 167
        9.2.3   Description of photo-induced light scattering in SBN 168
   9.3   Experimental 172
        9.3.1   Experimental setup 172
        9.3.2   Spatial distribution of the scattering intensity 173
        9.3.3   Investigating the relaxor-kind phase transition 174
        9.3.4   Determination of material parameters: gain factor Γ, effective electro-optic coefficient ($\zeta \cdot r_{\text{eff}}$) and effective trap density N_{eff} 177
        9.3.5   Investigating ferroelectric properties 179
        9.3.6   Investigating the polar structure 180
   9.4   Summary 186
   Bibliography 187

10 Ferroelectric Domain Breakdown: Application to Nanodomain Technology
(G. Rosenman, A. Agronin, D. Dahan, M. Shvebelman, E. Weinbrandt, M. Molotskii, Y. Rosenwaks) **189**
   10.1 Introduction 190
   10.2 Nanodomain size limitations 191
        10.2.1 Technological demands for FE domain-based devices 191
        10.2.2 Physical limit of domain dimensions in FE 192
   10.3 AFM nanodomain tailoring technology 193
        10.3.1 Nanoscale switching electrode 193
        10.3.2 Low and high voltage AFM for nanodomain reversal in FE bulk crystals 195
        10.3.3 Indirect electron beam induced ferroelectric domain breakdown . . . 198
   10.4 Ferroelectric domain breakdown 202
        10.4.1 Domain shapes under FDB 202
        10.4.2 The domain shape invariant 206
        10.4.3 Theory and experimental data of FDB effect 208
        10.4.4 Ferroelectric domain breakdown mechanism 208
   10.5 Nanodomain superlattices tailored by multiple tip arrays of HVAFM 210
   10.6 Conclusions 216
   Bibliography 217

11 Pyroelectric Ceramics and Thin Films:
Characterization, Properties and Selection   *(R. W. Whatmore)*   **221**
   11.1 Introduction 221
   11.2 The physics of pyroelectric detectors 222
        11.2.1 Pyroelectric response 222
        11.2.2 Comparison of noise and signal 225
        11.2.3 Other sources of noise 226
        11.2.4 The piezoelectric effect in pyroelectric detectors 226
   11.3 Measurement of physical parameters 227
        11.3.1 Dielectric properties 227
        11.3.2 Pyroelectric properties 228
        11.3.3 Electrical resistivity 231
        11.3.4 Thermal properties 231
        11.3.5 Piezoelectric property determination 231
   11.4 Pyroelectric materials and their selection 232
   11.5 Pyroelectric ceramics and thin films 234
   11.6 Conclusions 238
   Bibliography 238

12 Nano-inspection of Dielectric and Polarization Properties at Inner and Outer Interfaces in PZT Thin Films   *(L. M. Eng)*   **241**
   12.1 Introduction 241
   12.2 Methods 242
        12.2.1 Piezoresponse force microscopy (PFM) 242
        12.2.2 Kelvin Probe Force Microscopy (KPFM) 242

	12.2.3 Pull-off force spectroscopy (PFS)	243
12.3	Materials	244
12.4	Results	244
	12.4.1 Polarization profile across the PZT film	244
	12.4.2 Relaxation dynamics within the PZT film	247
	12.4.3 Local dielectric constant at the PZT surface	247
12.5	Conlusion	248
Bibliography		249

13 Piezoelectric Relaxation and Nonlinearity investigated by Optical Interferometry and Dynamic Press Technique   (*D. Damjanovic*)   **251**

13.1	Introduction	251
13.2	Measurement techniques	252
	13.2.1 Optical techniques for measurements of the converse effect	252
	13.2.2 Dynamic press for the measurements of direct effect	254
13.3	Investigation of the piezoelectric nonlinearity in PZT thin films using optical interferometry	255
13.4	Investigation of the piezoelectric relaxation in ferroelectric ceramics using dynamic press	257
	13.4.1 Maxwell-Wagner piezoelectric relaxation and clockwise hysteresis . .	257
	13.4.2 Piezoelectric relaxation and Kramers-Kronig relations in a modified lead titanate composition	258
	13.4.3 Evidence of creep-like piezoelectric response in soft PZT ceramics .	259
Bibliography		261

14 Chaotic Behavior near the Ferroelectric Phase Transition
(*H. Beige, M. Diestelhorst, R. Habel*)   **263**

14.1	Introduction	263
14.2	Dielectric nonlinear series-resonance circuit	263
14.3	Nonlinear nature of the resonant system	264
14.4	Tools of the nonlinear dynamics	264
14.5	Experimental representation of phase portraits	265
14.6	Comparison of calculated and experimentally observed phase portraits	266
14.7	Controlling chaos	269
14.8	Summary	273
Bibliography		274

15 Relaxor Ferroelectrics – from Random Field Models to Glassy Relaxation and Domain States
(*W. Kleemann, G. A. Samara, J. Dec*)   **275**

15.1	Introduction	275
15.2	Polar nanoregions	279
15.3	Cubic relaxors	283
15.4	Role of pressure	285
15.5	Dynamics of the dipolar slowing-down process	288

15.6 Uniaxial relaxors 291
15.7 Domain dynamics in uniaxial relaxors 292
Bibliography 299

16 Scanning Nonlinear Dielectric Microscope (Y. Cho) 303
16.1 Introduction 303
16.2 Nonlinear dielectric imaging with sub- nanometer resolution 304
    16.2.1 Principle and theory for SNDM 304
    16.2.2 Nonlinear dielectric imaging 306
    16.2.3 Comparison between SNDM imaging and piezo-response imaging . . 308
    16.2.4 Observation of domain walls in PZT thin film using SNDM 310
16.3 Higher order nonlinear dielectric microscopy 312
    16.3.1 Theory for higher order nonlinear dielectric microscopy 313
    16.3.2 Experimental details of higher order nonlinear dielectric microscopy . 314
16.4 Three-dimensional measurement technique 316
    16.4.1 Principle and measurement system 316
    16.4.2 Experimental results 317
16.5 Ultra High-Density Ferroelectric Data Storage Using Scanning Nonlinear Dielectric Microscopy 319
    16.5.1 SNDM domain engineering system 320
    16.5.2 Nano-domain formation in LiTaO$_3$ single crystal 320
16.6 Conclusions 324
Bibliography 327

17 Electrical Characterization of Ferroelectric Properties in the Sub-Micrometer Scale (T. Schmitz, S. Tiedke, K. Prume, K. Szot, A. Roelofs) 329
17.1 Introduction 329
17.2 Sample preparation 330
17.3 Contact problems 332
17.4 Parasitic capacitance 336
17.5 In-situ compensation 337
Bibliography 341

18 Searching the Ferroelectric Limit by PFM
(A. Roelofs, T. Schneller, U. Böttger, K. Szot, R. Waser) 343
18.1 Introduction 343
18.2 Polycrystalline ferroelectric PTO thin films on platinized silicon substrates . . 344
18.3 Separated lead titanate nano-grains 348
18.4 Conclusion 352
Bibliography 353

19 Piezoelectric Studies at Submicron and Nano Scale
(E. L. Colla, I. Stolichnov) 355
19.1 Introduction 355
    19.1.1 Fatigue in FeRAM: macroscopic results invoking nano scale features . 355

        19.1.2 Piezoelectric characterization at nano scale of ferroelectric thin films . 359
   19.2 Investigating cycling induced suppression of switchable polarization in FeCaps 361
        19.2.1 Appearance of frozen polarization nano domains 361
        19.2.2 Nano scale hysteresis loops of fatigued FeCaps 364
   19.3 Size effect on the polarization patterns in μ-sized ferroelectric film capacitors 367
        19.3.1 Downscaling of ferroelectric capacitors 367
        19.3.2 Size induced polarization instability 368
   19.4 Direct observation of inversely-polarized frozen nanodomains in fatigued Fe-
        Caps 371
        19.4.1 Removable electrodes 371
        19.4.2 Inversely-polarized nanodomains 372
   Bibliography 377

Authors 379

Index 381

1 Dielectric Properties of Polar Oxides

Ulrich Böttger

Institut für Werkstoffe der Elektrotechnik, RWTH Aachen, Germany

Abstract

This chapter gives an introduction to the class of polar oxides. Basic principles about symmetry classification, dielectric and ferroelectric polarization, phase transitions as well as electrical and piezoelectric properties are included. Landau-Ginzburg-Devonshire theory and the soft mode concept for the phase transition from an unpolar to a polar phase are also topic of this chapter. Specially addressed are the most relevant ferroelectric materials as $BaTiO_3$, $Pb(Zr,Ti)O_3$ and $SrBi_2Ta_2O_9$, and how the microstructure, modificatitions or doping will influence their properties. Ferroelectric domains evoked by the reduction of the electric and elastic energy are decisive for the dielectric and piezoelectric properties and the switching behavior. These effects are discussed for bulk ceramics as well as for thin films.

1.1 Introduction

Among the 32 crystallographic point groups describing all crystalline systems, 11 are centrosymmetric and contain an inversion center. In that case polar properties become not possible because any polar vector may be inverted by an existing symmetry transformation. All other 21 point groups without an inversion center (except the point group *432*) can exhibit piezoelectricity which describes the coupling between mechanical and electrical energies in a material. An external mechanical stress X leads to a change in the electric polarization P or dielectric displacement D respectively or an external electric field E causes an elastic strain x. The relation is given by the piezoelectric coefficient d_{ijk} being a third rank tensor (see Tutorial "Piezoelectric Characterization"):

$$D_i = d_{ijk} X_{jk} \qquad x_{ij} = d_{ijk} E_k \tag{1.1}$$

There are 10 polar groups with a unique polar axis among the 21 point groups without an inversion center. This class of crystals may show a spontaneous polarization parallel to the polar axis. E.g. barium titanate (in its tetragonal phase) is such a material (see Figure 1.1). However in the cubic phase (perovskite structure), the central titanium atom serves as an inversion center - then spontaneous polarization is not possible. Only with the occurrence of a tetragonal deformation, where the positively charged barium and titanium ions are displaced with respect to the six negatively charged oxygen ions, a polar axis is formed in the direction of the tetragonal deformation, which marks the direction of the spontaneous polarization [1].

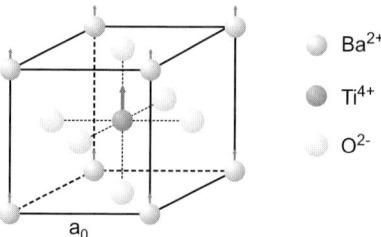

Figure 1.1: Unit cell of cubic BaTiO$_3$. The arrow schematically indicates one of the possible displacement of the central Ti^{4+} ion at the transition to the tetragonal ferroelectric structure that leads to a spontaneous polarization, in reality all ions are displaced against each other.

Following Maxwell's equations, the spontaneous polarization is connected with surface charges $P_s = \sigma$. The surface charges in general are compensated by charged defects. A temperature change changes the spontaneous polarization. This effect is called the pyroelectric effect.

If it is possible to reorient the spontaneous polarization of a material between crystallographically equivalent configurations by an external electric field, then in analogy to ferromagnetics one speaks about ferroelectrics. Thus, it is not the existence of spontaneous polarization alone, but the "switchability" by an external field which defines a ferroelectric material. Figure 1.2 displays a characteristic hysteresis loop occurring during the reversal of the polarization in a ferroelectric.

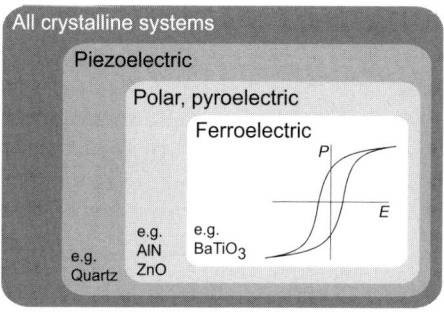

Figure 1.2: Classification of the crystallographic groups by their electrical properties

The class of ferroelectric materials have a lot of useful properties. High dielectric coefficients over a wide temperature and frequency range are used as dielectrics in integrated or in SMD (surface mounted device) capacitors. The large piezoelectric effect is applied in a variety of electromechanical sensors, actuators and transducers. Infrared sensors need a high pyroelectric coefficient which is available with this class of materials. Tunable thermistor properties in semiconducting ferroelectrics are used in PTCR (positive temperature coefficient

1.2 Dielectric polarization

resistors). The significant non-linearities in electromechanical behavior, field tunable permittivities and refractive indices, and electrostrictive effects open up a broad field of further different applications. In addition, there is growing interest in ferroelectric materials for memory applications, where the direction of the spontaneous polarization is used to store information digitally.

1.2 Dielectric polarization

1.2.1 Macroscopic and microscopic view

In accordance to the Poisson equation, the source of the dielectric displacement $\vec{D}$ is given by the density of free (conducting) charges ρ:

$$\mathrm{div}\vec{D} = \rho_{\mathrm{free}} \tag{1.2}$$

The overall charge neutrality of matter in an external field is described by:

$$\vec{D} = \epsilon_0 \vec{E} + \vec{P} \tag{1.3}$$

The vacuum contribution caused by the externally applied electric field is represented by the term $\epsilon_0 \vec{E}$, and the electrical polarization of the matter in the system is described by $\vec{P}$, e.g. [2]. This relation is independent of the nature of the polarization which could be pyroelectric polarization, by piezoelectric polarization or dielectric polarization (by an external electric field).

Considering a simple parallel plate capacitor filled with matter (see Figure 1.3), two cases have to be distinguished: (i) If the applied voltage is kept constant ($E = \mathrm{const}$, short circuit condition), additional free charges need to flow into the system to increase D according to Equation (1.2). If the charges on the plates are kept constant ($D = \mathrm{const}$, open circuit condition), the electric field E and, hence, the voltage between the plates will decrease according to Equation (1.3).

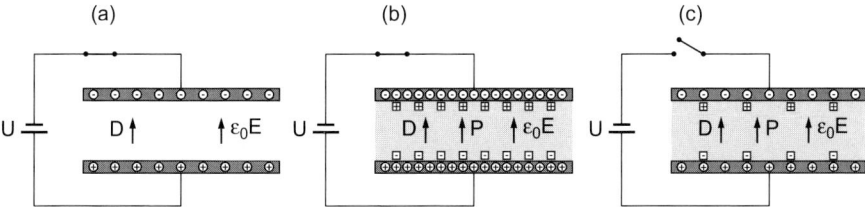

Figure 1.3: Parallel plate capacitor (a) without any dielectric, (b) filled with dielectric under short circuit condition ($E = \mathrm{constant}$) and (c) filled wtih dielectric under open circuit condition ($D = \mathrm{constant}$).

For a pure dielectric response of the matter the polarization is proportional to the electric field in a linear approximation by

$$P = \epsilon_0 \chi_e E \quad \text{or} \quad D = \epsilon_0 \epsilon_r E \tag{1.4}$$

The dielectric susceptibility χ is related to the relative dielectric constant ϵ_r by $\chi = \epsilon_r - 1$. Equations (1.4) are only valid for small fields. Large amplitudes of the ac field lead to strong non-linearities in dielectrics, and to sub-loops of the hysteresis in ferroelectrics. Furthermore, the dielectric response depends on the bias fields as shown in Figure 1.4. From the device point of view this effect achieves the potential of a tunable dielectric behavior, e. g. for varactors.

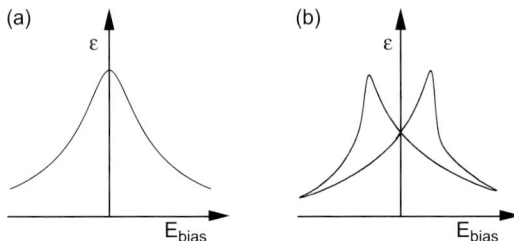

Figure 1.4: Bias field dependence of the dielectric constant of (a) dielectric and (b) ferroelectric material

Equations (1.3) and (1.4) describe the mean properties of the dielectric. This macroscopic point of view does not consider the microscopic origin of the polarization [3]. The macroscopic polarization P is the sum of all the individual dipole moments p_j of the material with the density N_j.

$$P = \sum_j N_j p_j \tag{1.5}$$

In order to find a correlation between the macroscopic polarization and the microscopic properties of the material a single (polarizable) particle is considered. A dipole moment is induced by the electric field at the position of the particle which is called the local electric field E_{loc}

$$p = \alpha E_{loc} \tag{1.6}$$

where α is the polarizability of an atomic dipole. If there is no interaction between the polarized particles, the local electric field is identical to the externally applied electric field $E_{loc} = E_0$, resulting in a simple relation between the susceptibility and the polarizability $\epsilon_0 \chi = N_j \alpha_j$.

In condensed matter, the density and therefore the electrostatic interaction between the microscopic dipoles is quite high. Hence, the local field E_{loc} at the position of a particular dipole is given by the superposition of the applied macroscopic field E_0 and the sum

1.2 Dielectric polarization

of all other dipole fields. For cubic structures and for induced dipoles (ionic and electronic polarization), the calculation reveals a relation between the atomic polarizability α and the macroscopic permittivity $\epsilon = \epsilon_0 \epsilon_r$ which is referred to the Clausius-Mossotti equation [4].

$$\epsilon = \frac{\epsilon_0 + 2N_j\alpha_j}{\epsilon_0 - N_j\alpha_j} \tag{1.7}$$

1.2.2 Mechanisms of polarization

In general, there are five different mechanisms of polarization which can contribute to the dielectric response [3].

- **Electronic polarization** exists in all dielectrics. It is based on the displacement of the negatively charged electron shell against the positively charged core. The electronic polarizability α_{el} is approximately proportional to the volume of the electron shell. Thus, in general α_{el} is temperature-independent, and large atoms have a large electronic polarizability.

- **Ionic polarization** is observed in ionic crystals and describes the displacement of the positive and negative sublattices under an applied electric field.

- **Orientation polarization** describes the alignment of permanent dipoles. At ambient temperatures, usually all dipole moments have statistical distribution of their directions. An electric field generates a preferred direction for the dipoles, while the thermal movement of the atoms perturbs the alignment. The average degree of orientation is given by the Langevin function $\langle \alpha_{or} \rangle = p^2/(3k_BT)$ where k_B denotes the Boltzmann constant and T the absolute temperature.

- **Space charge polarization** could exist in dielectric materials which show spatial inhomogeneities of charge carrier densities. Space charge polarization effects are not only of importance in semiconductor field-effect devices, they also occur in ceramics with electrically conducting grains and insulating grain boundaries (so-called Maxwell-Wagner polarization).

- **Domain wall polarization** plays a decisive role in ferroelectric materials and contributes to the overall dielectric response. The motion of a domain wall that separates regions of different oriented polarization takes place by the fact that favored oriented domains with respect to the applied field tends to grow.

The total polarization of dielectric material results from all the contributions discussed above. The contributions from the lattice are called intrinsic contributions, in contrast to extrinsic contributions.

$$\epsilon = \underbrace{\epsilon_{elec} + \epsilon_{ion}}_{intrinsic} + \underbrace{\epsilon_{or} + \epsilon_{dw} + \epsilon_{sc}}_{extrinsic} \tag{1.8}$$

Each contribution stems from a short-range movement of charges that responds to an electric field on different time scales and, hence, through a Fourier transform, in different frequency regimes. If the oscillating masses experience a restoring force, a relaxation behavior is found (for orientation, domain walls, and space charge polarization). Resonance effects are observed for the ionic and electronic polarization. The dispersion of the dielectric function is shown in Figure 1.5, and holds the potential to separate the different dielectric contributions.

The space charge polarization is caused by a drift of mobile ions or electrons which are confined to outer or inner interfaces. Depending on the local conductivity, the space charge polarization may occur over a wide frequency range from mHz up to MHz. The polarization due to the orientation of electric dipoles takes place in the frequency regime from mHz in the case of the reorientation of polar ligands of polymers up to a few GHz in liquids such as water. It is often possible to distinguish between space charge and orientation because of the temperature dependence of α_{or}. In the infrared region between 1 and 10 THz, resonances of the molecular vibrations and ionic lattices constituting the upper frequency limit of the ionic polarization are observed. The resonance of the electronic polarization is around 10^{15} Hz. It can be investigated by optical methods.

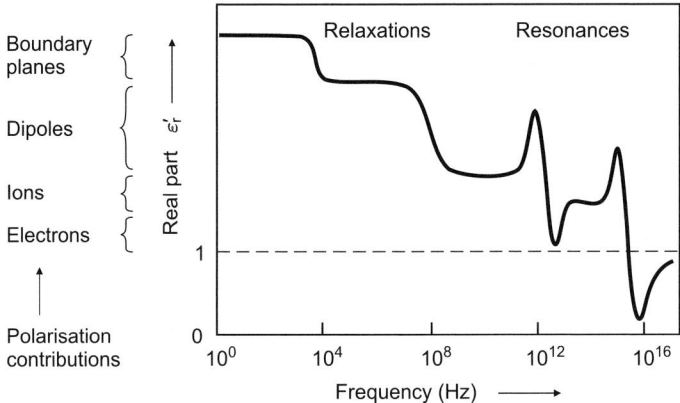

Figure 1.5: Frequency dependence of real part of the dielectric function.

The dispersion of the dielectric response of each contribution leads to dielectric losses of the matter which can be mathematically expressed by a complex dielectric permittivity:

$$\epsilon = \epsilon' + i\epsilon'' \tag{1.9}$$

Dielectric losses are usually described by the loss tangent:

$$\tan \delta = \frac{\epsilon''}{\epsilon'} \tag{1.10}$$

It should be taken into account that the general definition of the $\tan \delta$ is related to the ratio of loss energy and reactive energy (per periode), i.e. all measurements of the loss tangent also include possible contributions of conductivity σ of a non-ideal dielectric given by $\tan \delta = \sigma/\omega\epsilon'$.

1.3 Ferroelectric polarization

An ideal single crystal shows a $P(E)$ behavior as depicted in Figure 1.6. The non-ferroelectric dielectric ionic and electronic polarization contributions are clearly linear, and are suposed by the spontaneous polarization P_s (dashed curve in Figure 1.6). To reverse the polarization an electrical field with an amplitude $E > E_c$ is required. In opposite to single crystals in poly-domain ferroelectric ceramics, the remanent polarization P_r is smaller than the spontaneous one P_s due to backswitching even for opposite fields as shown in Figure 1.6. In that case P_s can be estimated by extrapolation of (non-switching) P-values to $E \to 0$.

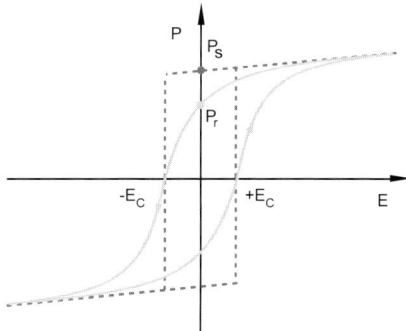

Figure 1.6: Ferroelectric hysteresis of single domain single crystal (dashed line) and polycrystalline sample (full line)

A ferroelectric "model" material is barium titanate BaTiO$_3$. On cooling from high temperatures, the permittivity increases up to values well above 10,000 at the phase transition temperature T_C. The inverse susceptibility as well as the dielectric permittivity follows a Curie-Weiss law $\chi^{-1} \approx \epsilon^{-1} \propto (T - \Theta)$. The appearance of the spontaneous polarization is accompanied with a spontaneous (tetragonal) lattice distortion.

The phase transition in barium titanate is of first order, and as a result, there is a discontinuity in the polarization, lattice constant, and many other properties, as becomes clear in Figure 1.7. It is also clear in the figure that there are three phase transitions in barium titanate having the following sequence upon cooling: rhombohedral, orthorhombic, tetragonal and cubic.

There is a small thermal hysteresis of the transition temperature, which depends on many parameters such as the rate of temperature change, mechanical stresses or crystal imperfections. From a crystal chemical view, the Ba-O framework evokes an interstitial for the central Ti^{4+} ion which is larger than the actual size of the Ti^{4+} ion. As a result, the serie of phase transformations takes place to reduce the Ti cavity size. Certainly, the radii of the ions involved impact the propensity for forming ferroelectric phases; thus both PbTiO$_3$ and BaTiO$_3$ have ferroelectric phases, while CaTiO$_3$ and SrTiO$_3$ do not [5].

The optical properties of ferroelectric materials are characterized by birefringence. Barium titanate is isotropic only in the cubic phase. The tetragonal and the rhombohedral phases are

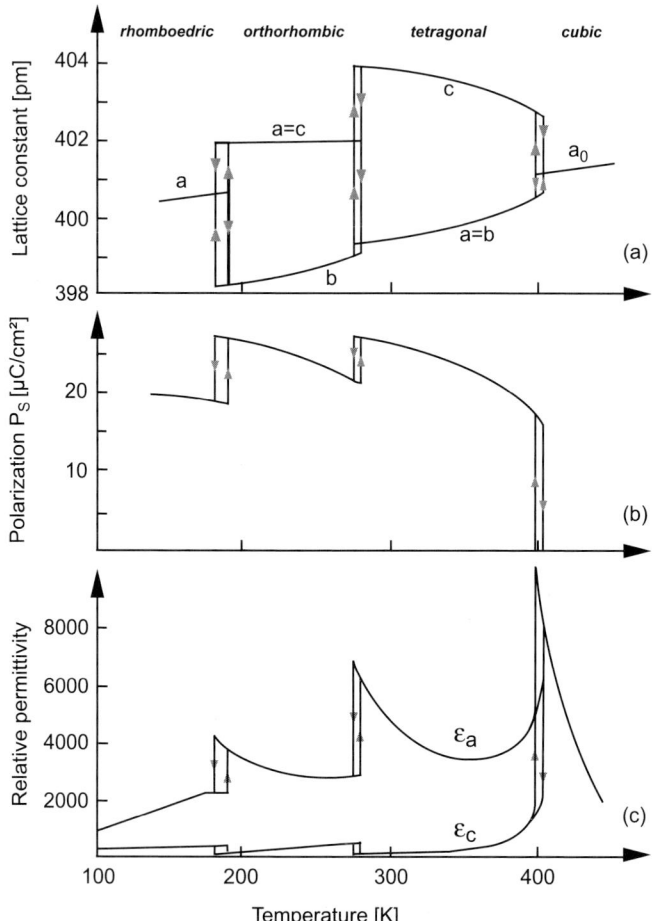

Figure 1.7: Various properties of barium titanate as a function of temperature. Anisotropic properties are shown with respect to the lattice direction. (a) Lattice constants, (b) spontaneous polarization P_s and (c) relative permittivity ϵ_r.

uniaxially birefringent while the orthorhombic phase exhibits birefringent behavior with two axes. Figure 1.7 (c) displays the temperature dependence of the permittivity in BaTiO$_3$.

1.4 Theory of Ferroelectric Phase Transition

1.4.1 Ginzburg-Landau Theory

The Ginzburg-Landau theory is equivalent to a mean field theory considering the thermodynamic entity of the dipoles in the mean field of all the others. It is reasonable if the particular

1.4 Theory of Ferroelectric Phase Transition

dipole interacts with many other dipoles. The theory introduces an order parameter P, i.e. the polarization, which for a second order phase transition diminishes continuously to zero at the phase transition temperature T_c [3]. Close to the phase transition, therefore, the free energy may be written as an expansion of powers of the order parameter. All the odd powers of P do not occur because of symmetry reasons.

$$F(P,T) = \frac{1}{2}g_2 P^2 + \frac{1}{4}g_4 P^4 + \frac{1}{6}g_6 P^6 \tag{1.11}$$

The highest expansion coefficient (here g_6) needs to be larger than zero because otherwise the free energy would approach minus infinity for large P. All coefficients depend on the temperature and in particular the coefficient g_2. Expanding g_2 in a series of T around the Curie temperature Θ which is equal to or less than the phase transition temperature T_c, we can approximate:

$$g_2 = \frac{1}{C}(T - \Theta). \tag{1.12}$$

Stable states are characterized by minima of the free energy with the necessary and sufficient conditions:

$$\frac{\partial F}{\partial P} = P(g_2 + g_4 P^2 + g_6 P^4) = 0 \tag{1.13}$$

$$\text{and} \quad \frac{\partial^2 F}{\partial P^2} = \frac{1}{\chi} = g_2 + 3g_4 P + 5g_6 P^3 > 0 \tag{1.14}$$

Two cases are to distinguish: (i) $g_4 > 0 \Rightarrow g_6 \approx 0$ which corresponds to a phase transition of second order, and (ii) $g_4 < 0 \Rightarrow g_6 > 0$ which is related with a phase transition of first order. In both cases, the trivial solution $P = 0$ exists, representing the paraelectric phase. Inserting Equation (1.12) into (1.14) it becomes obvious that above T_c the coefficient g_2 needs to be larger than zero in order to obtain stable solutions. A comparison of Equation (1.12) and (1.14) shows that g_2 is expressed by the susceptibility χ, for which a Curie-Weiss law is found.

$$\chi = \frac{C}{T - \Theta} \tag{1.15}$$

Second order phase transition

For $T < \Theta$ a spontaneous polarization exists. It can easily be shown that the Curie temperature Θ is equal to the phase transition temperature T_C. The spontaneous polarization depends on the distance from the phase transition temperature with a square root law.

$$P_s = \sqrt{\frac{T_c - T}{C g_4}} \tag{1.16}$$

Figure 1.8 schematically displays the free energy close to the second order phase transition for different temperatures as a function of the order parameter P_s. For $T > T_c$ a minimum

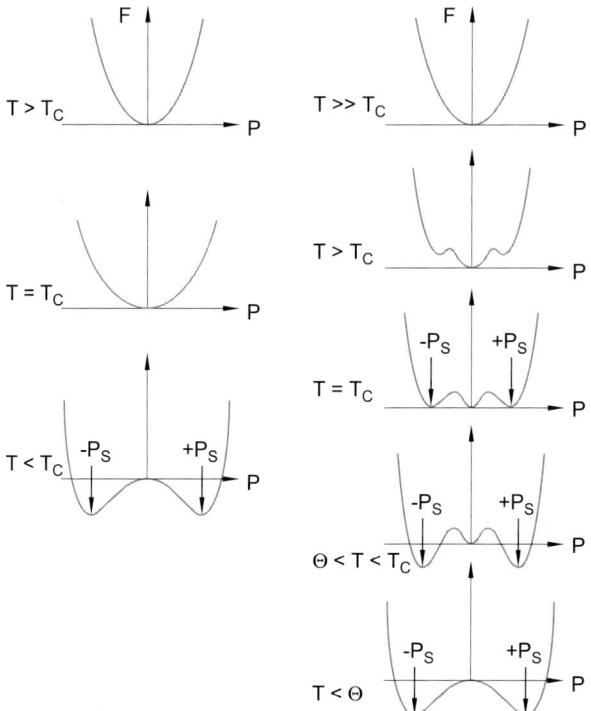

Figure 1.8: Free energy of a ferroelectric with a second-order phase transition (left) and with a first-order phase transition (right) at different temperatures. T_C is the phase transition temperature and Θ is the Curie temperature.

is found for $P^2 = 0$. At $T = T_c$, this minimum shifts continuously to final values of the polarization. The temperature dependence of the susceptibility in the ferroelectric phase is obtained by inserting Equation (1.16) into (1.14).

$$\left.\frac{1}{\chi}\right|_{T<T_c} = 2\frac{T_c - T}{C} \tag{1.17}$$

The slope of the inverse susceptibility below T_c is just twice of the slope above T_c. The theory is in good agreement to the experimental data, see Figure 1.9.

First order phase transition

If the expansion coefficients are chosen as $g_4 < 0$ and $g_6 > 0$, stable states will again be found from Equation (1.13):

$$P_s^2 = \frac{1}{2g_6}\left(|g_4| + \sqrt{g_4^2 - 4C^{-1}(T - \Theta)g_6}\right) \tag{1.18}$$

1.4 Theory of Ferroelectric Phase Transition

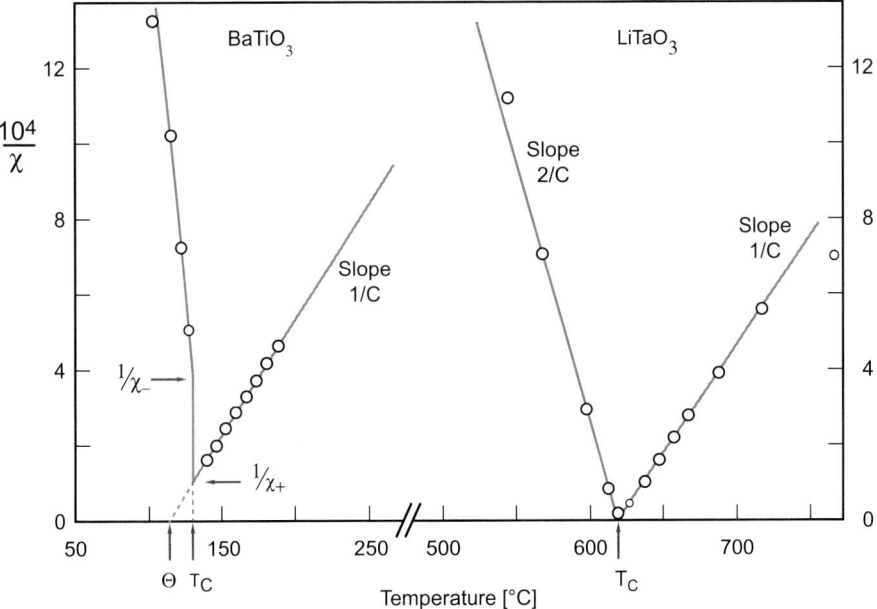

Figure 1.9: Reciprocal dielectric susceptibility at the phase transition of lithium tantalate (second order phase transition) and of barium titanate (first order phase transition).

Inserting Equation (1.18) into (1.11) results in the free energy as a function of polarization and of temperature. In Figure 1.8 the behavior of $F(P_s, T)$ is shown for some relevant temperatures. At high temperatures the free energy assumes a parabolic shape with a minimum corresponding to a stable paraelectric phase. During cooling, secondary minima at finite polarizations become visible. Their energy level at the beginning, however, is higher than that at $P = 0$. In this regime the paraelectric phase is stable and the ferroelectric phase metastable. Lowering the temperature further, at $T = T_C$ all three minima of the free energy are at the same level. Below T_C, F becomes negative and favors a finite spontaneous polarization. In the temperature regime between T_C and Θ the paraelectric phase coexists with the ferroelectric phase with the paraelectric phase being metastable. Somewhere during cooling through this regime, the first order phase transition to the ferroelectric state will occur with a corresponding jump of the spontaneous polarization from zero to a finite value.

The susceptibility in the ferroelectric phase is given by:

$$\frac{1}{\chi}\bigg|_{T<T_c} = \frac{3g_4^2}{4g_6} + 8\frac{T - T_c}{C} \tag{1.19}$$

The dielectric behavior closed to a phase transistion is displayed in Figure 1.9 for barium titanate (first-order transition with $T_c = 135°C$, $C = 1.8 \cdot 10^5\,°C$) and for lithium tantalate (second-order transition with $T_c = 618°C$ and $C = 1.6 \cdot 10^5\,°C$).

1.4.2 Soft Mode Concept

The atoms of a crystal vibrate around their equilibrium position at finite temperatures. There are lattice waves propagating with certain wavelengths and frequencies through the crystal [7]. The characteristic wave vector $\vec{q}$ can be reduced to the first Brillouin zone of the reciprocal lattice, $0 \leq q \leq \pi/a$, when a is the lattice constant.

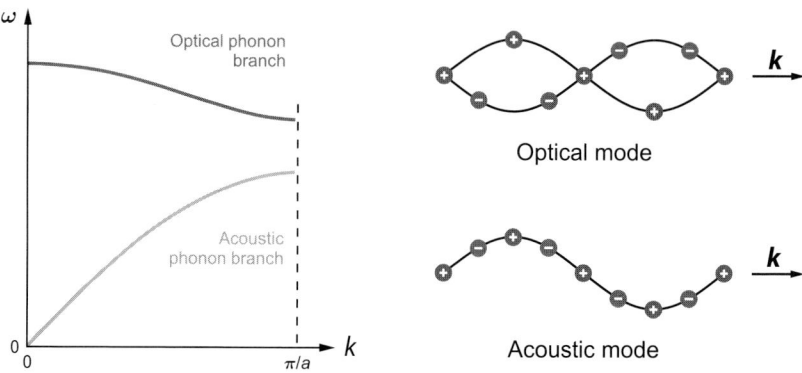

Figure 1.10: Transverse acoustic (TA) and optic mode (TO) of the phonon spectrum.

In every lattice there are three modes with different frequencies belonging to one longitudinal and two transverse branch of the acoustic phonons, as shown in Figure 1.10. A vibration of the atoms perpendicular to the propagation corresponds to a transverse wave, a vibration in the direction of the propagation corresponds to a longitudinal wave. The acoustic phonons have an elastic nature. All atoms vibrate as a linear chain independent of the number of different atoms per lattice cell. The wavelengths of the acoustic phonons are given by the sound velocity v_s, therefore, no coupling of acoustic phonons with electromagnetic waves exists ($v_s \ll v_{\text{Light}}$).

In case of non-primitive lattices with different atoms in the elementary cell, the sub-lattices can vibrate against each other (optical modes, see Figure 1.10). A vibration with a frequency $\omega \neq 0$ becomes possible even for $k = 0$. The opposite movement of neighboring atoms evokes large dipole moments allowing a coupling to electromagnetic waves.

In general, each mode of the phonon dispersion spectra is collectively characterized by the relating energy, i.e. the frequency and wave vector k, and is associated with a specific distortion of the structure.

The local electric field in ionic crystals leads to a splitting of the optical vibration modes. The longitudinal mode frequency is shifted to higher frequencies while the transverse mode frequency is shifted to lower frequencies. The softening of the transverse modes is caused by a partial compensation of the short-range lattice (elastic) forces on the one hand and the long-range electric fields on the other hand. This effect is strongest at the zone center [3]. If the compensation is complete, the transverse optic mode frequency becomes zero when the temperature is decreased, $\omega_{\text{TO}}(T \rightarrow T_c) \rightarrow 0$, and the soft phonon condenses out so that at

1.4 Theory of Ferroelectric Phase Transition

T_c a phase transition to a state with spontaneous polarization takes place (ferroelectric phase transition). The mechanism becomes clearer considering Figure 1.11 (b). At the zone center ($k = 0$) the wavelength of the TO mode is infinite ($\lambda \to \infty$), i.e the region of homogeneous polarization becomes infinite. In the case of the softening of the TO mode the transverse frequency becomes zero and no vibration exists anymore ("frozen in").

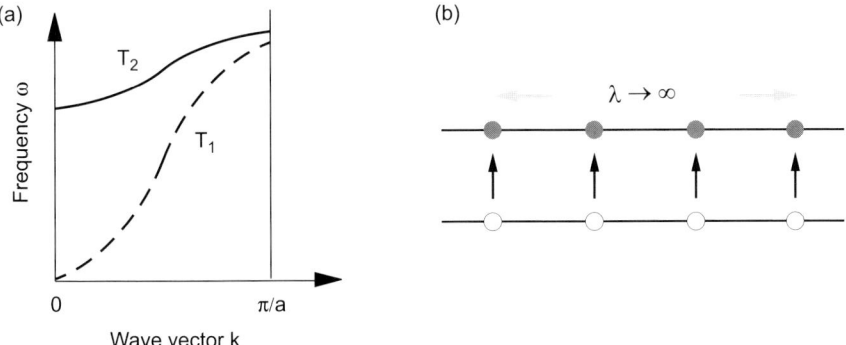

Figure 1.11: (a) Softening of the TO modes for $T_2 > T_1$ at the Brillouin zone center and (b) freezing of the TO modes for $T \to T_c$.

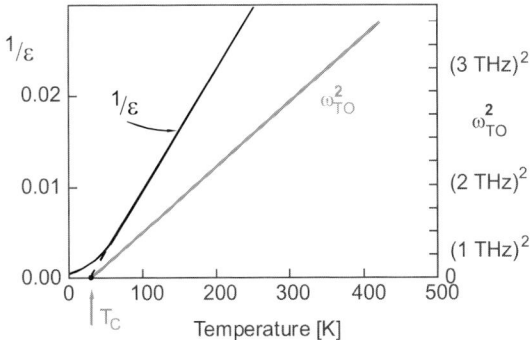

Figure 1.12: Frequency of the TO mode and dielectric behavior at the phase transition

A linear relation between ω_{TO}^2 and T at the zone center is found (see Figure 1.12) suggesting that the temperature dependence of the optic mode frequency relates to the phase transition. In accordance with the Lyddane-Sachs-Teller relation

$$\frac{\epsilon_s(T)}{\epsilon_\infty} = \frac{\omega_{LO}^2}{\omega_{TO}^2(T)} \tag{1.20}$$

ω_{TO} relates directly to the dielectric constant ϵ_s, i.e. the dielectric anomaly is associated with a soft mode condensation. From the extrapolation according to Equation (1.15) a phase transition at $T_c = 50$ K would be expected. This phase transition, however, does not really take place. It is dominated by a competing displacive phase transition at the zone boundary.

1.5 Ferroelectric Materials

1.5.1 Basic Compositions

Many ferroelectric materials were found in the past. However, there is a limited number of structures that are adopted by the majority of the commercially important ferroelectric materials. In each of these structures, the ferroelectricity is tied to distortion of the coordination polyhedra of one or more of the cations in the structure. One example is the perovskite structure. Cations that seem to be especially susceptible to forming such distorted polyhedra include Ti, Zr, Nb, Ta, and Hf. All of these ions lie near crossover points between the stability of different electronic orbitals, and so may be likely to form distorted coordination polyhedra [5]. Polarizable cations such as Pb and Bi are also common to many ferroelectric materials. In this case, it has been suggested that the lone pair electrons may play an important role in stabilizing ferroelectric structures. Thus the ferroelectric transition temperature and spontaneous distortion of PbTiO$_3$ is much larger than that of BaTiO$_3$.

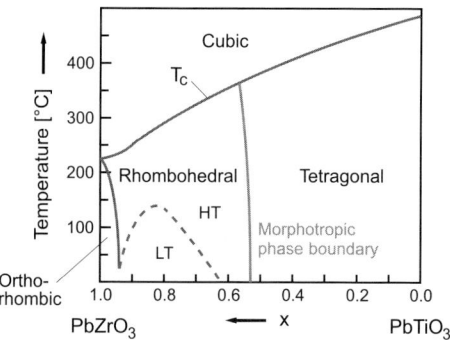

Figure 1.13: Phase diagram for PZT showing the morphotropic phase boundary between rhombohedral and tetragonal phases.

Solid solutions of PbTiO$_3$-PbZrO$_3$ (PZT) are one of the most important ferroelectric and piezoelectric materials [6]. Over the entire solid solution range, PZT adopts distorted versions of the perovskite structure, as shown in Figure 1.13. At the morphotropic phase boundary (MPB), the tetragonal phase and the rhombohedral phase co-exists leading to a higher polarizability due to the presence of a larger number of possible polarization directions (6 from tetragonal phase, and 8 from the rhombohedral phase). Thus, the dielectric and piezoelectric properties show strong maxima near this composition. The high piezoelectric coefficients (i.e. d_{33} values of 250 to 400 pC/N), coupled with a high transition temperature are the main reason that PZT ceramics are so widely used as piezoelectric sensors and actuators. The orthorhombic phase near the PbZrO$_3$ side of the PZT phase diagram corresponds to an antiferroelectric distortion of the perovskite structure, in which the polarization is cancelled on a unit cell level. For ferroelectric memory applications utilizing PZT, mostly Ti-rich compositions are used because of their large spontaneous polarization available, in accordance to Equation (1.16), and because the hysteresis loops of ferroelectric thin films are squarer for tetragonal than for rhombohedral compositions. This typically results in remanent polarizations in excess of 30 μC/cm^2. Today, low density memories based on PZT are in commercial production, and there is a good prospect for scaling to high densities.

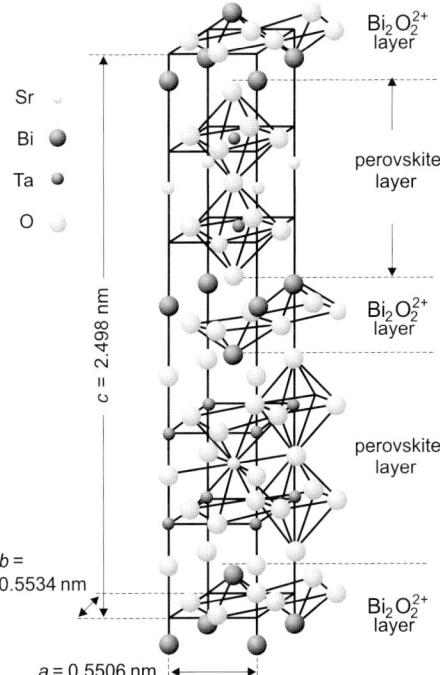

Figure 1.14: Bismuth layer structure SBT

An alternative structure that has also been widely investigated both for high temperature piezoelectric, as well as for ferroelectric memory applications is the bismuth layer structure family as shown in Figure 1.14 for SrBi$_2$Ta$_2$O$_9$ (SBT), e.g. [8]. The structure consists of perovskite layers of different thicknesses, separated by Bi$_2$O$_2^{2+}$ layers. It has been shown that when the perovskite block is an even number of octahedra thick, the symmetry imposes a restriction on the polarization direction, confining it to the a-b plane. In contrast, when the perovskite block is an odd number of octahedra thick, it is possible to develop a component of the polarization along the c axis (nearly perpendicular to the layers). This could be used in order to enhance the remanent polarization of ferroelectric thin films for memory applications; since there are comparatively few allowed directions for the spontaneous polarization, the remanent polarization is rather small for many film orientations as shown in comparison to PZT. However, the SBT materials are less susceptible to fatigue (a reduction in switchable polarization on repeated cycling).

It is important to realize that thin films may differ in some substantial ways from bulk ceramics or single crystals of the same composition. One source of these differences is the substantial in-plane stresses that thin films are typically under, ranging from MPa to GPa [9]. Because many ferroelectric materials are also ferroelastic, imposed stresses can markedly affect the stability of the ferroelectric phase, as well as the ease with which polarization can be reoriented in some directions. The phase diagram becomes considerably complicated by the presence of a dissimilar substrate [10]. It is obvious that the material coefficients are drastically changed.

1.5.2 Grain Size effects

In bulk $BaTiO_3$ ceramics the grain size has a strong effect on the low frequency permittivity for grain sizes below approx. 10 μm as shown in Figure 1.15 (a). The permittivity is rising at decreasing grain sizes up to a maximum at $g_m \approx 0.7\mu m$ [11]. The increase of ϵ could be caused by internal stresses because each grain is clamped by its surrounding neighbors or by the increase of the number of domain walls contributing to the dielectric constant. Below this size g_m, the permittivity sharply decreases again in conjunction with a reduction of the tetragonality and of the remanent polarization. The drop in permittivity may be interpreted by the effect of a low permittivity interfacial layer of 0.5 to 2 nm thickness at the grain boundaries. This layer shows no difference of the composition and crystal structure in comparison to the bulk and is believed to be of photonic nature.

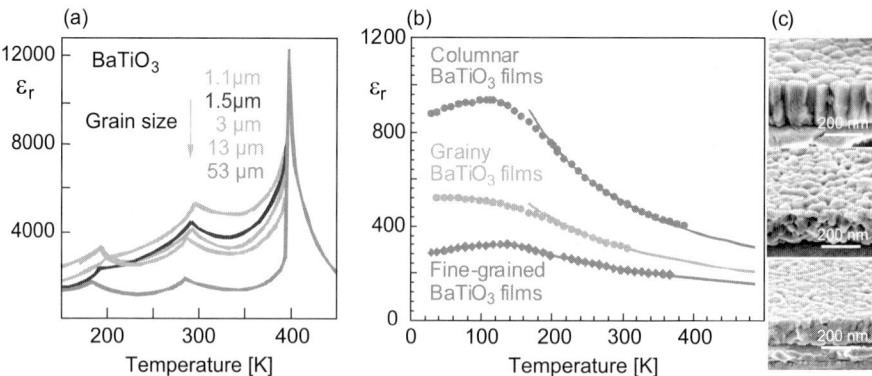

Figure 1.15: $BaTiO_3$: Temperature dependence of the permittivity for (a) bulk ceramics with different grain sizes and (b) thin films with different grain sizes and (c) microstructure of thin films.

For BaTiO$_3$ thin films a significant increase in the room temperature permittivity from 500 to 900 is observed which was induced by the change in the morphology from a granular to a columnar microstructure (Figure 1.15 (b) and (c)). In contrast to BaTiO$_3$ bulk ceramics, which exhibit a paraelectric to ferroelectric phase transition with decreasing temperature accompanied by a sharp peak of the permittivity at around 123° C, only a broad maximum in the permittivity vs. temperature curve is observed for polycrystalline thin films. Additionally, the BaTiO$_3$ thin films do not show a ferroelectric hysteresis at room temperature. While the absence of a remanent polarization is typical for paraelectric material, the grain size dependence indicates a superparaelectric behavior of BaTiO$_3$ thin films. Compared to bulk ceramics, thin BaTiO$_3$ films of the same average grain size show a significantly lower permittivity, although the grain size dependence is still observed. The difference in the absolute permittivity values may be explained by a combination of thin film effects, which result in a further decrease of the permittivity, as for example film-electrode interfaces and stress effects.

1.5.3 Influence of Substitutes and Dopants

The compositions of most dielectric materials used for ceramic capacitors are based on ferroelectric barium titanate. As discussed in detail in Pragraph 1.3 the permittivity of ferroelectric perovskites shows marked changes with temperature, particularly close to the phase transition. From the device point of view a high dielectric permittivity with stable properties over a wide temperature range is required. There are various specifications which have to be fulfilled (e.g. X7R: $\Delta C/C(T = 25°C) < \pm 0.15$ in a range between $-55°C$ and $125°C$).

By substitution or doping it becomes possible to taylor the ferroelectric materials to different properties. If an ion of the perovskite structure ABO$_3$ is replaced by a different ion of the same valence, isovalent doping is present, e.g. on the A^{2+}-site $(Ba_{1-x}Sr_x)TiO_3$, $(Ba_{1-x}Pb_x)TiO_3$, ... or on the B^{4+}-site $Ba(Sn_xTi_{1-x})O_3$,... . Aliovalent doping exists when donors like La^{3+} on A-site or Nb^{5+} on B-site as well as acceptors like Ni^{2+} or Mn^{3+} on B-site are incorporated in the crystal. At least the site occupancy is determined by the size and the valence of the dopant, e.g. Ca^{2+} may be isovalent A-site or acceptor-type on B-site.

Diffuse Phase Transitions

Doping generally leads to a shift of the phase transition temperature. In Figure 1.16 the change of the transition temperatures is plotted in dependence of the dopant concentration for different dopants in BaTiO$_3$. It is found that Pb-doping stabilizes the tetragonal phase in the sense of their existence over a wide temperature range whereas Sr-doping destabilized the tetragonal phase. This behavior could be understood by the fact that the larger Pb-ions on the A-site form a larger cavity giving the central ion more space for off-center positions than the Sr-ions. In case of high dopant concentrations sequence of the phase transitions in BaTiO$_3$ tends to be supressed.

For $Ba(Zr_xTi_{1-x})O_3$ all phase boundaries meet at a Zr content x≈0.18, see Figure 1.16. Because of the superposition of the particular phase transitions the resulting transition becomes diffuse with a broad maximum of the dielectric permittivity as shown in Figure 1.16. Therefore, this composition has the potential as suitable temperature-stable dielectric for ceramic capacitors.

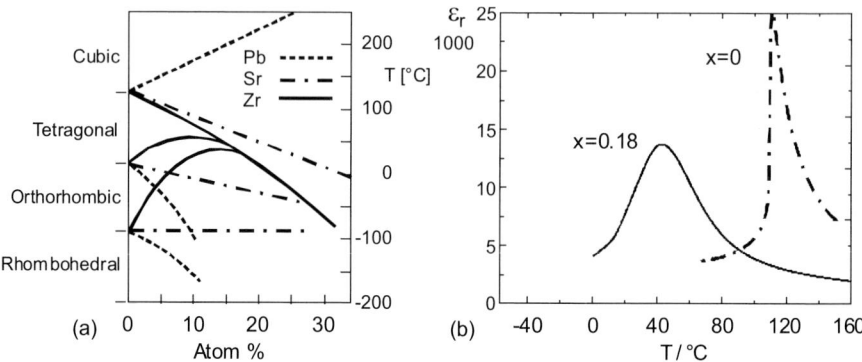

Figure 1.16: (a) Change of the phase transition tempreratures for different dopants and (b) change of a sharp phase transition (BaTiO3) to a diffuse phase transition observed in a broadening of the dielectric peak

Core-Shell Structures

A different approach to overcome the temperature instabilities of the dielectric coefficient is the addition of more complex compositions (than doping with atoms) as $CdBi_2Nb_2O_9$ [12]. By controlling the reaction kinetic of the sintering process the microstructure exhibits a grain core-grain shell structure. The core consists of $BaTiO_3$, and the perovskite material in the shell shows a mixture of $BaTiO_3$ with the complex perovskites $Ba(Bi_{1/2}Nb_{1/2})O_3$ and $Ba(Cd_{2/3}Nb_{2/3})O_3$ having an approximate Curie point at $T_c \approx -80°C$. Figure 1.17 displays schematically the ferroelectric core and the paraelectric shell. The material fulfills the X7R specification for its dielectric behavior. The chemical inhomogeneity emerges during a process of reactive liquid-phase sintering. Application of too-high sintering temperatures leads to uniform distributions of the additives via solid-state diffusion and to the loss of the X7R characteristic.

Relaxors

Relaxor materials as $Pb(Mg_{1/3}Nb_{2/3})O_3$ (PMN), $Pb(Zn_{1/3}Nb_{2/3})O_3$ (PZN), and $(Pb_{0.92}La_{0.08})(Zr_{0.7}Ti_{0.3})O_3$ (PLZT) are a subgroup of ferroelectrics with diffuse phase transistions. Characteristic behavior of this class are the strong dielectric dispersion related with high dielectric losses, see Figure 1.18.

The disturbance of the long-range dipolar interactions in relaxor materials by translational lattice inhomogeneities, defects or segregation of chemical order-disorder leads to micropolar regions (clusters) with typical diameters of $\approx$ 10 nm to 100 nm, which are surrounded by a paraelectric phase. Over a large temperature regime this metastable phase exists. By mechanical stresses or electric fields the ferroelectric state is induced because of the growth of the micropolar regions.

The dynamic of the cluster boundaries explains also the strong dielectric response under small ac fields. The naturally diffuse behavior (temperature stability) with high dielectric

1.5 Ferroelectric Materials

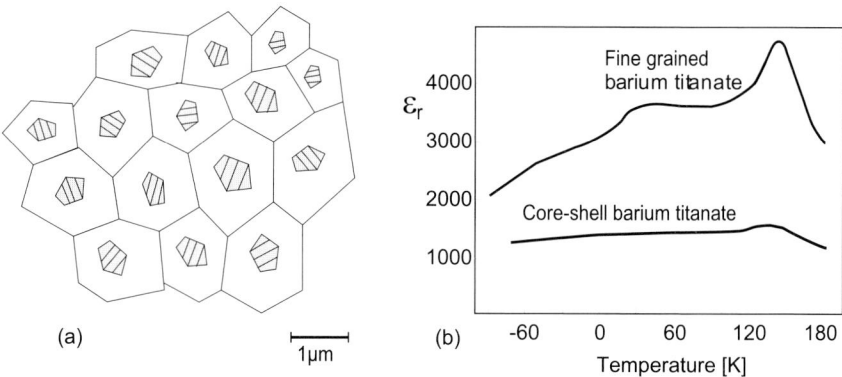

Figure 1.17: (a) Schematic view of a core-shell structure and (b) Temperature behavior of the dielectric coefficient of BaTiO$_3$ doped with CdBi$_2$Nb$_2$O$_9$ and hot-pressed fine-grained BaTiO$_3$ (grain size ≈ 0.5 μm)

coefficient as well as long-term stability make the class of relaxors promising candidates as dielectric in capacitors. Furthermore, relaxors have great technical importance because of their electro-mechanical properties. Especially in single crystals extremely high strains are achieved with small hysteresis.

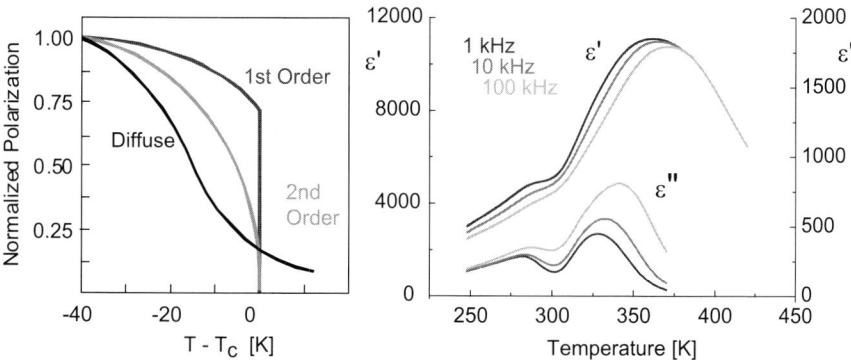

Figure 1.18: (a) Normalized polarization for first-order, second-order and diffuse phase transition in ferroelectric and relaxor materials and (b) dielectric behavior of relaxor-type (Pb$_{0.92}$La$_{0.08}$)(Zr$_{0.7}$Ti$_{0.3}$)O$_3$ (PLZT)

1.6 Ferroelectric Domains

When a ferroelectric single crystal is cooled below the phase transition temperature the electrical stray field energy caused by the non-compensated polarization charges is reduced by the formation of ferroelectric domains, see Figure 1.19. The configuration of the domains follows a head-to-tail condition in order to avoid discontinuities in the polarization at the domain boundary, $\nabla \vec{P} = \sigma$. The built-up of domain walls, elastical stress fields as well as free charge carriers counteract the process of domain formation. In addition, an influence of vacancies, dislocations and dopants exists.

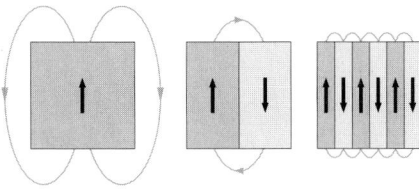

Figure 1.19: Reduction of electrical stray field energy by domain formation

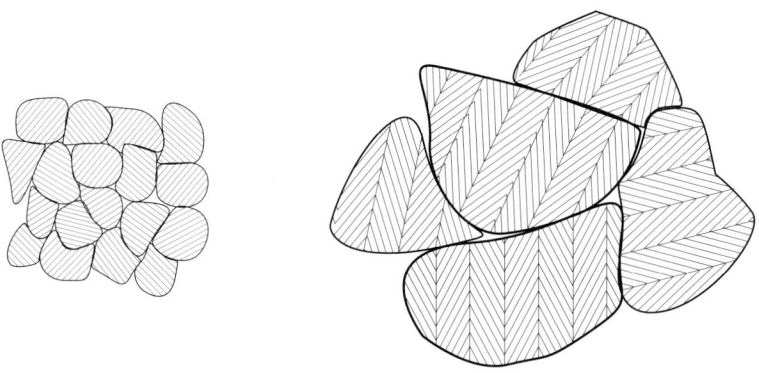

Figure 1.20: Scheme of domain pattern of fine grained BaTiO$_3$ ceramic (left) and coarse grained BaTiO$_3$ ceramic (right)

In polycrystalline bulk ceramics the pattern of domains is quite different because the domain structure of each grain is formed under elastic clamped conditions by its surrounding neighbors, whereas a single crystal is free [11]. It should be noted that only non-180° domains, i.e. 90° domains (for tetragonal structures) or 71° and 109° domains (for rhombohedral structures), have the potential to reduce elastic energy. There exists two types in coarse grained BaTiO$_3$, called herringbone and square net pattern. The first one is by far the most common in

1.6 Ferroelectric Domains

unpoled ceramics. As shown in Figure 1.20, by decreasing the grain size the domain pattern changes from a banded to a laminar structure [13].

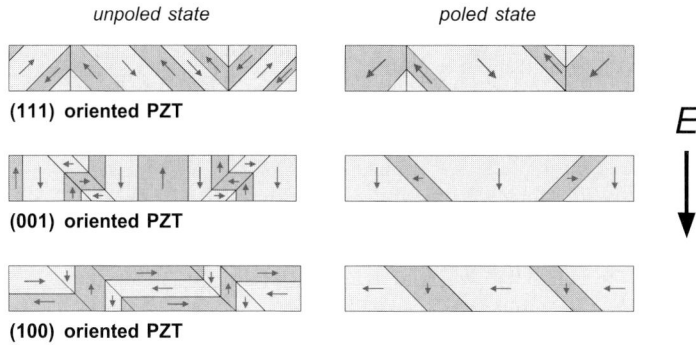

Figure 1.21: Domain structures of tetragonal PZT with different orientations.

Real ferroelectric thin films show polydomain patterns. In Figure 1.21 possible domain configurations of different textures of tetragonal films of PbZr$_{1-x}$Ti$_x$O$_3$ with x > 0.48 are depicted. For compressive stress the polarization is predominantly out-of-plane, (001), oriented. 90° as well as 180°-domains are expected. Such orientation could be realized by deposition of tetragonal PZT on magnesium oxide substrates [15]. Under the influence of an electric field the number of 180°-domains is decreased. The resulting pattern predominantly consists in 90° domains. A (100)-orientation, i. e. in-plane orientation of the polarization, is caused by tensile stress and is achieved by using a buffer layer of yttrium stabilized zirconium and an oxide electrode of lanthanum strontium cobaltate or by depositing on a (100)-SrTiO$_3$-substrate with SrRuO$_3$ electrode [14]. The change of the domain structure by poling is similar to the (001)-orientation but, the a-axis orientation is still preferred. In standard systems for ferroelectric thin films, e.g. PZT with platinum electrodes on oxidized silicon wafers, the orientation of the crystallographic axes of PZT is in (111)-direction. Poling should evoke the single domain state while the "head-to-tail" configuration is required. However, there are a lot of indications that non-180° domain walls are generally immobile. It could be shown that these ferroelastic domain walls become able to move, when the 2-D clamping of the film on the substrate is annuled by patterning the ferroelectric film into discrete islands using a focused ion beam [29]. Further indications are discussed in the next section.

A tool to visualize the domains in thin films is the 3-D piezoresponse force microscope (PFM, see also Chapter 12). Figure 1.22 shows an epitaxial PZT thin films grown on a (001) single crystalline SrTiO$_3$ substrate coated with La$_{0.5}$Sr$_{0.5}$CoO$_3$ oxide layer. A detailed analysis point out that a self-polarized polarization mechanism is observed because the out-of-plane polarization in c-domains are preferentially orientated towards the bottom electrode, i. e. a remanent polarization exists without applying an external field. The domain configuration is always of the "head-to-tail" type.

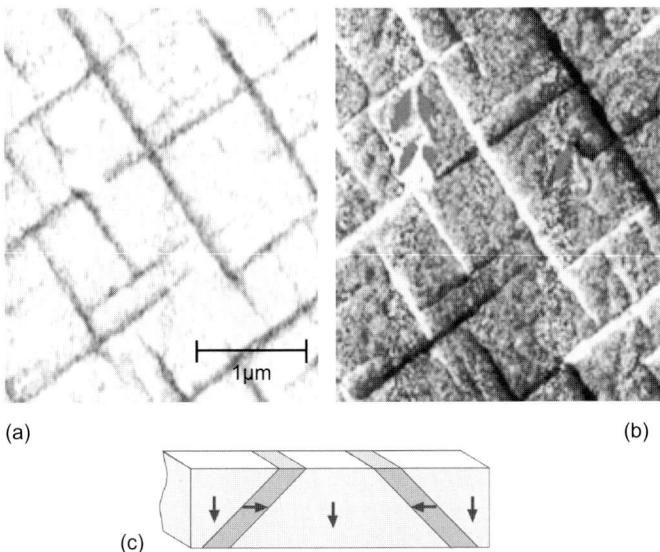

Figure 1.22: (a) Out-of-plane and (b) in-plane piezoresponse images of as-grown epitaxial PZT. The contrast corresponds to different orientations of polarizations in the domains.

1.6.1 Reversible and Irreversible Polarization Contributions

To characterize ferroelectric materials usually the dependence of the polarization on the applied voltage is measured by means of a Sawyer-Tower circuit or by recording the current response to a voltage step. The $P(V)$-hysteresis curve is used to determine the remanent polarization and coercive voltage, respectively coercive field. These two parameters are of critical importance to the design of external circuits of FeRAMs.

The ferroelectric hysteresis originates from the existence of irreversible polarization processes by polarization reversals of a single ferroelectric lattice cell (see Section 1.4.1). However, the exact interplay between this fundamental process, domain walls, defects and the overall appearance of the ferroelectric hysteresis is still not precisely known. The separation of the total polarization into reversible and irreversible contributions might facilitate the understanding of ferroelectric polarization mechanisms. Especially, the irreversible processes would be important for ferroelectric memory devices, since the reversible processes cannot be used to store information.

For ferroelectrics, mainly two possible mechanisms for irreversible processes exist. First, lattice defects which interact with a domain wall and hinder it from returning into its initial position after removing the electric field that initiated the domain wall motion ("pinning") [16]. Second, the nucleation and growth of new domains which do not disappear after the field is removed again. In ferroelectric materials the matter is further complicated by defect dipoles and free charges that also contribute to the measured polarization and can also interact with domain walls [17]. Reversible contributions in ferroelectrics are due to ionic and electronic

1.6 Ferroelectric Domains

displacements and to domain wall motions with a small amplitude. These mechanisms are very fast. The reorientation of dipoles and/or defect or free charges also contributes to the total polarization. These mechanisms are usually much slower, but they also might be reversible (relaxation).

A domain wall under an external electric field moves in a statistical potential generated by their interaction with the lattice, point defects, dislocations, and neighboring walls. Reversible movement of the wall is regarded as a small displacement around a local minimum. When the driven field is high enough, irreversible jumps above the potential barrier into a neighboring local minimum occur (see Figure 1.23).

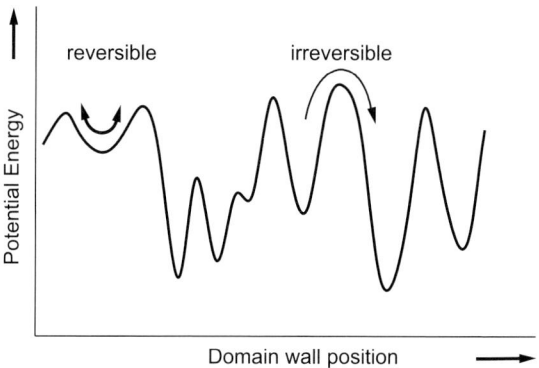

Figure 1.23: Movement of a domain wall in the lattice potential.

Based on these assumptions the measurement of the large signal ferroelectric hysteresis with additional measurements of the small signal capacitance at different bias voltages are interpreted in terms of reversible and irreversible parts of the polarization. As shown for ferroelectric thin films in Figure 1.24, the separation is done by substracting from the total polarization the reversible part, i. e. the integrated $C(V)$-curve [18].

$$P_{irr}(V) = P_{tot}(V) - \frac{1}{A} \int_0^V C(V')\, dV' \tag{1.21}$$

Analogous $C(V)$ curves were recorded on PZT bulk ceramics with compositions around the morphotropic phase boundary (MPB). Figure 1.25 displays the relative permittivity as a function of DC-bias for a tetragonal ($x = 0.48$), a morphotropic ($x = 0.52$) and a rhombohedral ($x = 0.58$) sample. In contrast to thin films additional "humps" observed in the $\varepsilon(E)$ curves. This could be explained by different coercive fields for 180° and non-180° domains [31]. Their absence in ferroelectric thin films could be taken as evidence for suppressed non-180° domain switching in thin films [30].

A further approach to separate the reversible and irreversible 90° and non-90° contributions is the investigation of the piezoelectric small and large signal response of the ferroelectric

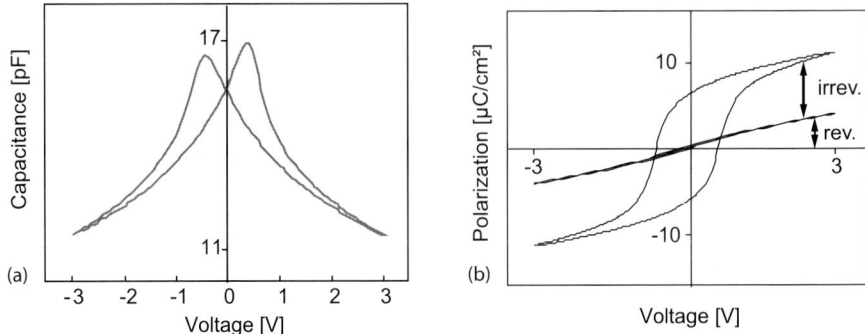

Figure 1.24: (a) $C(V)$ - curve and (b) reversible and irreversible contribution to the polarization of a ferroelectric SBT thin film.

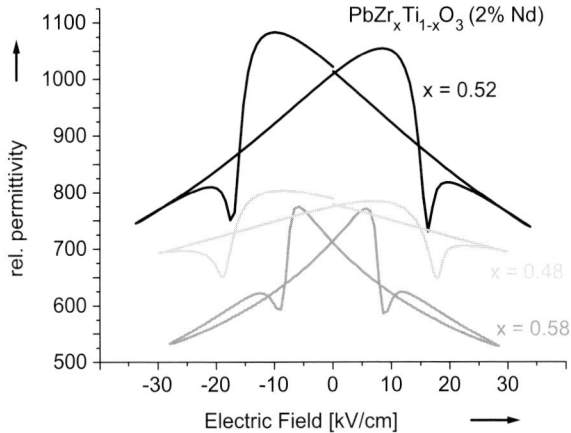

Figure 1.25: Relative permittivity of 2% Nd-doped PbZr$_x$Ti$_{1-x}$O$_3$ bulk ceramics.

material. While both 180° and non-180° walls contribute to the permittivity, only non-180° walls affect the piezoelectric properties. A displacement of a 180° wall does not change the strains and thus yields no piezoelectric response.

Analogue to the dielectric case, the reversible contribution to the strain (see Equation 1.1) can be determined by the integration of the piezoelectric small signal coefficient d_{3j} over the applied bias field.

$$x_{3j,rev}(E_{\text{bias}}) = \int_0^{E_{\text{bias}}} d_{3j}(E_3') \, dE_3' \tag{1.22}$$

Figure 1.26 displays the results of the d_{33} coefficient as function of the bias field, the large signal x_{33} response ("butterfly loop") as well as constructed $x_{33,rev}$ curve of the same PZT

1.6 Ferroelectric Domains

thin film. The curves were taken by double-beam laser interferometry having a resolution better than 1 pm [33]. In contrast to bulk ceramics almost the complete strain response of the film appeared to originate from reversible processes. Since the polarization response of the film was determined by both 180° and non-180° domain wall motion and the piezoelectric response was solely due to non-180° boundaries, the presented results are evidence that most reversible domain wall motions in ferroelectric thin films are due to reversible motion of non-180° domain walls. The clamping effect of the substrate which entails rather stringent mechanical boundary conditions apparently only allows for minute motions of the non-180° walls, which immediately return to their initial positions when the external electric field that initiated the motion is returned to zero.

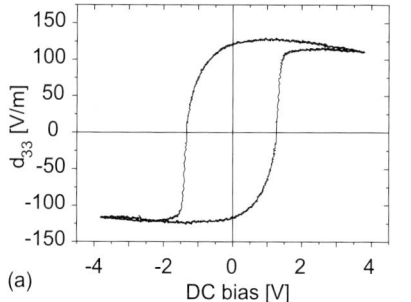
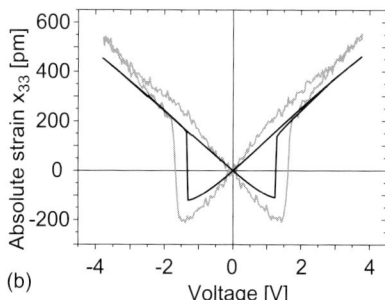

Figure 1.26: (a) Piezoelectric coefficient d_{33} and (b) "Butterfly" loop and integrated d_{33} response of a PZT 45/55 film.

1.6.2 Ferroelectric Switching

The polarization reversal in single crystals has been intensively investigated by direct observation of the formation and the movement of the domain walls. E.g., in $BaTiO_3$ single crystals it was found by Merz [19], and Fousek [20] that in response to a voltage step the process happens by forming of opposing 180° or orthogonal 90° domains in the shape of needles and wedges. Both, the resulting maximum displacement current i_{max} as well as the switching time t_s, which is the most significant quantity and describes the duration of the polarization reversal, were measured as a function of the applied field E and follow empirical laws

$$i_{max} = i_0 \cdot \exp(-\alpha/E), \tag{1.23}$$

$$t_s = t_0 \cdot \exp(\alpha/E), \tag{1.24}$$

where i_0 and t_0 are a constants. The constant activation field α in both equations is the same [21].

The above mentioned equations are only applicable when the applied field E is constant during the polarization reversal, i.e., the time constant of the dielectric charging τ_{RC} must be much smaller than the switching time t_s. The dielectric charging is determined by the capacitance of the sample and inevitable series resistors (source, lines etc.). The switching time is determined by many factors, the domain structure, the nucleation rate of opposite domains, the mobility of the domain walls, and many others. As shown in Figure 1.27 the condition is fullfilled and the dielectric charging is clearly separated from the polarization switching hump.

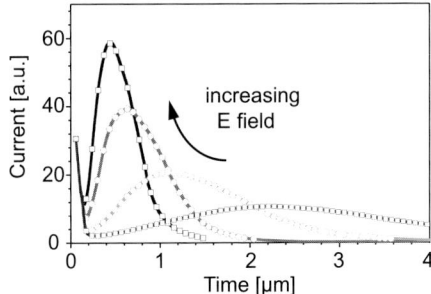

Figure 1.27: Polarization switching current of TGBF single crystal with decreasing electrical field.

Applying voltage steps on capacitors of PZT thin films up to voltages above the coercive voltage, the current response could behave differently, even if the switching is complete. In Figure 1.28 the responses of two $PbZr_{0.3}Ti_{0.7}O_3$ thin films (undoped and 1 % Nb-doped) are depicted. It depends on the squareness of the hysteresis loop whether the current shows the typical ferroelectric behavior with maximas $i_{max}(E)$ as known from single crystals or not.

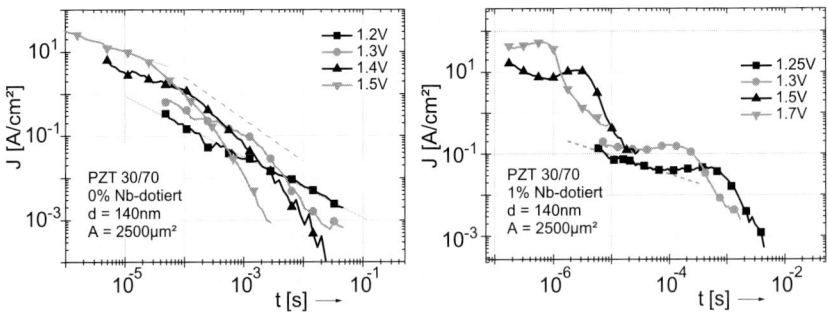

Figure 1.28: Current response of undoped and 1% Nb-doped $PbZr_{0.3}Ti_{0.7}O_3$ after different voltage steps.

During polarization switching, the current response of the undoped film clearly shows a typical Curie-von-Schweidler behavior

$$i(t) = i_o t^{-\kappa} \qquad (1.25)$$

where κ is a constant and has a value of less or equal to one.

The Curie-von-Schweidler behavior is generally found in all thin films, in dielectric [25] as well as in ferroelectric thin films [26]. The physical mechanism of the Curie-von-Schweidler relaxation could be based either on the fact that the hopping motion of a charged particle always affects the motion of the neighboring charges [27] or on the variation in charge transport barrier, e. g. at grain boundaries leading to a large distribution of relaxation times [28]. Independent of the physical origin the Curie-von-Schweidler behavoir leads to a different view of the polarization reversal.

Using a simplified picture of a real ferroelectric capacitor, consisting of an ideal ferroelectric and a non-ferroelectric interface layer with a strong dielectric dispersion, the current response can be modeled. It can be shown that the dielectric interface has a strong influence on the switching process and the delay of the polarization reversal is determined rather by dielectric dispersive polarization mechanisms than by the real ferroelectric switching. These interface effects could also provide an explanation for the frequency dependence of the coercive field in ferroelectric thin films which have a different origin compared with single crystals. Therefore, the frequency dependence of the coercive field depends on the ferroelectric itself and the electrode material. In $SrBi_2Ta_2O_9$ with Pt electrodes the exponential factor s of the empirical relation $E_C \propto f^{-\beta}$ [32] is about 0.1 in a raw approximation, while in $PbZr_{0.3}Ti_{0.7}O_3$ with Pt electrodes it is about an order of magnitude less.

Bibliography

[1] B. C. Fraser, H. Danner, R. Papinsky, *Phys. Rev.* **100**, 745 (1955).

[2] N. W. Ashcroft, N. D. Mermin, D. Mermin, *Solid State Physics, Holt*, Rinehart and Winston, New York, 1976.

[3] C. Kittel, *Introduction to Solid State Physics*, Wiley, New York, 1996.

[4] R. P. Feynman, *The Feynman Lectures on Physics 'Mainly Electromagnetism and Matter'*, Calif. Addison-Wesley, Redwood City, 1989.

[5] R. E. Newnham, *Structure-Property Relations*, Springer-Verlag, New York, 1975.

[6] B. Jaffe, W. Cook, and H. Jaffe, *Piezoelectric Ceramics*, Academic Press, London, 1971.

[7] L. Brillouin, *Wave Propagation in Periodic Structures*, Dover Publications, New York, 1953.

[8] R. E. Newnham, R. W. Wolfe, and J. F. Dorrian, *Mat. Res. Bull.* **6**, 1029 (1971).

[9] M. Shaw, S. Trolier-McKinstry, and P. C. McIntyre, *Annu. Rev. Mater. Sci.* **30**, 263 (2000).

[10] N. Pertsev, A. Zembilgotov, A. Tagantsev, *Phys. Rev. Lett.* **80**, 1988 (1998).

[11] G. Arlt, D. Hennings, and G. de-With, *J. Appl. Phys.* **58**, 1619 (1985).

[12] D. Hennings and G. Rosenstein, *Jour. Amer. Cer. Soc.* **67**, 249 (1984).

[13] G. Arlt, *J. Mat. Sci.* **25**, 2655 (1990).

[14] K. Nagashima, M. Aratani, and H. Funakubo, *J. Appl. Phys.* **89**, 4517 (2001).

[15] K. Nashimoto, D. K. Fork, and G. B. Anderson, *Appl. Phys. Lett.* **66**, 822 (1995).

[16] T. J. Yang, V. Gopalan, P. J. Swart, and U. Mohideen, *Phys. Rev. Lett.* **82**, 4106 (1999).

[17] O. Boser and D. N. Beshers, M*at. Res. Soc. Symp. Proc.* **82**, 441 (1987).
[18] O. Lohse, D. Bolten, M. Grossmann, R. Waser, W. Hartner, and G. Schindler, *Ferroelectric Thin Films VI. Symposium. Mater. Res. Soc. 1998*, 267 (1998).
[19] W. J. Merz, *Phys. Rev.* **95**, 690 (1954).
[20] J. Fousek and B. Brezina, *Czech. J. Phys., Sect. B* **10**, 511 (1960).
[21] C. F. Pulvari and W. Kuebler, *J. Appl. Phys.* **29**, 1742 (1958).
[22] O. Lohse, M. Grossmann, U. Böttger, D. Bolten, and R. Waser, *J. Appl. Phys.* **89**, 2332 (2001).
[23] Y. Ishibashi and H. Orihara, *Integr. Ferroelectr.* **9**, 57 (1995).
[24] M. Avrami, *J. Chem. Phys.* **7**, 1103 (1939).
[25] M. Schumacher, G. W. Dietz, and R. Waser, *Integr. Ferroelectr.* 10, 231 (1995).
[26] X. Chen, A. I. Kingon, L. Mantese, O. Auciello, and K. Y. Hsieh, *Integr. Ferroelectr.* 3, 355 (1993).
[27] L. A. Dissado and R. M. Hill, *J. Mater. Sci.* 16, 1410 (1981).
[28] H. Kliem, *IEEE Trans. Electr. Insul.* **24**, 185 (1989).
[29] V. Nagarajan et al., *Nature Materials* **2**, 43 (2003).
[30] D. Damjanovic, *Rep. Prog. Phys.* **61**, 1267 (1998).
[31] N. Bar-Chaim, M. Brunstein, J. Grünberg, and A. Seidman, *J. Ap. Phys.* **45**, 2398 (1974).
[32] J. F. Scott, *Ferroelectrics Review* **1**, 1 (1998).
[33] P. Gerber et al., *Rev. Sci. Instr.* **74**, 2613 (2003).

2 Piezoelectric Characterization

Susan Trolier-McKinstry

Materials Science and Engineering Department and Materials Research Institute, Penn State University, USA

Abstract

Piezoelectric coefficients need to be measured accurately over a wide range of temperature, drive field amplitude, and frequency, in order to predict device performance appropriately. There are multiple methods available for such characterization in bulk materials and thin films. This paper overviews some of the standard characterization tools, with an emphasis on the methods utilized in the IEEE Standard on Piezoelectricity. In addition, several of the evolving methods for making accurate piezoelectric coefficient measurements on thin films are reviewed. Some of the common artifacts in piezoelectric measurements, as well as means of avoiding them, are discussed.

2.1 Important piezoelectric constants

The piezoelectric effect entails a linear coupling between electrical and mechanical energies. Numerous piezoelectric coefficients are in use, depending on the electrical and mechanical boundary conditions imposed on the part under test. Each of the piezoelectric d, e, g, and h coefficients can be defined in terms of a direct and a converse effect; the two sets of coefficients are related by thermodynamics. For example, the piezoelectric charge coefficient, d_{ijk}, can be defined via [1]:

$$D_i = d_{ikl}\sigma_{kl} \tag{2.1}$$

where D is the induced dielectric displacement and σ is the applied stress. The same d_{ijk} coefficients are used in the converse piezoelectric effect,

$$x_{ij} = d_{kij}E_k \tag{2.2}$$

where E is the applied electric field and x is the induced strain. Similarly the other piezoelectric coefficients can be defined via [1]:

$$\begin{aligned}\sigma_{ij} &= -e_{kij}E_k & D_i &= e_{ikl}x_{kl} \\ x_{ij} &= g_{kij}D_k & E_i &= -g_{ikl}\sigma_{kl} \\ \sigma_{ij} &= -h_{kij}D_k & E_i &= -h_{ikl}x_{kl}\end{aligned} \tag{2.3}$$

Polar Oxides: Properties, Characterization, and Imaging
Edited by R. Waser, U. Böttger, and S. Tiedke
Copyright © 2005 WILEY-VCH Verlag GmbH & Co. KGaA, Weinheim
ISBN: 3-527-40532-1

The piezoelectric coefficients are third rank tensors, hence the piezoelectric response is anisotropic. A two subscript matrix notation is also widely used. The number of non-zero coefficients is governed by crystal symmetry, as described by Nye [2]. In most single crystals, the piezoelectric coefficients are defined in terms of the crystallographic axes; in polycrystalline ceramics, by convention the poling axis is referred to as the "3" axis.

Because the piezoelectric coefficients can each be expressed in two ways, there are in general two different approaches to measuring the piezoelectric response; approaches based on measurement of charge (or current), and those based on measurements of displacement (or strain). Choice of which coefficient to measure is often a matter of convenience.

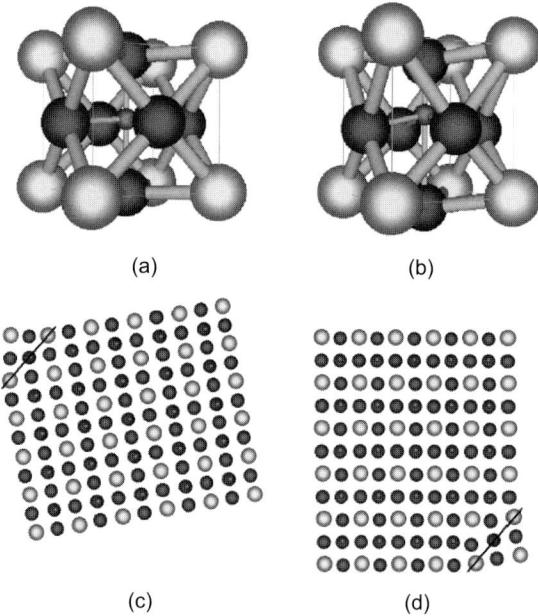

Figure 2.1: Schematic illustrations of intrinsic and extrinsic contributions to the piezoelectric constant of perovskite ferroelectrics. (a) and (b) correspond to the intrinsic unit cell shape (a) without and (b) with applied electric field. (c) and (d) correspond to the extrinsic response associated with the change in position of a non-180° domain wall (shown as a black line) (c) before and (d) after an electric field is applied. Note that both intrinsic and extrinsic responses lead to a change in shape of the material due to application of an electric field (and hence to a piezoelectric response). In both cases, the actual distortions are significantly exaggerated to make visualization easier.

In many ferroelectric materials, the net piezoelectric effect is a result of both intrinsic and extrinsic responses. Here, intrinsic refers to the response that would result from an appropriately oriented single crystal (or ensemble thereof, in a polycrystalline sample). The extrinsic response is typically the result of motion of non-180° domain walls. The principle of these

2.1 Important piezoelectric constants

two contributions is illustrated in Figure 2.1. Because there is a finite spontaneous strain in ferroelectrics, when non-180° domain walls are moved, they carry with them a strain component that contributes to the net sample distortion.

It is also important to realize that piezoelectricity implies a *linear* coupling between dielectric displacement and strain, for example. However, in many ferroelectric materials, this response is linear only over a relatively limited field range (See for example, Figure 2.2). Non-linearity is especially important in ferroelectric materials which show a strong extrinsic contribution to the piezoelectric response [5]. In addition, it is quite common for the response to be hysteretic. The amount of hysteresis that is observed depends strongly on the measurement conditions. Larger amplitude excitations often result in larger extrinsic contributions to the coefficients, and more non-linearity and hysteresis in the response.

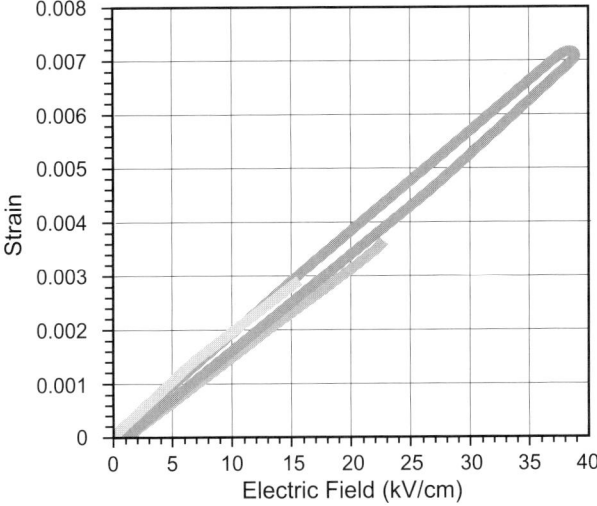

Figure 2.2: Unipolar strain–field curves for a 0.955 PbZn$_{1/3}$Nb$_{2/3}$O$_3$–0.045 PbTiO$_3$ single crystal, shown for different field amplitudes.

Piezoelectric coefficients are also temperature dependent quantities. This is true for both the intrinsic and the extrinsic contributions. Typically, the piezoelectric response of a ferroelectric material increases as the transition temperature is approached from below (See Figure 2.3) [3]. Where appropriate thermodynamic data are available, the increase in intrinsic d_{ijk} coefficients can be calculated on the basis of phenomenology, and reflects the higher polarizability of the lattice near the transition temperature. The extrinsic contributions are also temperature dependent because domain wall motion is a thermally activated process. Thus, extrinsic contributions are lost as the temperature approaches 0 K [4]. As a note, while the temperature dependence of the intrinsic piezoelectric response can be calculated on the basis of phenomenology, there is currently no complete model describing the temperature dependence of the extrinsic contribution to the piezoelectric coefficients.

It should be noted that in practice, the piezoelectric response will typically not continue to rise all the way to the transition temperature, as elevated temperatures induce depoling of the ferroelectric, unless appropriate care is taken to insure that the material remains polarized (e. g. by application of a bias electric field). Depoling of this type is often important at temperatures of $\sim 1/2$ of the Curie temperature, making high transition temperature materials interesting both for the decreased temperature dependence in the response, and the wider use range that can be achieved.

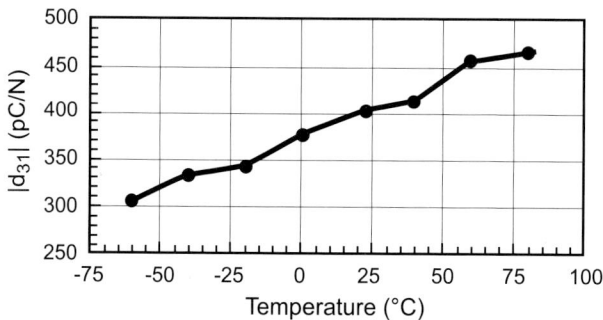

Figure 2.3: The magnitude of d_{31} as a function of temperature in a PZT-5H ceramic measured using a large amplitude electric field, after [3].

Piezoelectric coefficients are also a strong function of the domain state in ferroelectric materials. This is shown in Figure 2.4, where the strain-field curve for two single crystals with different domain states is shown. When the domain state is altered by the applied electric field, large, but very hysteretic, strains can be induced. In general, since hysteresis results in temperature rises in an actuator, lower hysteresis is preferred in the piezoelectric. It should also be noted that in cases where the domain state is unstable, the piezoelectric coefficient can also change with time, resulting in aging of the coefficients. This is typical in ceramic samples. Typical practice is to report the magnitude of the small signal piezoelectric response 1 day after the sample was poled, to minimize the impact of aging.

Finally, while the piezoelectric d, e, g, and h constants are typically reported as real numbers, there is increasing use of the fact that the material response is not always in phase with the applied field. This can be due to a variety of factors, including domain wall motion in ferroelectrics [5]. Thus, coefficients can be described as complex quantities. Discussions of how to measure these constants are given in [6–10].

In summary, piezoelectric coefficients are complex numbers that depend on the measurement frequency, excitation field, temperature, and time (e. g. time after poling in samples that show finite aging rates). Consequently, in reporting piezoelectric data, it is important to specify how the property was measured.

2.2 Measurements in bulk materials

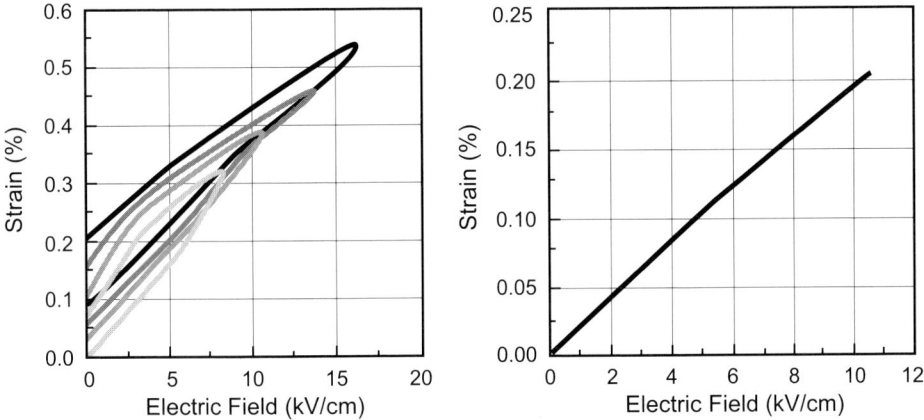

Figure 2.4: Strain-field curves for <001> oriented $0.91PbZn_{1/3}Nb_{2/3}O_3$–$0.09PbTiO_3$ single crystals. The sample in (a) was poled at room temperature, where the resulting domain state is unstable (due to induction of tetragonal material associated with the curved morphotropic phase boundary), yielding substantial hysteresis. In (b) the crystal was poled at low temperatures to keep it in the rhombohedral phase. When measured at room temperature, the piezoelectric response is much more linear and non-hysteretic, due to the improved stability of the ferroelectric domain state. Data courtesy of S. E. Park.

2.2 Measurements in bulk materials

An excellent reference describing appropriate ways of measuring the piezoelectric coefficients of bulk materials is the IEEE Standard for Piezoelectricity [1]. In brief, the method entails choosing a sample with a geometry such that the desired resonance mode can be excited, and there is little overlap between modes. Then, the sample is electrically excited with an alternating field, and the impedance (or admittance, etc.) is measured as a function of frequency. Extrema in the electrical responses are observed near the resonance and antiresonance frequencies. As an example, consider the length extensional mode of a vibrator. Here the elastic compliance under constant field can be measured from

$$s_{11}^E = \frac{1}{4}\rho f_s^2 l^2 \qquad (2.4)$$

where s_{11}^E is the elastic compliance measured at constant electric field, ρ is the sample density, f_s is the frequency of maximum conductance, and l is the length. Then, the coupling coefficient can be determined from Equation (2.5), where f_p is the frequency or maximum resistance and Δf is the difference between the $f_p - f_s$. Once both of those parameters are known, the d_{31} coefficient can then be calculated from Equation (2.6), where ε_{33}^T is the dielectric constant measured at constant stress.

$$\frac{(\hat{k}_{31}^t)^2}{1-(\hat{k}_{31}^t)^2} = \frac{\pi}{2}\frac{f_p}{f_s}\tan\frac{\pi}{2}\frac{\Delta f}{f_s} \qquad (2.5)$$

$$(\hat{k}^t_{31})^2 = \frac{d^2_{31}}{\varepsilon^T_{33} s^E_{11}} \tag{2.6}$$

The IEEE Standard on Piezoelectricity [1] describes these measurements in considerable detail, and gives the necessary sample geometries for determination of a number of the piezoelectric constants. For example, the relations that enable determination of the other common piezoelectric moduli in bulk ceramics are given in Equation (2.7).

$$k_{15} = \frac{d_{15}}{\sqrt{s^E_{44}\varepsilon^T_1}} \qquad k_{33} = \frac{d_{33}}{\sqrt{s^E_{33}\varepsilon^T_3}} \qquad k_{31} = \frac{d_{31}}{\sqrt{s^E_{11}\varepsilon^T_3}} \tag{2.7}$$

In most cases, resonance measurements are made at low excitation levels, so that small signal numbers are derived. Methods of extending resonance methods to higher powers are discussed in [11].

It is important to note that measurement of the complete set of piezoelectric, elastic, and dielectric constants thus requires a number of samples. In order for the data to be self-consistent, each of these samples must have comparable poling states, something which may or may not be easy to achieve in practice. Moreover, difficulties can arise with conventional resonance-based methods for materials with unusually high piezoelectric constants and coupling coefficients [12]. These difficulties can be ameliorated by introduction of additional characterization tools, e. g. ultrasonic measurements of the elastic and piezoelectric constants [13]. Ultrasonic measurements typically entail launching a sound wave into the material of interest, and measuring the flight time required for the sound wave to propagate across the sample. The elastic constants can then be extracted from the measured wave velocities. Combined data sets of this type enable the self-consistency of the measurements to be verified, and the data weighted appropriately to improve the accuracy [12].

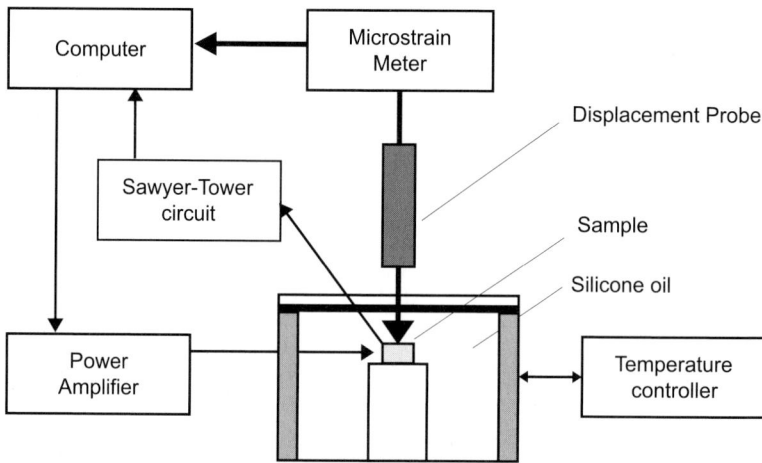

Figure 2.5: Schematic measuring system for simultaneous measurements of sample polarization and displacement. After [15].

2.2 Measurements in bulk materials

Quasi-static measurements of the piezoelectric coefficient can be made a number of different ways. Figure 2.5 shows a typical schematic for simultaneous characterization of the converse piezoelectric coefficient and polarization of a bulk specimen. Measurements of the piezoelectric coefficient are made by applying an electric field to an electroded sample and determining the field-induced displacement of the surface using one of a variety of sensors (i.e. an LVDT, a DVRT, or an optical probe). Alternatively, in some cases the strain is measured by a strain gauge. Because the displacements involved are typically rather small, the ability to accurately track deflection is important. Charge amplification electronics can be used to simultaneously detect the charge associated with the electric field-induced polarization. In this way, the observed strain can be correlated unambiguously to the polarization change. Numerous manifestations of such equipment have been reported in the literature, some of which are fixtured to allow the measurements to be made as a function of temperature or applied stress levels [14–16].

In such a measurement, the sample is clamped as lightly as possible, and the displacement of the surface in monitored. The amount of sample clamping is important, because the mechanical constraints can impact the ferroelastic response of the sample. That is, in samples where the mechanical coercive stress is low, it is possible to change the domain state of the material by improperly clamping it in the sample fixture. This is especially important in elastically soft piezoelectrics, such as many of the relaxor ferroelectric $PbTiO_3$ single crystals.

Optical methods can also be utilized to track the displacement of samples surfaces. Various types of interferometers have been employed for measurements of this type, yielding accuracy in the surface displacement between ~0.1 and 10^{-5} Å [6, 9, 15]. Both single and double beam versions have been developed for measuring the d_{33} coefficient of bulk samples. Cubic samples can avoid some of the difficulties associated with sample flexure on actuation [9]. In some cases, one of the probing beams can also be brought to the side of the sample, so that lateral and tangential displacements can be tracked simultaneously [15].

Fundamentally, interferometric measurements entail bouncing a beam of coherent light from the sample surface, and then interfering the reflected beam with a reference beam. The light intensity is then a function of the displacement of the sample surface. When an alternating electric field is applied to the sample, then a lock–in amplifier can be used in conjunction with the detector to yield a precise measurement of the field–induced deflection of the sample surface.

For a single beam interferometer, the light intensity around the quarter wave point (where the accuracy of the measurement is high) is given by [9]

$$I = \frac{1}{2}(I_{\max} + I_{\min}) + \frac{1}{2}(I_{\max} - I_{\min}) \sin \frac{4\pi \Delta d}{\lambda} \qquad (2.8)$$

where I is the measured light intensity, I_{max} and I_{min} are the light intensities for complete constructive and destructive interference of the light, Δd is the sample displacement, and λ is the wavelength of the light utilized.

In such measurements, it is very important to specify the excitation level and the frequency at which the measurement was made. As can be seen in Figure 2.2, even when the sample hysteresis is low, the piezoelectric coefficients of materials are field-dependent quantities. Thus, there is some curvature of the strain–field response. Typically the "low–field" response is defined as the piezoelectric coefficient measured near the origin (often at field levels $< 1/10^{\text{th}}$

of the coercive field of ferroelectric compositions). Clearly, however, these low field coefficients are not necessarily the appropriate ones to describe high power actuators, where a more substantial part of the curve is traced out. The calculated piezoelectric coefficient will also depend on the manner in which it was derived; instantaneous slopes of the strain–field curves differ from linear fits through the data endpoints. The measurement frequency is impacted by factors such as power supply limitations, mechanical resonances of the samples or fixturing, and whether or not "free" or "clamped" coefficients are desired [1].

Piezoelectric measurements in bulk samples can also be made by applying an alternating stress and measuring the charge developed. Recent work in this area has been reported by Damjanovic and co–workers using a dynamic press apparatus [17]. In that work, the importance of measurement frequency and stress amplitude was also clearly pointed out.

Other common means for determining the direct d_{33} coefficients of bulk samples include Berlincourtstyle approaches. Berlincourt meters are available commercially from several sources. In most cases, the sample to be measured is mechanically clamped between jaws with pressures on the order of a few N. The charge output due to a small mechanical oscillation (forces $\sim$0.1–0.3 N) is then determined. It is important to note that this technique is appropriate for measurements of bulk samples with stable domain states, only. Measurement accuracy is also better when highly resistive samples are used.

There is less consensus in the literature about the best means of reporting the hysteresis in the sample response, especially at high field drive levels. For relatively modest levels of hysteresis, Rayleigh and modified Rayleigh approaches have been utilized [17, 18]. The Rayleigh treatment gives the amplitude dependence of the piezoelectric coefficients as

$$d_{33}(\sigma_o) = d_{init} + \alpha\sigma_o \qquad (2.9)$$

This is valid under relatively small signal excitation conditions, and describes the motion of domain walls in local random fields. α describes the irreversibility of the domain wall motion. Under the conditions where the Rayleigh model holds, the hysteresis in the piezoelectric response is then given by

$$Q = (d_{init} + \alpha\sigma_o)\sigma \pm \frac{\alpha}{2}(\sigma_o^2 - \sigma^2) \qquad (2.10)$$

In other cases, the large signal dielectric loss [19] (calculated as in Figure 2.6) is reported as a measure of the energy loss per cycle. Work is currently on–going on preparation of an IEEE standard in this area.

2.3 Measurements in thin films

There are several methods that have been implemented for measurements of thin film piezoelectric coefficients. In looking at the relative accuracy of the techniques, it is important to consider the differences between thin film and bulk samples. Firstly, thin films are, by definition, thin. Since the maximum strain that can typically be generated is $\leq 1\,\%$, the film displacements, themselves are quite small. The second, rather obvious, difference is that thin films are not amenable to fabrication of unsupported samples. A limited amount of information is available on the piezoelectric coefficients of thin films released by dissolving away the

2.3 Measurements in thin films

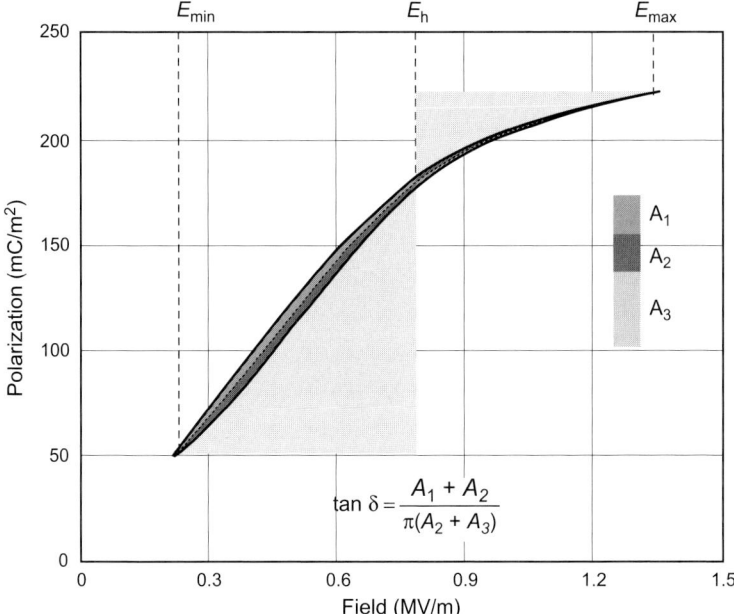

Figure 2.6: Calculation of large signal loss tangent, after [19].

substrate. In some cases, the resulting structures show piezoelectric coefficients comparable to bulk materials [20]. However, measurements of this type are rather rare, and the resulting piezoelectric coefficients should differ from those of the same films on a substrate. Lefki and Dormans also pointed out that one consequence of this is that the effective piezoelectric coefficients for the direct and converse effects may be different in films [23].

Far more frequently, however, measurements are made with a relatively thin film (often $< 5\,\mu m$) on a comparatively massive substrate (often $> 400\,\mu m$ in thickness). In principle, the piezoelectric and electromechanical properties of the films can be extracted by fitting conventional resonance measurements. In practice, multiple resonances due to the film/substrate system are measured; it is then necessary to deconvolute the data in order to extract information on the films itself. In practice, this is difficult to do unambiguously [21, 22].

As a result of the presence of the supporting substrate, the piezoelectric coefficients reported for films are usually effective numbers, rather than the free piezoelectric coefficients typically reported for bulk samples [23]. The two coefficients most widely measured are the effective $d_{33,f}$ coefficient and the $e_{31,f}$ coefficients given in Equation (2.11).

$$e_{31,f} = \frac{d_{31}}{s_{11}^E + s_{12}^E} \qquad d_{33,f} = d_{33} - \frac{2s_{13}^E}{s_{11}^E + s_{12}^E}d_{31} \qquad (2.11)$$

Here the subscript f denotes the property measured for the film. It is clear from an inspection of the equations that $d_{33,f}$ is lower than the value which would be obtained from an unconstrained d_{33}, due to the substrate clamping. To illustrate this, consider a thin film electroded on both surfaces, on a substrate. When an electric field is applied parallel to the spontaneous polarization in a perovskite thin film, the film attempts to expand out–of–plane, and contract laterally. The motion is, however, partially clamped by the massive (and passive) substrate. Thus, the d_{33} response of the film is reduced relative to that of a free sample.

In many microelectromechanical systems (MEMS) based on piezoelectric thin films, flexure is deliberately used to amplify the available displacements (or alternatively to increase the sensitivity of a sensor). For simplicity (and to keep poling and actuation voltages low), films are often poled and driven by electrodes at the top and bottom surfaces. As a result, the critical piezoelectric coefficient is often $e_{31,f}$, rather than $d_{33,f}$ [24]. For the direct effect, the effective film coefficient, $e_{31,f}$ can be defined by

$$D_i = e_{ijk} x_{jk} \qquad (2.12)$$

where the relevant strains are those in the film plane. Use of the e_{ijk} coefficients, rather than the d_{ijk} coefficients is also easier since the elastic moduli of thin film piezoelectrics are rarely known accurately. In contrast to $d_{33,f}$, which tends to be comparatively small in films, the $e_{31,f}$ value can be comparable to that in bulk materials of the same composition.

A number of techniques have emerged as methods for measuring the d_{33} coefficients of thin films on substrates. Among these, double beam interferometry is the best established [25]. While in principle measurements of the field–induced displacement typically be made from a single surface only, in practice, thin films on passive substrates act like unimorph structures when driven by an electric field. Thus, the combination of the in–plane strain in the piezoelectric and the passive substrate results in flexure of the sample. Since the flexural structure acts as a mechanical amplifier, the displacement of the surface can be quite large, even though the dilation of the film is modest. Consequently, measurements of converse piezoelectric $d_{33,f}$ constants from a single surface only are prone to errors in films [25]. This is one reason for a number of anomalously high reported piezoelectric coefficients in thin film samples. While this error can be minimized by actuating with electrodes that are very small in lateral extent in displacement measurements from a single surface, measurement of the relative motion of both surfaces is preferable (providing good alignment is maintained).

Unfortunately, errors can also arise in direct measurements of the piezoelectric response of thin films. As an example of this, consider the following. Measurements of the direct piezoelectric coefficients require that a known stress level is applied to the sample. In principle, then, $d_{33,f}$ could be measured by applying a uniaxial compressive stress to the film, and monitoring the charge output. However, in making measurements of $d_{33,f}$ by this approach, it is important to consider that the wafer may not be completely flat. Such curvature of film–substrate systems is nearly universal in ferroelectric thin films because of the finite values of residual stress in the film. Thus, in attempting to apply a compressive stress, it is typical for there to be a small change in the existing radius of curvature. This is problematic, as even a small change in radius of curvature results in an in–plane stress. This in–plane stress is often large relative to the intended uniaxial stress. The net result is that the induced charge arises both from the applied uniaxial compressive stress, and an in–plane tensile stress. Given that d_{33} and d_{31} are of opposite sign in most useful piezoelectric films, the resulting charge is amplified by the in–plane stress, and artificially high "d_{33}" values are obtained [27]. Errors up to an order of magnitude can be introduced in this way.

2.3 Measurements in thin films

Alternative methods for measuring the $d_{33,f}$ coefficients of thin films include pneumatic loading methods, in which the direct effect is made by mechanically loading the film with a uniaxial stress, and measuring the resulting charge [26]. An example of the fixturing required to make this measurement, as well as the charge output from a $PbZr_{1-x}Ti_xO_3$ thin film is shown in Figure 2.7. It is important to note that, here again, unintended application of an in–plane stress component can bend the specimen under test, and artificially inflate the measured properties [27]. Appropriate correction of the in–plane component of the stress can minimize this artifact, yielding excellent agreement with double beam interferometry [26].

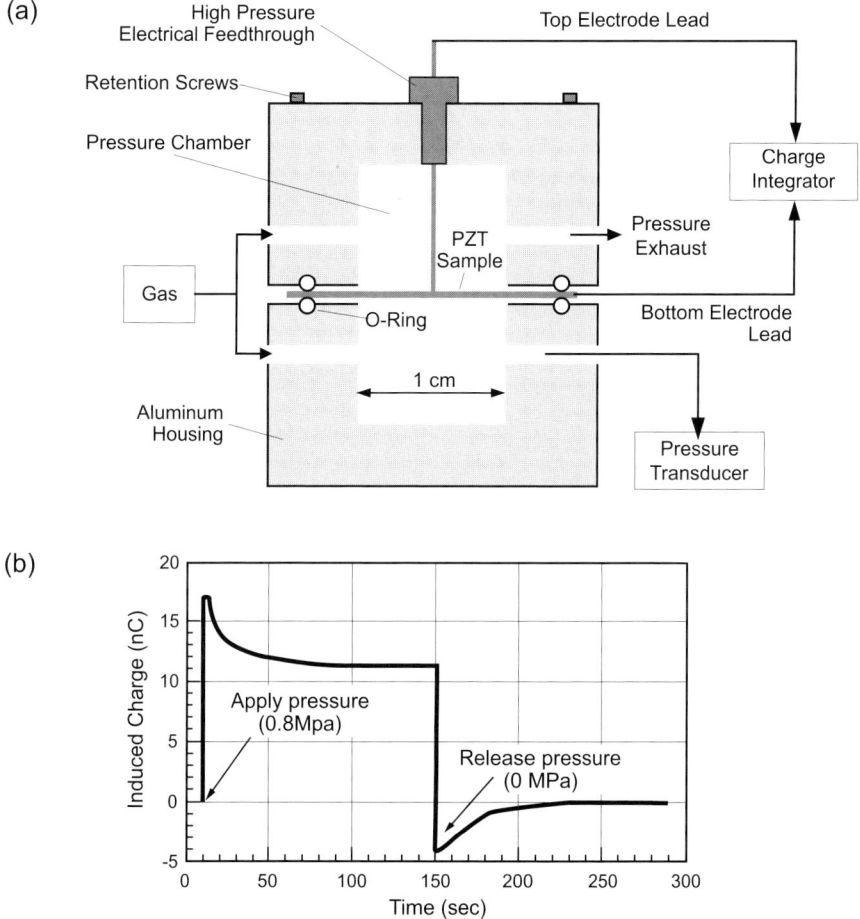

Figure 2.7: (a) Apparatus for charge–based measurement of thin film $d_{33,f}$ coefficient, (b) measured electrical signal on pressurizing and depressurizing cavities, after [26].

A third tool that is now in quite widespread use is scanning probe microscopy (SPM). Here, the sample is electrically driven either with a conductive tip or with electrodes on the sample, and the SPM tip is used to follow the sample displacement. Domain orientation can be determined by noting whether the displacement is in or out of phase with the applied electric field. This approach provides exceptional spatial resolution (in the range of tens of nm laterally) [28]. It also provides an excellent means of following the relative changes in piezoelectric response, i. e. as a function of time. However, in general, it is not as quantitative as a $d_{33,f}$ measurement as the other tools [29]. This is because the measured response is dependent on a wide variety of factors, including field drop near tip, long–range electrostatic surface–tip interactions, the force used to hold the tip in contact with the film, and the tip quality [29, 30]. Good work on improving the accuracy of the technique has been reported by Kalinin [30]. Errors in measurement of the $d_{33,f}$ coefficient by this technique can be reduced by applying the electric field through top and bottom electrodes, rather than with the tip, and by using small diameter top electrodes.

Several methods have been described to make accurate measurements of $e_{31,f}$. Two of these are shown schematically in Figure 2.8 [31, 32]. In general, what is required is a sample geometry in which known levels of strain can be applied. This can be accomplished by flexing an end–clamped, cantilever–shaped specimen, or by bending a film–coated diaphragm. For example, the measurement shown in Figure 2.8 (a) is achieved by machining a cantilever from a film–coated wafer. The sample is clamped at one end in a fixture and excited in bending by a piezoelectric actuator. The resulting piezoelectrically–generated charge can then be measured. The method shown in part (b) entails clamping a film–coated wafer over a cavity, and modulating the air pressure in the cavity to induce an in–plane stress. Again, the resulting charge can be measured using a charge amplifier coupled with a lock–in amplifier to improve accuracy. In the case where full wafers are available, small deflection plate theory can be used to calculate the induced stress. In other cases (e. g. where a small sample is glued to a carrier wafer) the applied strain can be measured using a strain gauge in order to calculate $e_{31,f}$. Measurements by these methods typically yield reproducibility within ~10 %.

2.4 Conclusions

This paper described a number of the means for measuring the piezoelectric coefficients of bulk materials and thin films. In bulk materials, excellent references are available. Numerous means have been used over the years to measure the piezoelectric coefficients, which can be loosely grouped as charge-based and displacement–based. Accurate data can be obtained by many of the techniques, and agreement between measurement types is usually reasonable, provided that comparable excitation levels are utilized. In contrast, for thin films attached to substrates, the mechanical boundary conditions differ in charge and displacement based techniques. As a result, the direct and converse coefficients are not identical. In addition, perhaps because of the relative immaturity of the field, the numerous possible artifacts are not always accounted for, which can lead to erroneous results in thin film measurements.

2.4 Conclusions

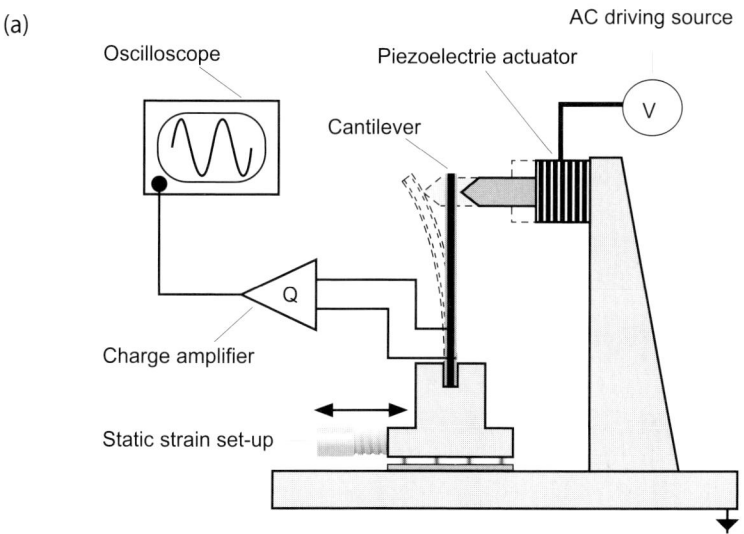

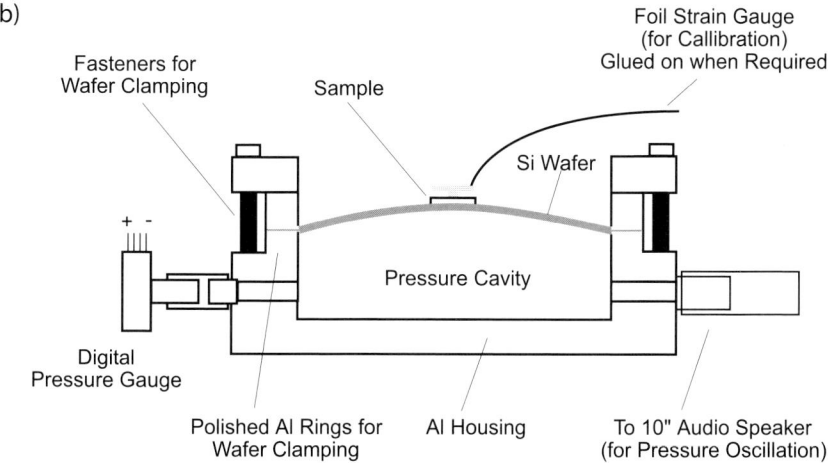

Figure 2.8: Two methods for measuring $e_{31,f}$. (a) Cantilever based method, (b) wafer flexure method. After [31, 32].

Bibliography

[1] IEEE *Standard on Piezoelectricity,* ANSI/IEEE Std 176-1987.
[2] J. F. Nye, *Physical Properties of Crystals: Their Representation by Tensors and Matrices*, Clarendon Press, Oxford, 1979.
[3] D. Wang, Y. Fotinich, and G. P. Carman, *J. Appl. Phys.* **83**, 5342 (1998).
[4] X. L. Zhang, Z. X. Chen, L. E. Cross, and W. A. Schulze, *J. Mater. Sci.* **18**, 968 (1983).
[5] G. Arlt, and H. Dederichs, *Ferroelectrics* **29**, 47 (1980).
[6] Q. M. Zhang, W. Y. Pan, and L. E. Cross, *J. Appl. Phys.* **63**, 2492 (1988).
[7] T. Yamaguchi, K. Hamano, and K. Katsumi, *J. Appl. Phys.* **18**, 927 (1979).
[8] D. Damjanovic, T. R. Gururaja, S. J. Jang, and L. E. Cross, *Mat. Lett.* **4**, 414 (1986).
[9] J. F. Li, P. Moses, and D. Viehland, *Rev. Sci. Instrum.* **66**, 215 (1995).
[10] X. H. Du, Q. M. Wang, and K. Uchino, IEEE TUFFC **50**, 312 (2003).
[11] S. Priya, D. Viehland, A. V. Carazo, J. Ryu, and K. Uchino, *J. Appl. Phys.* **90**, 1469 (2001).
[12] W. H. Jiang, R. Zhang, B. Jiang, and W. W. Cao, *Ultasonics* **41**, 55 (2003).
[13] H. F. Wang, and W. W. Cao, *J. Appl. Phys.* **92**, 4578 (2002).
[14] Q. M. Zhang, J. Z. Zhao, K. Uchino, and J. H. Zheng, *J. Mat. Res.* **12**, 226 (1997).
[15] W. Ren, A. J. Masys, G. Yang, and B. K. Mukherjee, *J. Phys. D Appl Phys.* **35**, 1550 (2002).
[16] http://utmr.npt.nuwc.navy.mil/facilities/facilities.htm.
[17] D. Damjanovic, *J. Appl. Phys.* **82**, 1788 (1997).
[18] G. Robert, D. Damjanovic, and N. Setter, *J. Appl. Phys.* **90**, 4668 (2001).
[19] H. C. Robinson, *Proceedings of the 7th* SPIE *Symposium on Smart Materials and Structures*, Newport Beach, CA, March 2000.
[20] I. Kanno, S. Fujii, T. Kamada, and R. Takayama, *Appl. Phys. Lett.* **70**, 1378 (1997).
[21] M. Lukacs, T. Olding, M. Sayer, R. Tasker, and S. Sherrit, *J. Appl. Phys.* **85**, 2835 (1999).
[22] M. Lukacs, M. Sayer, and S. Foster, IEEE TUFFC **47**, 148 (2000).
[23] K. Lefki, and G. J. M. Dormans, *J. Appl. Phys.* **76**, 1764 (1994).
[24] P. Muralt, IEEE TUFFC **47**, 903 (2000).
[25] A. L. Kholkin, C. Wutchrich, D. V. Taylor, and N. Setter, *Rev. Sci. Instrum.* **67**, 1935 (1996).
[26] F. Xu, F. Chu, and S. Trolier-McKinstry, *J. Appl. Phys.* **86**, 588 (1999).
[27] A. Barzegar, D. Damjanovic, N. Ledermann, and P. Muralt, *J. Appl. Phys.* **93**, 4756 (2003).
[28] A. Gruverman, O. Auciello, and H. Tokumoto, *J. Vac. Sci. Tech. B* **14**, 602 (1996).
[29] G. Zavala, J. H. Fendler, and S. Trolier-McKinstry, *J. Appl. Phys.* **81**, 7480 (1997).
[30] S. V. Kalinin, and D. A. Bonnell, *Phys. Rev. B* **65**, 125408 (2002).
[31] M.-A. DuBois, *Ph. D. Thesis*, Ecole Polytechnique Federale de Lausanne, Switzerland (1999).
[32] J. F. Shepard, P. J. Moses, and S. Trolier-McKinstry, *Sens. Actuators A* **71**, 133 (1998).

3 Electrical Characterization of Ferroelectrics

Klaus Prume, Thorsten Schmitz, Stephan Tiedke

aixACCT Systems GmbH, Aachen, Germany

3.1 Introduction

The electrical characterization of ferroelectric materials is crucial to investigate their suitability for ferroelectric memories, piezoelectric or pyroelectric sensor and actuator devices, and other applications. Substantial for electrical characterization of a sample is the measurement of its electrical response due to a mostly electrical excitation signal. So, the first part (3.2) of this chapter deals with *measurement methods* to record the electrical response signals. The second part (3.3) describes different *measurement types* with their typical excitation signals and helps to analyze and compare the received results. Besides the hysteresis loop as standard large signal measurement, also pulse testing, and various lifetime relevant measurements, e.g. fatigue, retention, and imprint are introduced. Additional information about the sample can be received from small signal response measurements.

3.2 Measurement methods

In general the sample response has to be measured on an applied excitation signal for material characterization. Due to the fact, that all ferroelectrics are also pyro- and piezoelectric, all permutations of electrical, thermal, or mechanical excitation and response signals are possible. However we focus on measurement methods exclusively with electrical signals, and therein with the most common approach of a voltage excitation of the sample and measuring the current or charge response. But Section 3.2.4 describes the alternative principle of a current excitation and measuring the voltage response and compares it with the methods mentioned before.

First of all, the range of signal magnitude of relevant parameters needs to be estimated to determine the requirements for an electrical amplifier to guarantee accurate measurements. Relevant parameters are charge, current, applied voltage, frequency, and signal to noise ratio. The total polarization charge is determined by the geometry of the capacitor, the material, and the applied voltage. The magnitude of the current I is given by the charge Q and the change in voltage per time, which is called the slew rate. Where A is the area of the capacitor, P the polarization of the material, V the applied voltage and dV/dt describes the slew rate of the signal form that is applied, e.g. a voltage pulse or a triangular voltage signal at different frequencies, which is a typical signal to record the hysteresis loop. The magnitude of the

Polar Oxides: Properties, Characterization, and Imaging
Edited by R. Waser, U. Böttger, and S. Tiedke
Copyright © 2005 WILEY-VCH Verlag GmbH & Co. KGaA, Weinheim
ISBN: 3-527-40532-1

current flow during a polarization reversal in a ferroelectric capacitor can be calculated as follows:

$$Q = D \cdot A \approx P \cdot A \Rightarrow I = \frac{dQ}{dt} = A \cdot \frac{dP}{dV} \cdot \frac{dV}{dt} \qquad (3.1)$$

Based on a typical thin film sample capacitor geometry with an area $A = 0.25 \, \text{mm}^2$, an operating voltage of 3 V, and a remanent polarization value $P_r = 30 \, \mu\text{C/cm}^2$ of the material, we calculate a switching charge of $Q = P_{sw} \cdot A = 2P_r \cdot A = 150 \, \text{nC}$. If we choose a triangular excitation signal with 100 Hz frequency, the average current magnitude will be roughly calculated by:

$$I = \frac{60 \, \mu\text{Ccm}^{-2} \cdot 0.25 \, \text{mm}^2}{3 \, \text{V}} \cdot \frac{3 \, \text{V}}{2.5 \, \text{ms}} \approx 60 \, \mu\text{A} \qquad (3.2)$$

Because switching occurs in a shorter time interval around the coercive voltage V_c, i.e. dP/dV is not constant and the peak switching current will increase easily one or two orders of magnitude. Besides the current magnitude we need to know the bandwidth of the current in order to decide how the current can be recorded. The frequency spectrum of the current will define the bandwidth requirements for the recording amplifier and this information can help to select a certain type of amplifier. Furthermore, the noise and ground bouncing play an important role, which will finally result in the estimation of the signal to noise ratio (SNR). These parameters all together decide how well the electrical properties of a ferroelectric capacitor can be recorded. Figure 3.1 shows the electrical excitation signal and the current response of the hysteresis loop measurement of a PZT material. The frequency spectrum of the corresponding excitation voltage and current response is shown in Figure 3.2. The current response shows higher order harmonics, which have dependent on the slope of the hysteresis loop a 30 to 50 times higher bandwidth than the voltage excitation signal base frequency. The spectrum is hysteresis shape dependent, which means a highly rectangular shape of the hysteresis loop requires more bandwidth in comparison to a more slanted loop. An amplifier offering less bandwidth than required will cause an increase of the coercive voltage due to the phase shift between input and output. Therefore, it is very important to separate this measurement error from the substantial material inherent frequency dependency of the coercive voltage.

The signal of interest has been calculated above. Now, the noise needs to be estimated. Noise is defined as all the signal sources that are not of interest, but superpose the signal of interest. In our case the thermal noise and related sources, which is system immanent to any circuit is almost negligible even for small sample sizes, as the following calculation will show. The thermal noise of a $10 \, \text{k}\Omega$ resistor R at room temperature with a frequency bandwidth Δf of 100 kHz can be calculated to

$$U = \sqrt{4kTR\Delta f} \approx 4 \, \mu\text{V} \qquad (3.3)$$

where k is the Boltzmann constant and T the temperature in Kelvin. In contrast, a current of 100 nA causes a voltage drop across a $10 \, \text{k}\Omega$ resistor of 1 mV. The noise induced from the power lines is a more serious problem and requires good shielding against electromagnetic waves and electrodynamic fields, which can be achieved by good shielding connected to a stable ground point. But besides these problems other effects play a more decisive role for precise measurements.

3.2 Measurement methods

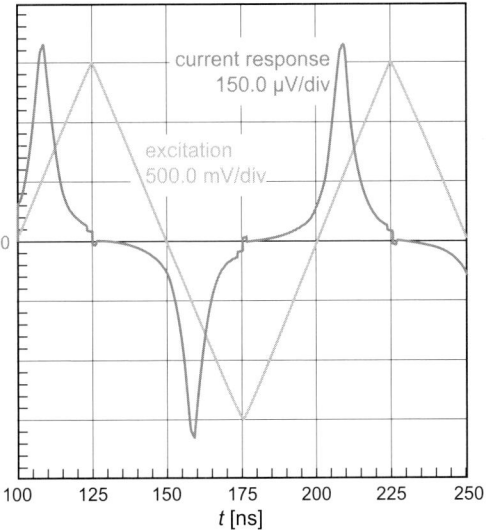

Figure 3.1: Voltage excitation and current response for hysteresis measurement.

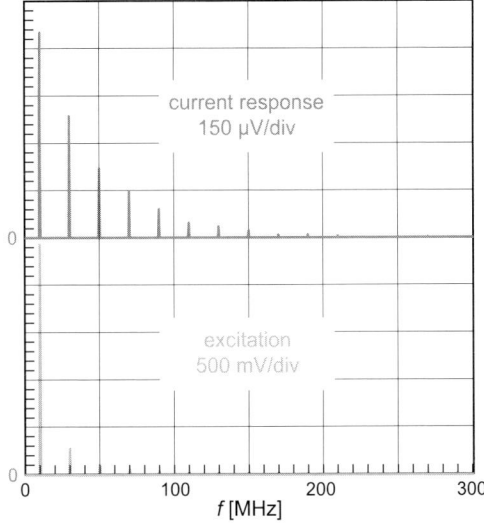

Figure 3.2: Frequency spectra of current (upper graph) and voltage (lower graph)

The estimation of the capacitance of the ferroelectric capacitor shows that the set-up itself can significantly affect the recorded results. An idea about the equivalent circuit of the set-up is given in Figure 3.3, which shows the influence of the cabling in a simplified manner (not showing the propagation delays), but also parasitic capacitors and resistors of the set-up. Typical parasitic capacitance of a set-up in parallel to the sample is in the range of a few picofarad. But cabling capacitance can easily reach a few hundred picofarad [1].

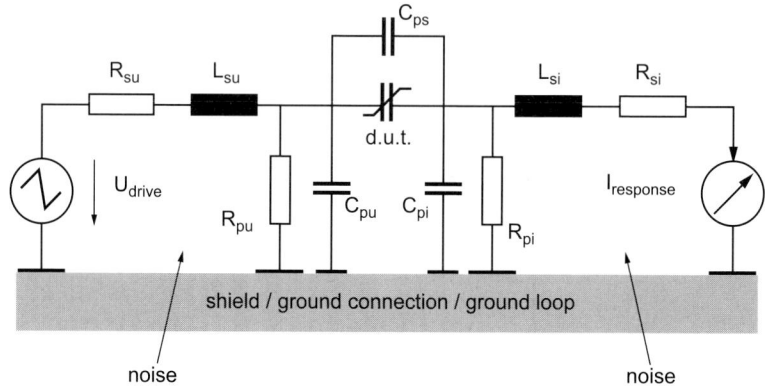

Figure 3.3: Schematic of the parasitic components of the measurement set-up.

With respect to the equivalent circuit in Figure 3.3, an evaluation of the known methods for hysteresis measurements will be given, in view of the effective parasitic capacitance and the influence of reflection. Well known methods to record the hysteresis loop of ferroelectric capacitors by measuring the current response are Sawyer Tower, Virtual Ground, and Shunt measurement as shown in Figure 3.4.

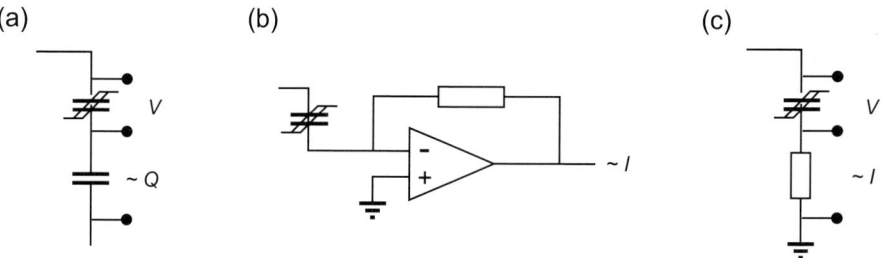

Figure 3.4: Schematic comparison of (a) Sawyer Tower, (b) Virtual Ground, and (c) Shunt hysteresis measurement method.

3.2.1 Sawyer Tower method

The Sawyer Tower measurement circuit is based on a charge measurement method which relies on a reference capacitor in series with the ferroelectric capacitor [2]. The voltage drop across the reference capacitor is proportional to the polarization charge as defined by $V = Q/C$. But if the voltage on the reference capacitor increases, the voltage across the sample decreases (back voltage effect). So the reference capacitor is chosen much larger than the measured capacitor, e.g. if the reference capacitor is 100 times larger, the voltage drop is about 1 %. This means the reference capacitor has to be adapted to each sample. The Sawyer Tower method can be used up to high speeds which is primarily limited by cable reflections. As parasitic effects, cabling capacitances of the wiring between sample, reference capacitor, and the recording amplifier are in parallel to the reference capacitor. Typical cable capacitance values are between 33 pF and 100 pF per meter. For small capacitors, the total measured capacitance is increased over the capacitance of the ferroelectric material. Furthermore, it is difficult to get precise reference capacitors which typically have several percent tolerance, and additionally the cable capacitance adds to this capacitance. Furthermore the input resistance of the voltage measurement device is in parallel to the reference capacitor and discharges it with a corresponding time constant, therefore the Sawyer Tower is less suitable for slow measurements.

3.2.2 Shunt method

The Shunt measurement substitutes the reference capacitor of the Sawyer Tower circuit by a reference resistance (shunt resistor). This measurement method is a current based method, i.e. the switching current is measured as a voltage drop across the shunt resistor ($V = RI$), and later integrated (e.g. numerically) to get the polarization charge $Q = \int I dt$. Similar difficulties as in case of the Sawyer Tower set-up appear. Though it is easier to get precision resistors, the resistance value of choice depends not only on the sample capacitance but also on the excitation frequency. Thus, the voltage drop increases with increasing frequency, and the time constant between ferroelectric capacitor and shunt resistor influences the result at higher speed. Additionally, the cable capacitance as well as the input capacitance for the voltage measuring device are in parallel to the reference resistor. Therefore accurate measurements for large pads are possible but become increasingly difficult in case of small pads, where the parasitic capacitances come into effect.

3.2.3 Virtual ground method

The Virtual Ground method uses a current to voltage converter which is based on current measurement using a feedback resistor across an operational amplifier. The output of the current to voltage converter is connected to the inverting input of the operational amplifier via the feedback resistance, the non inverting input is connected to ground. The voltage difference between both inputs is ideally zero and in reality a few microvolts. So, the inverting input is virtually on ground level. This is helpful for the measurement, especially of small capacitors as the cable capacitance is physically in place but electrically ineffective, because both electrodes of the capacitor are kept on the same potential. Also the sample always is applied to the full excitation voltage, since there is no back voltage. For high speed measurements or voltage pulses, the inductance of the set-up and the impedance mismatch caused by the sample and sample holder will lead to reflection on the cable and thus to a measurable change of the results related to the set-up. Also the stability, bandwidth, and phase shift of the operational amplifier must be taken into account. But as a whole the Virtual Ground method enables the highest precision for ferroelectric measurements.

3.2.4 Current step method

By comparison to all of the above mentioned measurement methods the current step method uses an applied current as excitation signal instead. This current induces a directly measurable voltage drop over the sample. Therefore, as advantage compared to the Sawyer-Tower or shunt method no additional reference component is needed and the resulting limitations and approximations are not in effect any more. The second advantage of this method is the smaller bandwidth of the voltage signal which has to be measured. A rectangular current signal will cause a triangular voltage response and vice versa across an ideal capacitor. As can be seen in Figure 3.2 the bandwidth requirements for measuring the voltage response are much lower than for measuring the current. Table 3.1 summarizes the advantages and disadvantages of the different measurement methods.

Method	Measured quantity	Reference component	Integration necessary	Bandwidth requirement	Influence of parasitics
Sawyer Tower	charge Q	capacitor	no	moderate	high
Virtual Ground	current I	no	yes	high	low
Shunt	current I	resistor	yes	high	high
Current Step	voltage V	no	no	moderate	moderate

Table 3.1: Comparison of different measurement methods for hysteresis measurements of ferroelectrics.

3.3 Measurement types

3.3.1 Hysteresis loop and characteristic values

After consideration of the different measurement methods we now focus on typical measurement types which are all based on one of the voltage excitation methods but differ in the excitation and the electrical *treatment signal* in between repeated measurements.

Typical result of all the measurements is the hysteresis curve of the polarization respectively the development of one of its characteristics. In the following the nomenclature of characteristic values for the evaluation of the measured data is introduced ([1], see also Figure 3.5):

P_{r+} positive state of **remanent polarization** of the dynamically measured hysteresis loop

P_{rrel+} positive state of **relaxed remanent polarization**, relaxed for one second in the P_{r+} state. Equal to the positive state of remanent polarization of the quasi statically measured loop (see Section 3.3.4)

3.3 Measurement types

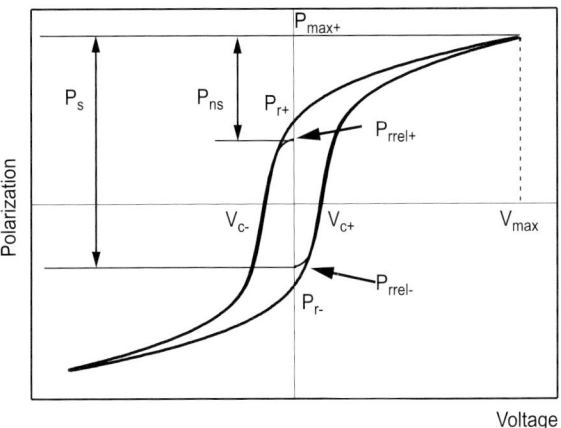

Figure 3.5: Nomenclature used.

P_{max+} state of polarization when the stimulating signal reaches its maximum value - **positive saturation**

V_{c+} positive **coercive voltage**, voltage where the polarization crosses the x-axis by increasing voltage values

$P_{r-}, P_{rrel-}, P_{max-}, V_{c-}$ are the corresponding values for the negative field and polarization direction

P_s $(P_{max+} - P_{rrel-})$ change of polarization when the sample is switched from the negative state of the relaxed remanent polarization into the positive saturation - **switching case**

P_{ns} $(P_{max+} - P_{rrel+})$ change of polarization when the sample is driven into the positive saturation from the positive state of the relaxed remanent polarization - **non-switching case**

ΔP_s $(P_s - P_{ns})$ detectable polarization difference between switching and non-switching case

3.3.2 Dynamic hysteresis measurement

To obtain the dynamic hysteresis loop of a ferroelectric capacitor the polarization is measured versus the applied voltage. Since the hysteresis is neither a linear nor a time invariant property, the hysteresis loop is dependent on the sample history and on the measurement method. To have a standardized and comparable hysteresis loop, certain parameters are commonly fixed. One is the absolute position of the loop on the polarization axis, since the initial (virgin) state of the polarization is unknown in almost all cases, the hysteresis loop is balanced to a reference value. Most commonly the positive and negative saturation polarization are set to equal absolute values to center the loop, however in some cases (e.g. asymmetric voltage excitation or initial loop) it can be centered around its positive and negative P_r values, or simply start at $P = 0$. Furthermore the relaxation of P_r is by standard measured after 1 second, and as excitation signal a triangular shaped bipolar symmetric voltage signal is used. The voltage excitation signal to obtain a complete hysteresis loop including the relaxed polarization P_{rrel} is shown in Figure 3.6 (a).

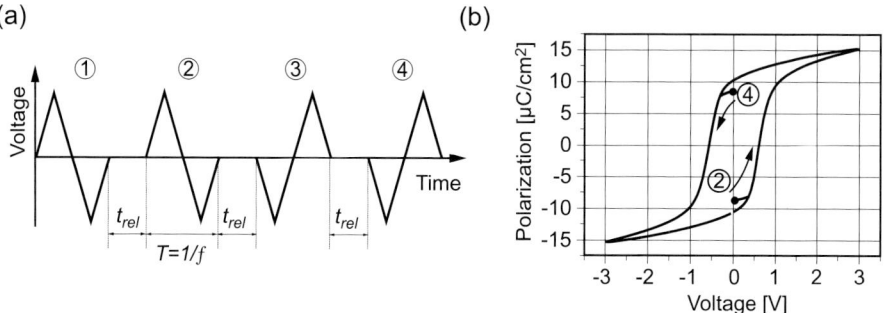

Figure 3.6: Excitation signal for hysteresis measurement.

The prepolarization pulse no. 1 establishes a defined polarization state, the negative state of relaxed remanent polarization after 1 second. The prepolarization pulse is followed by three consecutive bipolar excitation signals, each signal separated by 1 second relaxation time. The corresponding hysteresis loops to the bipolar excitation signal are shown in Figure 3.6 (b). The hysteresis loop corresponding to pulse no. 2 starts in the negative relaxed remanent polarization state (P_{rrel-}) and turns into the positive saturation (P_{max+}). When the voltage returns to zero the polarization reaches the positive remanent polarization state (P_{r+}), afterwards it continues into the negative saturation (P_{max-}) and back to the negative remanent polarization state (P_{r-}). This point is normally not equal to the starting point (P_{rrel-}), because of the polarization loss (relaxation) over time.

The third loop establishes the sample into the positive remanent polarization state without sampling data. The fourth loop now starts in the positive relaxed remanent polarization state (P_{rrel+}), turns into the negative saturation (P_{max-}), then crosses the polarization axis at zero volts excitation signal in the negative remanent polarization state (P_{r-}). Afterwards the sample is driven into the positive saturation (P_{max+}) and ends up in the positive remanent polarization state (P_{r+}) when the voltage is zero again. Subsequently, the hysteresis loop is balanced respectively to the values $P(+V_{max})$ and $P(-V_{max})$. From the data of the second loop the parameters V_{c-}, P_{r-}, P_{rrel-} are determined and from the data of the fourth loop the parameters V_{c+}, P_{r+}, P_{rrel+}. The closed hysteresis loop (continuous loop) can be calculated from the second half of the second loop and the second half of the fourth loop.

It is of particular importance to emphasize that the shape of the hysteresis curve changes with frequency, amplitude, shape, and relaxation time between prepolarization and recording pulses of the excitation signal. Therefore the extracted characteristic values differ for two

3.3 Measurement types

consecutive measurements on the same sample e.g. one with sine wave and one with a triangular excitation signal shape. The change of the hysteresis loop with increasing measurement frequency can be seen e.g. in Figure 3.7.

Furthermore, the history of a hysteresis loop plays an important role in the determination of lifetime and reliability of ferroelectric capacitors, especially for applications in ferroelectric memories. Three main effects are characterized in particular as changes in the hysteresis loop under various conditions, which are described later in this chapter as fatigue, retention, and imprint with the corresponding ways to measure these effects.

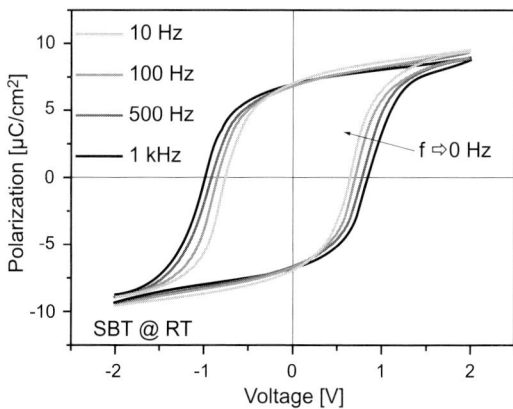

Figure 3.7: Hysteresis loops measured at different frequencies.

3.3.3 Pulse measurement

Conventional characterization of ferroelectric materials have been done in the past to estimate the suitability of these materials in view of ferroelectric memory (FeRAM) applications. The requirements of the industry increased when the development of the materials proceeded. Therefore testing methods for the material using excitation signals close to real world applications like pulse switching or opposite state retention tests have been developed. With pulse rise times in the nanosecond regime and pulse width down to a few ten nanoseconds, the operation of a true FeRAM cell can be emulated in good approximation. Coercive field and remanent polarization as the characteristic values of the material for the design of an integrated FeRAM device should be known at speeds equal to the speed in a real memory device due to the frequency dependence of the hysteresis loop [3, 4].

The investigation of the switching behavior contributes to the knowledge of values like access time of a FeRAM device that could be reached with a certain material and of loss mechanisms in the short time regime [5]. In Figure 3.7 hysteresis curves measured at different frequencies are displayed. It can clearly be seen that with increasing measuring frequency the coercive voltages also increases. Now, it is of interest how the growth of the coercive voltage proceeds with increasing frequency, especially at the speed of an integrated memory device.

To emulate the operation of the FeRAM cell of the integrated circuit the measurement setup has to generate pulses of both polarities. The Shunt method as it is described in Section 3.2.2 is useful to exclude the influence of the sense capacitor and to reach high speed.

Figure 3.8 (a) shows a simplified setup with an ultra fast digital control unit. The bridge design of the setup manages to emulate pulse trains of real memory devices as displayed in Figure 3.9.

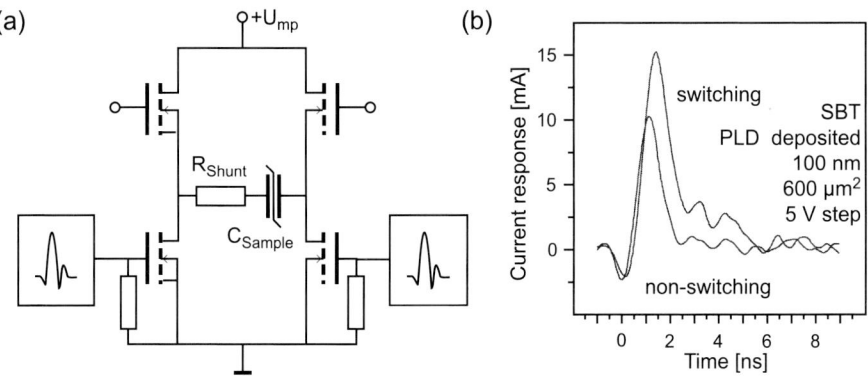

Figure 3.8: (a) Simplified pulse measurement setup and (b) typical signal rise times.

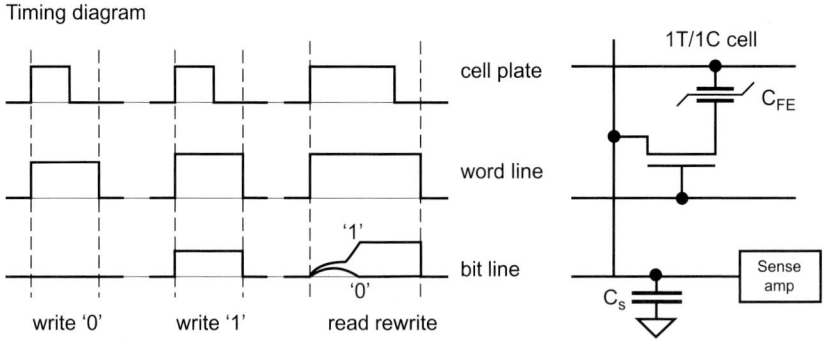

Figure 3.9: Timing diagram of a FeRAM cell [6].

Each switch is controlled separately. This guarantees a maximum of flexibility in pulse sequence generation. Switching and non switching case can be investigated for different amplitudes, pulse widths, and delay times. Figure 3.8 (b) shows the typical rise time of the setup (1 ns).

3.3 Measurement types

The impact of the pulse switching measurement can be demonstrated by comparison of measured and simulated data. The hysteresis loop in general gives a description of the course of the polarization as a function of the applied voltage when the voltage is continuously increased. One would expect that the pulse switching current can be simulated by means of these data when the equivalent network consists of the ferroelectric capacitor and a serial resistance as in Figure 3.10. Starting from the hysteresis loop recorded at frequencies of 100 Hz the polarization switching current and the correlated voltage can be calculated numerically using Equation 3.4. It has been assumed that the polarization is not a function of time.

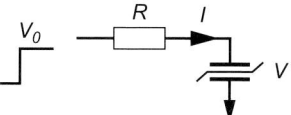

Figure 3.10: Equivalent circuit diagram.

$$I + R \cdot A \cdot \frac{\delta P(V)}{\delta V} \cdot \frac{\delta I}{\delta t} = 0 \tag{3.4}$$

Figure 3.11 (a) shows the results of the simulation of the current and the voltage across the ferroelectric capacitor using Equation 3.4 and assuming an ideal voltage step of 2 V and a resistance in series of 120 Ω. The course of the current is determined by the change of the slope with the changing voltage across the capacitor. If the simulated current is compared to the measured current response of the same capacitor a distinct difference in the course can be observed. Then, the polarization, determined by integrating the current, is plotted versus the voltage drop across the polarization course and is shifted parallel to the hysteresis curve (Figure 3.11 (b)).

3.3.4 Static hysteresis measurement

As described in Section 3.3.2 the dynamic hysteresis measurement applies signals of a certain frequency to record the hysteresis loop of a ferroelectric material. The shape of the hysteresis loop recorded by different excitation frequencies shows a frequency dependence. Therefore, it would be of interest to exclude the influence of the frequency of the excitation signal. But, this so called static hysteresis curve is determined by an infinite slow excitation signal. This would require an extremely sensitive current meter, a high precision low noise signal generator, and a leakage free capacitor. This is circumvented by the static hysteresis measurement method which allows to record the quasi-static hysteresis loop. This means, in each data point of the quasi-static hysteresis curve the sample relaxes for a definite time. This relaxation time can be varied to investigate the irreversible and reversible parts of polarization [8].

The ferroelectric sample is pre-polarized by measuring a complete hysteresis. The excitation voltage is then kept constant for the relaxation time at a particular voltage. The relaxed

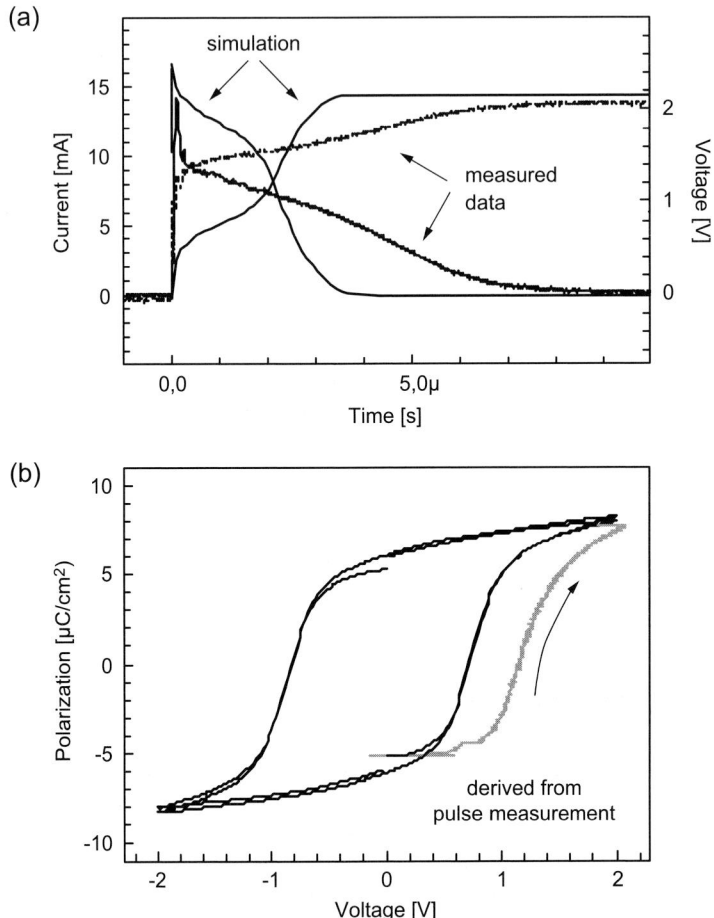

Figure 3.11: Simulated and measured data [7].

quasi-static polarization is determined by driving the capacitor into saturation and recording the current response. By integrating the current response the desired relaxed polarization is calculated.

After cycling once again through a complete hysteresis to ensure the same initial condition the excitation signal is stopped at the next voltage for the relaxation time. The relaxed polarization is again determined as described above. The whole procedure is repeated for each point until a complete quasi static hysteresis loop has been recorded.

In Figure 3.12 a static hysteresis loop is shown in correlation to a dynamic hysteresis loop. Two clippings of the excitation signal of the static hysteresis loop are shown. Each of them is used to measure one data point of the curve. To record 20 points for one loop, 20 of these excitation signals have to be applied to the sample, each one stopping at a different voltage.

3.3 Measurement types

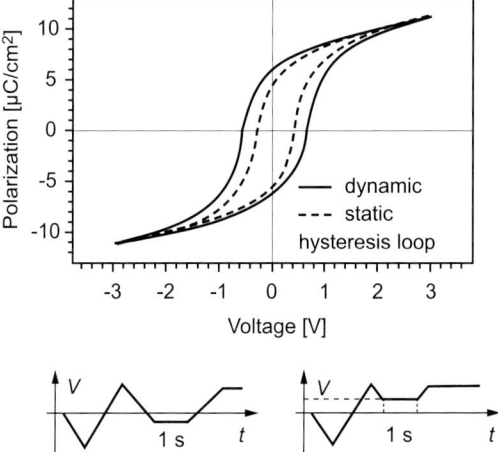

Figure 3.12: Typical measurement result of the static hysteresis and curve of the excitation signal to measure two consecutive data points of one and the same static hysteresis loop (below).

3.3.5 Leakage measurement

Conductivity is an important characteristic value of electroceramic materials. It depends on the mobility of free electronic or ionic charges in the material under the given conditions (e.g. temperature and applied electric field) [9–13]. An applied voltage results in a (leakage) current flow through the sample. The typical leakage current measurement is performed by applying a special step shaped voltage waveform to the sample and measuring the current response e.g. by a virtual ground amplifier as described in Section 3.2.3. Since a ferroelectric capacitor can be considered as a non-linear capacitor, and a voltage dependent resistor in parallel, the current response due to an applied step waveform has to be analyzed to extract the leakage current information, because the pure leakage current results only from the voltage dependent resistive part of the sample [14].

To determine polarization dependent leakage effects, the waveform should have a triangular step shape (Figure 3.13), similar to the hysteresis measurement. The maximum amplitude, the step height, and duration depend on the sample properties and the time constant.

A special measurement method is used to distinguish the leakage current from other currents, like the capacitive load current and relaxation currents in the sample, as resulting from the theoretical model. Thus the current is monitored during each voltage step, and after the exponential decay of the load current, the current reading is averaged in the region from 70% to 90% of the step time, giving a precise value for the leakage current. Averaging of the whole waveform would give inaccurate results, since it includes the load current and relaxation currents, too. Typically, the step width is around a second but it can be extended to measure capacitors with high time constants or very low currents.

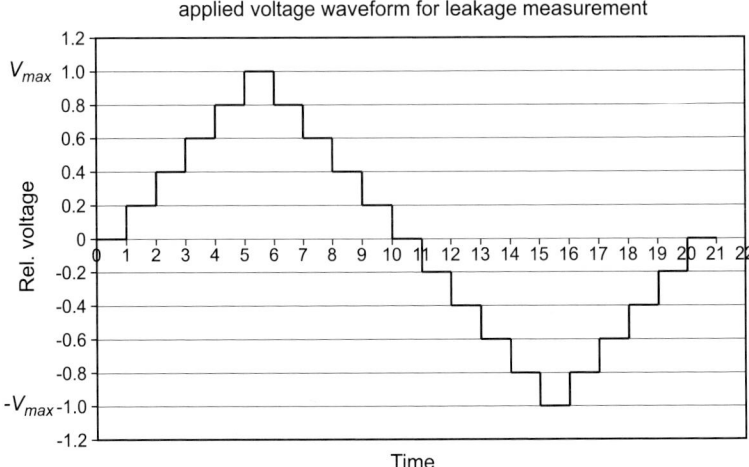

Figure 3.13: Waveform for leakage measurement.

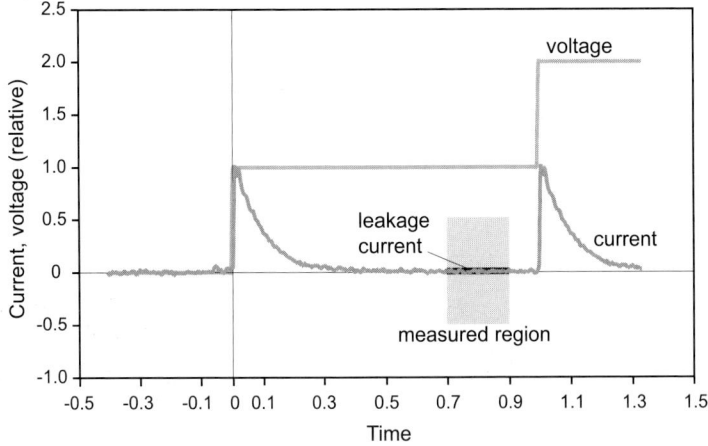

Figure 3.14: Current response and integration range for a leakage measurement.

3.3.6 Fatigue measurement

Fatigue describes the change of the hysteresis loop dependent on the number of switching cycles the ferroelectric capacitor has gone through [15–23]. The main effect of fatigue is a loss in the saturation polarization (P_{max}) and in the remanent polarization values (P_r, P_{rrel}) correspondingly, which is seen as a flattening of the hysteresis loop as shown in Figure 3.15. In a memory device switching cycles are performed during write and destructive read operations, state, thus imprint only arises from the internal polarization field within the capacitor. The imprint effect can be increased by applying an additional DC or pulsed unipolar bias field. Both effects are accelerated at elevated temperatures [30], which is of interest also for the lifetime prediction for memory devices to reach 10 years life time, but be predicted from short term measurements [31, 32]. A typical excitation signal is displayed in Figure 3.20. The first rectangular shaped signals represent the deaging sequence. The following triangle period is used for the initial and the in-between triangles for the following hysteresis measurement. Finally, the constant section of the voltage curve refers to the imprint condition. The applied imprint condition and the subsequent hysteresis measurement are repeated for the duration of the measurement in regular intervals depending on the desired number of measurement points and total measurement time.

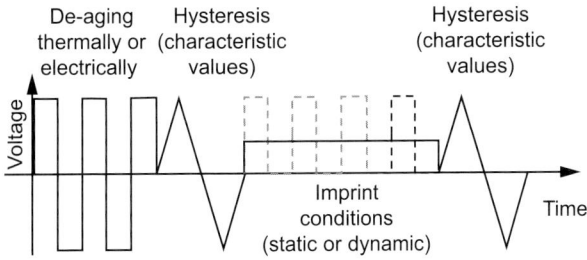

Figure 3.20: Measurement procedure of the imprint measurement.

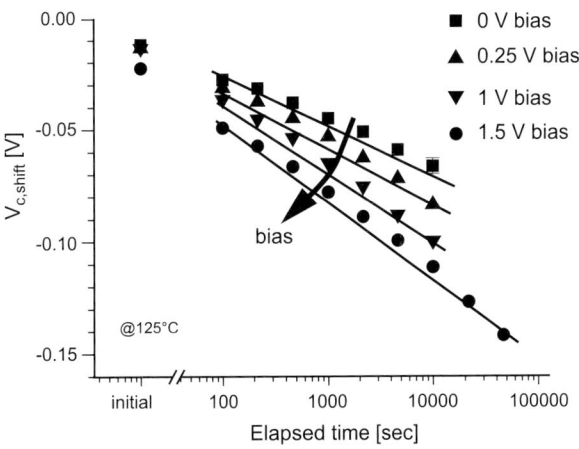

Figure 3.21: Typical result of an imprint measurement.

3.3 Measurement types

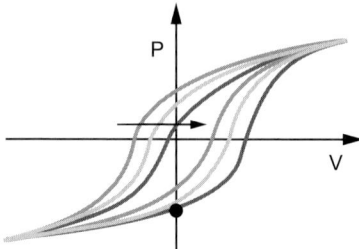

Figure 3.18: Shift of polarization of a sample in the negative state of remanent polarization due to imprint.

Furthermore, the remanent polarization values P_r and P_{rrel} also change due to the shift of the loop. A correlation between voltage shift and polarization change can be given with the static hysteresis loop (see Section 3.3.4).

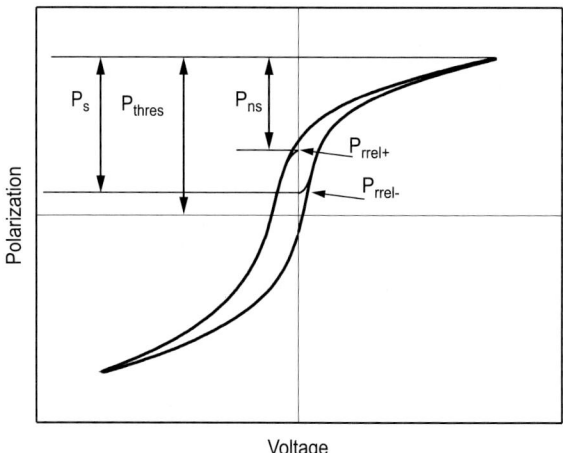

Figure 3.19: Failure mechanism related to the voltage shift in the hysteresis loop. P_{thres} is used as the threshold polarization to detect the stored information.

Since a ferroelectric capacitor usually shows imprint after processing, or after storage in a poled state, the initial imprint can be reduced to achieve a better defined starting point. This reduction of the initial imprint, so called de-aging procedure, can be performed by cycling the capacitor several times. For measuring the imprint effect, hysteresis measurements are taken between various stress conditions over time. These conditions can be shorted or open contact, DC bias field stress, unipolar voltage pulse stress, and additional temperature stress. In the simplest case the sample is shorted or not contacted during the imprint period in a poled way that in a logarithmic plot data points are depicted with an equal spacing. A typical result of a fatigue measurement combined with hysteresis is shown in Figure 3.17. It displays the remanent polarization of the in between recorded hysteresis loops versus the logarithm of the total number of applied cycles.

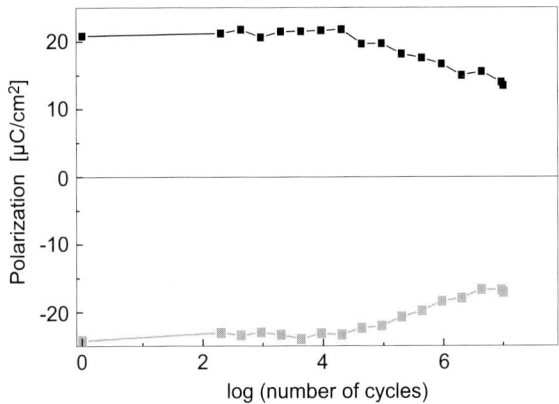

Figure 3.17: Typical result of a fatigue measurement in combination with hysteresis measurements is usually depicted as the remanent polarization versus log cycles of the excitation signal.

3.3.7 Imprint measurement

Imprint can be described as the preference of one polarization state over the other or the inability to distinguish between the two different polarization states [24–29]. Imprint affects the ferroelectric behavior of thin films in two ways. On the one hand, a shift of the ferroelectric hysteresis loop on the voltage axis is observed (Figure 3.18). On the other hand, imprint also leads to a loss of remanent polarization. Establishing and maintaining a negative state of polarization leads to a shift of the hysteresis loop on the voltage axis to the right (i.e., to positive voltages) and, additionally, to a loss of the positive state of polarization, which is the state opposite to the established one (see Figure 3.19). This effect is refered to as opposite state retention.

The failure mechanism within a memory cell is either due to the inability of the programming voltage to switch the ferroelectric material because of an increase of the coercive voltage (write failure) or due to a decrease in the difference of P_s and P_{ns}. This means the two different states of remanent polarization cannot be distinguished by the memory sense amplifier (read failure). This case is shown in Figure 3.19.

As typical parameter the voltage shift of a hysteresis loop is recorded over time under various stress conditions. Voltage shift means a change in V_{c+} and V_{c-}, the curve is not symmetrical any more. $V_{c,shift}$ is defined as followed:

$$V_{c,shift} = \frac{V_{c+} + V_{c-}}{2} \tag{3.5}$$

3.3 Measurement types

so fatigue would lead to a loss in the signal margin to determine between positive and negative polarized states and thus lead to a later memory cell failure.

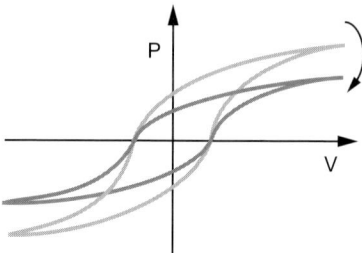

Figure 3.15: Polarization drop due to fatigue.

After measuring an initial hysteresis loop as described above, a fatigue signal sequence (see e.g. Figure 3.16) is applied to the capacitor. This signal is intended to switch the ferroelectric capacitor multiple times, thus it has to be sure that the amplitude of the fatigue signal is high enough to switch the capacitor completely within the given fatigue cycle time. Typically a rectangular waveform is used to achieve the highest amount of switching. The total number of switching cycles is related to the desired operation condition, e.g. in a ferroelectric memory device, and thus is desired to reach up to 10^{16} cycles which is comparable with a memory operation lifetime of 10 years. Since this number cannot be reached within a reasonable time frame even at high fatigue signal frequencies, the results have to be extrapolated.

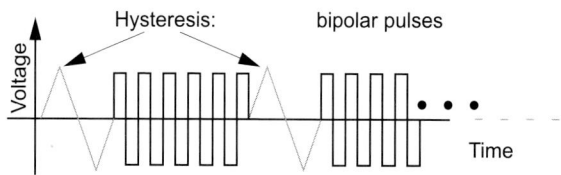

Figure 3.16: Typical fatigue excitation signal.

If fatigue switching is performed at high frequencies, it has to be ensured that the sample is still completely switched, because sample relaxation as well as time constant and driving power of the set up decrease the switched polarization at increasing frequency. This effect can be compensated by increasing the excitation voltage, but this also increases stress on the sample and requires higher driving power.

The fatigue treatment signal is interrupted regularly for hysteresis measurements to monitor the development of the hysteresis shape, or at least the polarization values. These intermediate hysteresis measurements are performed after increasing time intervals, e.g. in such a

3.3 Measurement types

The results of imprint measurements under different bias voltage conditions are shown in Figure 3.21. The voltage shift of the in between measured hysteresis loops is plotted versus the logarithm of the elapsed time, which can be closely fitted to a straight line that can be easily extrapolated to make a lifetime prediction. The values of polarization show a nonlinear change on the same scale which is more difficult to extrapolate, but can be transformed into an equivalent coercive voltage shift using the static hysteresis loop (see Section 3.3.4).

3.3.8 Retention measurement

Retention loss describes the drop in remanent polarization of a ferroelectric material with time (see Figure 3.22 and [33–35]). This can lead to a failure of a memory cell if after poling a sample into a known polarization state during a write pulse the polarization slightly decreases and drops below a limit which can be distinguished by the sense amplifier during read operation. Retention can be measured as a function of time by poling a sample into a known state and reading out the remaining remanent polarization after an increasing and defined time period. The signal sequence which is used to acquire the retention data is shown in Figure 3.23. This measurement is performed by a pulse measurement as described above. It consists of a write and five read pulses. The first pulse (x) is the most important one, since this pulse reads the charge left on the sample electrode after the retention time period. The measured change of polarization is compared to the change of polarization measured by the following four reference pulses. These reference pulses are used to determine the absolute value of the polarization. As another method, the polarization can also be determined by measuring the hysteresis loop, but leaving out the initial polarization pulse (pulse No. 1 in Figure 3.6).

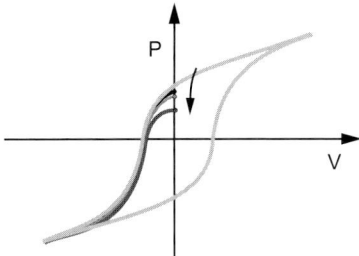

Figure 3.22: Polarization drop due to retention.

A typical result of a retention measurement is shown in Figure 3.24 which displays the remanent polarization of the recorded data versus the logarithm of the delay time between poling the sample into a defined polarization state and the read pulse. After reference delays the unknown polarization state is measured again. The change in polarization dependent on the delay time between write and read pulses gives the result of the retention measurement. During the delay period between write and read the capacitor can face various conditions, e.g. it can be shorted, or measured at elevated temperature, or a disturb signal can be applied,

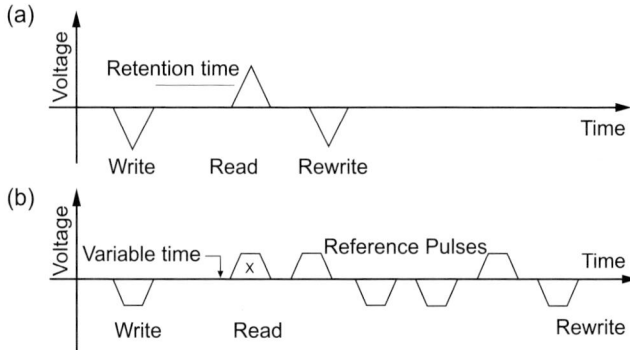

Figure 3.23: Principal (a) and realized (b) signal sequence to acquire retention data.

which simulates the behavior of a ferroelectric memory circuit that can also have crosstalk voltage noise applied to the capacitor.

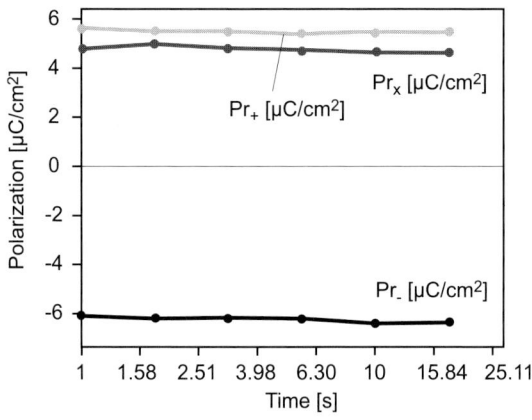

Figure 3.24: Typical result of a retention measurement of an almost retention free sample.

The time intervals between the intermediate measurements are here chosen in such a way that on a logarithmic plot the data points are depicted with an equal spacing (see dots in Figure 3.24).

The pulses have trapezoid shape or triangular shape with similar rise times of pulses to measure a closed hysteresis loop. The retention is measured by pulses instead of a standard hysteresis loop since the excitation has to be modified to get the unknown initial polarization state, but compare it to a reference value, e.g. polarization with 1 second delay. In principle this could also be measured using the hysteresis measurement as described above, and monitoring the polarization during the prepolarization pulse (pulse no. 1 in Figure 3.6). The

3.3 Measurement types

reference pulses are used to determine the absolute value of the polarization. The trapezoid waveform is used, because the polarization increases after the plateau of the applied pulse is reached. The polarization comes up to a characteristic value, after the relaxation faded. By this, a reference polarization is defined. The absolute values of the polarization are determined by the assumption that both characteristic values of the polarization in the plateau of the reference pulses are equal. Now, the change of polarization of the first pulse of the read sequence can be compared to the reference polarization and can be determined clearly.

3.3.9 Small signal measurements

With small signal capacitance vs. voltage $C(V)$ and loss tangent $tan(\delta)(V)$ measurements, further important parameters of ferroelectric capacitors can be determined. Especially it gives information about the reversible parts of polarization, and furthermore the switching process. Together with additional hardware to simultaneously measure the samples displacement even the piezoelectric coefficient $d_{33}(V)$ can be determined. A typical voltage excitation signal for $C(V)$ and $d_{33}(V)$ measurements is shown in Figure 3.25.

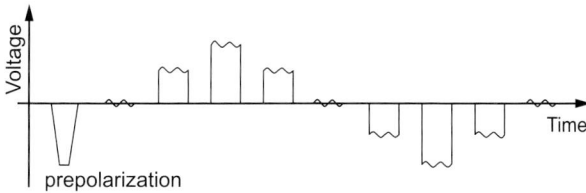

Figure 3.25: Typical waveform of a $C(V)$ measurement.

The prepolarization pulse establishes a defined polarization state, usually the negative state of relaxed remanent polarization after reaching $-V_{max}$. This pulse is followed by a number of consecutive unipolar excitation signal pulses, depending on the entered input parameters. The unipolar pulses are DC biased voltage pulses with an added sine wave AC small signal. The DC bias starts at zero volts, then increases with each pulse up to the desired maximum excitation voltage V_{max}, goes down to $-V_{max}$ and back to zero. This ensures cycling through the whole hysteresis loop during the measurement. The capacitance and loss tangent are then derived from the AC small signal current response of the sample. The resulting $C(V)$ and $tan(\delta)(V)$ curves corresponding to the small signal excitation are shown in Figure 3.26.

As an alternative a different test waveform with a continuous triangular DC bias voltage with overlayed AC signal can be used for the $C(V)$ measurement too. Additionally, impedance measurements are very important and common small signal measurements. They give the frequency dependency of capacitance and loss tangent. By using samples of certain geometries it is possible to extract a whole set of small signal parameters from the resonances in the impedance spectra. This method is descibed in detail in the standard DIN EN 50324-2.

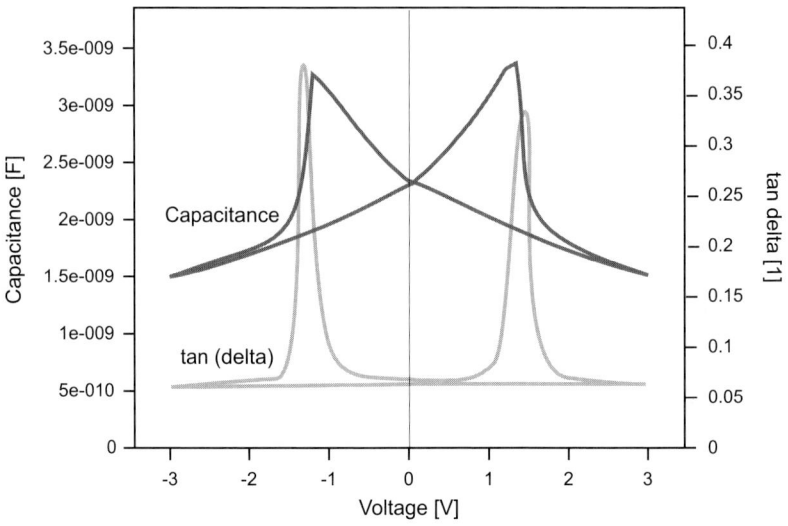

Figure 3.26: Typical $C(V)$ measurement.

Bibliography

[1] S. Tiedke, and T. Schmitz, Tutorial Session: Testing and Characterization. *Int. Symp. Integr. Ferroel. Conference*, Aachen, (2000).
[2] C. B. Sawyer, and C. H. Tower, *Physical Review* **35**, 269 (1930).
[3] J. Cillessen, M. Prins, and R. Wolf, *J. Appl. Phys.* **81**, 2777 (1997).
[4] O. Lohse, M. Grossmann, D. Bolten, U. Boettger, and R. Waser, *Mat. Res. Soc. Symp. Proc.* **655**, (2000).
[5] A. Tagantsev, M. Landivar, E. Colla, and N. Setter, *J. Appl. Phys.* **78**, 2623 (1995).
[6] R. E. Jones, *Int. Ferroel.* **17**, (1997).
[7] O. Lohse, S. Tiedke, M. Grossmann, and R. Waser, *Integrated Ferroelectrics* **22**, 123 (1998).
[8] O. Lohse, D. Bolten, M. Grossmann, R. Waser, W. Hartner, and G. Schindler, *Mat. Res. Soc. Symp. Proc.* **493**, 267 (1998).
[9] C. Ohly, S. Hoffmann, K. Szot, and R. Waser, *Integrated Ferroelectrics* **33**, 363 (2001).
[10] K. Abe and S. Komatsu, *Jpn. J. Appl. Phys.* **32**, 4186 (1993).
[11] T. Mihara and H. Watanabe, *Jpn. J. Appl. Phys.* **32**, 5664 (1995).
[12] T. Mihara and H. Watanabe, *Jpn. J. Appl. Phys.* **32**, 5674 (1995).
[13] G. Dietz, M. Schumacher, R. Waser, S. Streiffer, Basceri, and A. Kingon, *J. Appl. Phys.*, **82**, 2359 (1997).
[14] G. Dietz and R. Waser, *Integrated Ferroelectrics* **9**, 317 (1995).

[15] A. Tagantsev, I. Stolichnov, E. Colla, and N. Setter, *J. Appl. Phys.* **90**, 1387 (2001).
[16] W. Schulze and K. Ogino, *Ferroelectrics* **87**, 361 (1988).
[17] T. Mihara, H. Yoshimori, H. Watanabe, and C. P. D. Araujo, *Integrated Ferroelectrics* **10**, 351 (1995).
[18] Z. Pajak and J. Stankowski, *Proc. Phys. Soc.* **72**, 1144 (1958).
[19] S. Ikegami and I. Ueda, *J. Phys. Soc. Jap.* **22**, 725 (1967).
[20] R. Bradt and G. Ansell, *J. Am. Ceram. Soc.* **52**, 192 (1969).
[21] H. Dederichs and G. Arlt, *Ferroelectrics* **68**, 281 (1986).
[22] A. M. Bratkovsky and A. P. Levanyuk, *Phys. Rev. Lett.* **84**), 3177 (2000).
[23] J. Lee, C. Thio, M. Bhattacharya, and S. Desu, *Mat. Res. Soc. Symp. Proc.* **261**, 241 (1995).
[24] G. Arlt and H. Neumann, *Ferroelectrics* **87**, 109 (1988).
[25] R. Lohkämper, H. Neumann, and G. Arlt, *J. Appl. Phys.* **68**, 4220 (1990).
[26] M. Grossmann, *Imprint: An Important Failure Mechanism of Ferroelectric Thin Films in View of Memory Applications*. Dissertation. RWTH-Aachen, 2001. published by VDI Verlag, Fortschritt-Berichte VDI: Reihe 9 Elektronik/Mikro- und Nanotechnik.
[27] M. Grossmann, O. Lohse, D. Bolten, R. Waser, W. Hartner, G. Schindler, C. Dehm, N. Nagel, V. Joshi, N. Solayappan, and G. Derbenwick, *Int. Ferroelectrics* **22**, 95 (1998).
[28] M. Grossmann, O. Lohse, D. Bolten, R. Waser, W. Hartner, G. Schindler, C. Dehm, and N. Nagel, *Mat. Res. Soc. Symp. Proc.* **541**, 269 (1999).
[29] H. Al-Shareef, D. Dimos, W. Warren, and B. Tuttle, *J. Appl. Phys.* **80**, 4573 (1996).
[30] J. Benedetto, M. Roush, I. Lloyd, R. Ramesh, and B. Rychlik, *Integrated Ferroelectrics* **10**, 279 (1995).
[31] R. Suizu and S. Chapman, *Integrated Ferroelectrics* **16**, 87 (1997).
[32] M. Grossmann, O. Lohse, D. Bolten, U. Boettger, R. Waser, W. Hartner, M. Kastner, and G. Schindler, *Appl. Phys. Lett.* **76**, 363 (2000).
[33] Y. Shimada, K. Nakao, A. Inoue, M. Azuma, Y. Uemoto, and E. Fujii, *J. Appl. Phys. Lett.* **71**, 2538 (1997).
[34] B. S. Kang, Jong-Gul Yoon, D. J. Kim, T. W. Noh, T. K. Song, Y. K. Lee, J. K. Lee, and Y. S. Park, *Appl. Phys. Lett.* **82**, 2124 (2003).
[35] Y. Watanabe, M. Tanamura, and Y. Matsumoto, *Jpn. J. Appl. Phys.* **32**, 1564 (1996).

4 Optical Characterization of Ferroelectric Materials

Christoph Buchal

Research Center Juelich, Germany

4.1 Introduction: Light propagation within anisotropic crystals

The propagation of light within ferroelectric materials is an interesting subject, which can be used to evaluate some properties of the ferroelectric, especially the domain structure and the homogeneity of the material. Together with x-ray analysis, an optical inspection is the first method to be used, if bulk ferroelectric crystals have been grown. Most ferroelectrics have very anisotropic optical properties. Unpolarized light is not transmitted unchanged through anisotropic media, and two perpendicularly polarized waves travel with different speeds. We can visualize some of the essential differences between isotropic and uniaxial anisotropic crystals in the following way:

If we could place a point source of light inside an isotropic medium, flash the source for an infinitesimally small interval and then record photographically the locus of all the rays diverging from the source the locus would be spherical in shape since in any direction in the isotropic body the light waves travel with the same speed. Also the light arriving at any point on the sphere would be unpolarized. If we could now carry out the same experiment with the source placed inside a birefringent crystal, the results would be strikingly different. The photographic plate would show two surfaces. One of these, corresponding to the ordinary disturbance, would be spherical in shape since in any direction the ordinary wave travels with the same speed. We should also see that there was a second surface enclosing the first. In the case of calcite, a well known birefringent material, this second surface has the form of an ellipsoid of revolution which touches the spherical surface at the opposite ends of one diameter. The ellipsoidal surface represents the locus of the diverging extraordinary rays after an interval of time. The extraordinary disturbance travels with different speeds in different directions. In one direction, the optic axis direction, there is no double refraction; this is represented by the diameter joining the points at which the two surfaces touch. In all other directions the extraordinary is faster than the ordinary disturbance, the form of the variation being elliptical to give the maximum difference in speeds perpendicular to the optic axis. In any given direction each of the disturbances would be linearly polarized.

The surfaces seen on these photographs are known as *ray velocity surfaces*. For all uniaxial anisotropic crystals, we have a double surface; when the crystal has a positive optic sign (like quartz), the ellipsoid of revolution is enclosed by the sphere, but if the crystal is optically negative (like calcite), the ellipsoid encloses the sphere. The assumption that the form of the variation of velocity for the extraordinary disturbance in uniaxial crystals is ellipsoidal was first made empirically to explain double refraction. Once the form of the ray velocity surface is known, it can be used to predict the directions of the wave fronts within the medium.

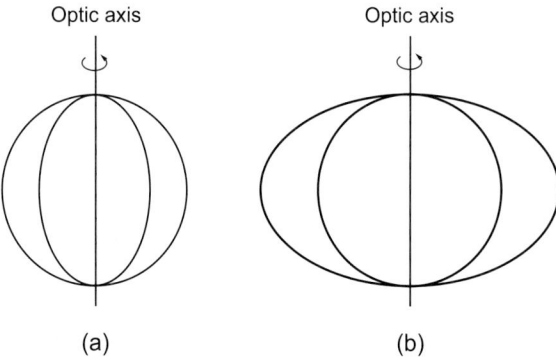

Figure 4.1: Sections of uniaxial ray velocity surfaces. (a) Optically positive crystal and (b) optically negative crystal.

4.1.1 Huyghens's construction for uniaxial crystals

We can illustrate this construction by considering the simple experiment showing the double refraction of an object dot viewed through opposite faces of a calcite rhomb. Let us take a section through the air-crystal interface through which light from the object enters the calcite. In Figure 4.2 the incident light beam, normal to the interface, is defined by the extreme rays PR and QS; in the Huyghenian construction the points R and S, on the wave front marked zero, act as sources for the secondary disturbances to simultaneously spread within the crystal, as do all intermediate points on the interface. During the interval of time between the wave fronts, the rays diverging from the secondary sources will have spread into the crystal to points on negative uniaxial ray velocity surfaces. The figure shows the section of the two double-surfaced figures arising from the limiting secondary sources R and S, although we must remember that there is an infinite number of such surfaces due to all the sources between R and S. We must now consider the implications of Figure 4.2.

The wave fronts transmitted within the crystal are the envelopes of all the surfaces representing the secondary wavelets; thus the +1 wave fronts in the crystal are given by the common tangents to extreme secondary wavelets. We see from the figure that there are two parallel wave fronts travelling in the crystal represented by TU and LM for the ordinary and extraordinary waves respectively, and that the wave normal direction is common both to them and the incident waves. It is also clear that the two parallel wave fronts travel with different speeds for they are at different positions within the crystal; the extraordinary wave fronts advance faster than the ordinary wave fronts (RN > RT). In order to locate the images of the dot formed by the two waves, we must now consider the direction of advance of a given point on the front; physically this is what is meant by the ray directions within the crystal.

4.1 Introduction: Light propagation within anisotropic crystals

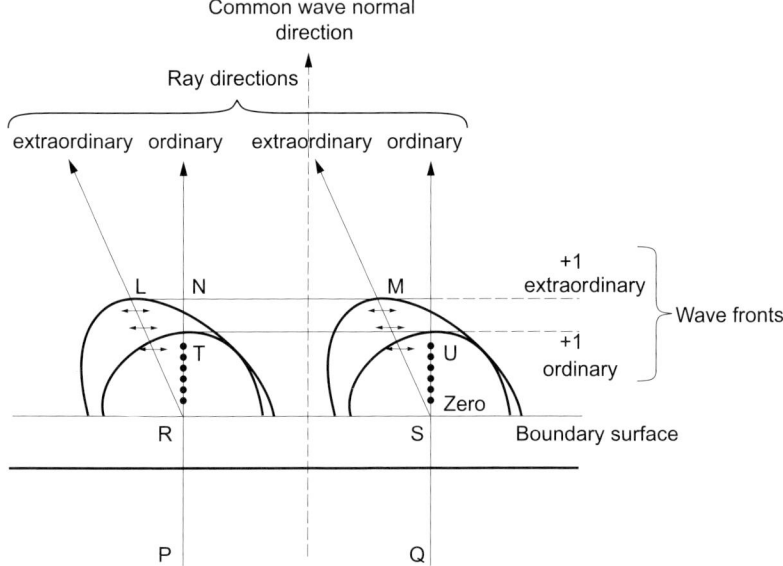

Figure 4.2: Huyghens's construction for light incident obliquely onto a rhombohedral face of a calcite crystal

Obviously, the ray directions associated with each wave front can be obtained by joining the source of the secondary wavelet to the point at which the surface of the wavelet is touched by the wave front. For ordinary waves, the ray directions are parallel to RT and SU, while the extraordinary ray directions are parallel to RL and SM. We see then that although the two transmitted waves have parallel fronts, i.e. they share a common wave normal direction, the ordinary and extraordinary ray directions diverge to give the double image of the object dot. The ordinary ray directions predict the undisplaced image, while the extraordinary ray directions give rise to the displaced image which rotates as the rhomb is rotated on the paper.

It is instructive to pursue the interpretation of the calcite rhomb experiment beyond the simple Huyghenian construction to learn something about the polarization of the transmitted light. The electromagnetic theory of light requires that the electric vector shall be contained in the plane of the wave front. The ordinary disturbances vibrate perpendicular to a principal section. Also the extraordinary disturbance must vibrate in the principal section plane.

It is apparent from this that the vibration direction for the transmitted extraordinary waves is not generally perpendicular to the ray direction.

We can now summarize the conclusions that have been reached for conditions of normal incidence about the light waves which pass through a general section of a uniaxial crystal. The two disturbances, ordinary and extraordinary, have parallel wave fronts but their velocities along the common wave normal direction are different; if the optic sign is positive the ordinary waves travel faster, and vice versa. The two transmitted waves are linearly polarized. For the ordinary disturbance the vibration direction is perpendicular to the principal section

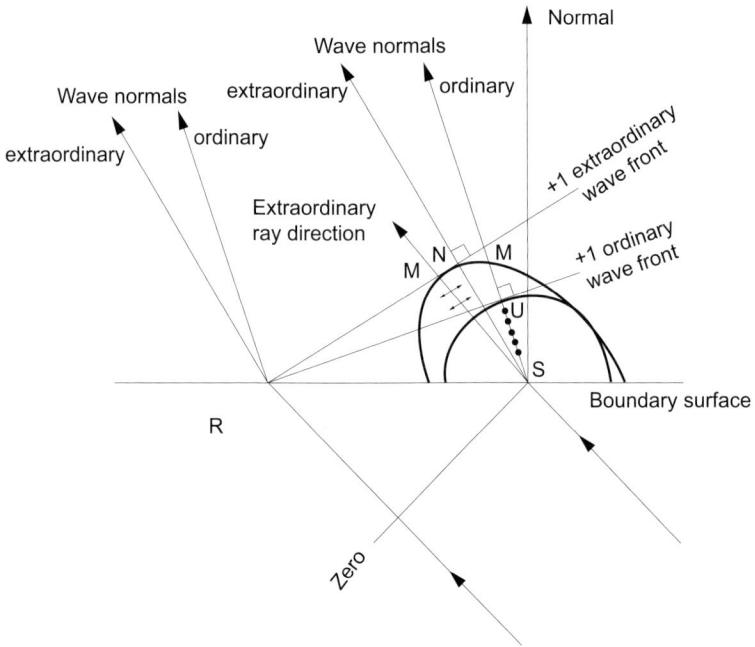

Figure 4.3: Huyghens's construction for light incident obliquely onto a rhombohedral face of a calcite crystal

containing the common wave normal and the optic axis; it lies in the plane of the wave front and perpendicular to the ray direction. For the extraordinary disturbance the vibration direction lies in the principal section containing the common wave normal and the optic axis; it lies in the plane of the wave front, but is inclined at a general angle to the ray direction.

4.1.2 The uniaxial indicatrix

For optically uniaxial crystals we know that the refractive index values for extraordinary waves are variable, with that for ordinary waves fixed. We can link this observation with that concerning the vibration directions for the two waves travelling along a general wave normal direction; the ordinary vibration direction is always perpendicular to the optic axis, while the extraordinary vibration is always in the plane containing the optic axis and wave normal direction. This suggests that we may connect the variation of the refractive index in the crystal with the vibration direction of the light. This concept allows a convenient representation of anisotropic optical properties in the form of a spatial plot of the variation of refractive index as a function of vibration direction. Such a surface is known as the *optical indicatrix*.

For uniaxial crystals, the optical indicatrix is a single-surfaced ellipsoid of revolution similar in shape to the extraordinary ray velocity surface. To construct the optical indicatrix for a particular example, say calcite, we construct the ellipsoid of revolution so that the radius

4.1 Introduction: Light propagation within anisotropic crystals

of its central circular section is directly proportional to the ordinary refractive index, and the length of the unique axis of revolution is directly proportional to the minimum value of the extraordinary refractive index; for this negative crystal, the indicatrix is an oblate ellipsoid (Figure 4.4 (a)). For an optically positive crystal, the radius of the central circular section is again proportional to the ordinary refractive index, while the length of the axis of revolution is proportional to the maximum value of the extraordinary refractive index; the indicatrix is a prolate ellipsoid (Figure 4.4 (b)). In both figures, we notice that the radius vector normal to the circular section is n_e, showing that a maximum (or minimum) refractive index is observed for light with this vibration direction; the axis of revolution must be the optic axis direction. Since the optic axis is perpendicular to the circular section of the indicatrix of radius n_o, light vibrating perpendicular to the optic axis always has this value of the refractive index. For any other radius vector of the figure, i.e. any general vibration direction, the refractive index n_e' is intermediate between the two principal values.

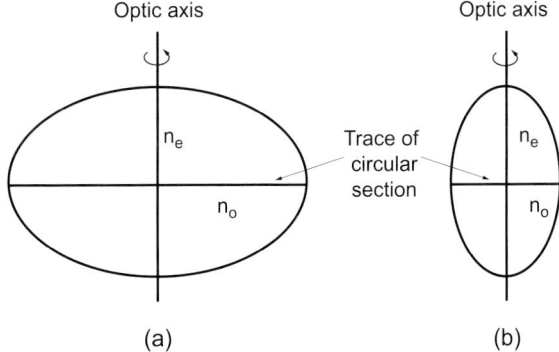

Figure 4.4: Sections of the optical indicatrix for uniaxial crystals. (a) Optically negative crystal and (b) optically positive crystal

We must now consider how the properties of such figures allow us to interpret the transmission of light in any direction in a crystal. Let us imagine that the transmitted waves are travelling within a negative uniaxial crystal at some general inclination to the optic axis, i.e. in the direction of the common wave normal of Figure 4.5. In order to find the optical properties of the crystal for light travelling in this direction, we cut a central elliptical section of the ellipsoid perpendicular to the wave normal, i.e. we take the central section of the indicatrix cut by the plane of the wave fronts in the crystal. The properties of the indicatrix are such that the two axes of this elliptical section define the two mutually perpendicular vibration directions permitted for this wave normal direction, while the length of the semiaxis is, in each case, therefore proportional to the appropriate refractive index (Figure 4.5). Consideration of the figure shows that every central section of the ellipsoid (except that perpendicular to the axis of revolution) is an ellipse with one of its axes perpendicular to the axis of revolution. One of the axes of the ellipse must always lie in the circular section of the indicatrix; this is the ordinary vibration direction with a corresponding refractive index n_o. The other axis of the ellipse must lie in the principal plane containing the wave normal; this is the extraordinary vibration direction with a corresponding refractive index n'_e; intermediate between the extreme values of n_e and n_o. For the special case when the wave normal is parallel to the axis of revolution, all vibration directions lie in the circular section of the ellipsoid and have equal values; they are not to be distinguished from each other, and there is only one refractive index value n_o so confirming that the optic axis direction corresponds to the axis of revolution. As the difference between n_e and n_o is reduced towards zero, the shape of the ellipsoid approaches a sphere, which is the indicatrix for an isotropic substance, in which there is only one value of the refractive index.

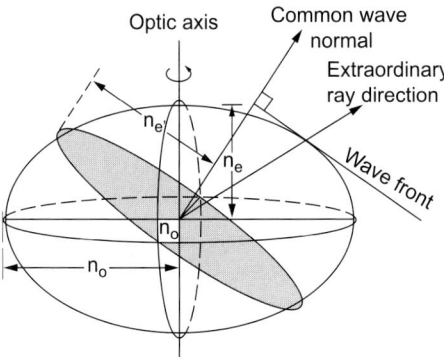

Figure 4.5: The properties of the negative uniaxial indicatrix

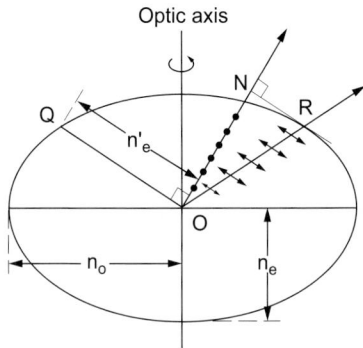

Figure 4.6: A principal section of the negative uniaxial indicatrix

4.1.3 The biaxial indicatrix

If we try to understand the transmission of light waves in biaxial crystals, we start from the concept of the indicatrix, and to attempt to visualize what shape this must have to show the variation of refractive index with vibration direction for such crystals. From our previous knowledge of the indicatrix for uniaxial crystals, an ellipsoid of revolution with two principal refractive indices, n_o and n_e, it is a simple step to see that the indicatrix for biaxial crystals will be a triaxial ellipsoid with three principal refractive indices, n_γ, n_β and n_α.

4.2 The electro-optic effect

Certain transparent materials change their index of refraction, if an electric field is applied to them. This may be used to control the propagation of light. We will first give the basic explanation of the phenomenon.

Light is an electromagnetic wave and its $\vec{E}$-vector interacts strongly with the electrons. Much of this was understood by H. A. Lorentz, who considered the electrons as negative charges, elastically bound to the nuclei. Their response to the oscillating electric field of the light wave is described by the physics of the driven harmonic oscillator. Because of the very high frequencies of the light wave, which amounts to 500 THz at a vacuum-wavelength of 0.6 µm, the heavy nuclei do not oscillate noticeably, but especially the valence electrons in their extended orbitals do respond to the periodic disturbance. This is the origin of the refractive index of matter

$$n(\omega) = \sqrt{\epsilon(\omega)} \tag{4.1}$$

On the other hand, the application of a static or "slowly" varying electric field will be able to displace ions and electrons away from their equilibrium positions and, as a consequence, the polarizability of the electrons will be modified. In the description of H. A. Lorentz's electronic oscillators, the small shifts in the ionic positions modify the "spring constants and restoring forces" of the electronic oscillators.

Keeping in mind that even the fast varying electromagnetic field of light has a wavelength, which is three orders of magnitude larger than the atomic dimensions, the optical response will be described by the average response from many atomic orbitals.

If crystals have a certain distinct arrangement of the ionic charges, the sign and orientation of the applied external $\vec{E}$- vector is important. Examples are the optical ferroelectrics like BTO or LNO and the III-V-semiconductors like GaAs (the charge distribution of GaAs may be seen as Ga^-As^+). In these cases, a reversal of the electric field $\vec{E}$ changes the sign of the index change Δn

$$n(\vec{E}) = n(0) + \Delta n(\vec{E}), \qquad \Delta n(+\vec{E}) = -\Delta n(-\vec{E}) \tag{4.2}$$

This is called the linear electro-optic effect, also called the "Pockels effect".

In non-polar, isotropic crystals or in glasses, there is no crystallographic direction distinguished and the linear electro-optic effect is absent. Nevertheless a static field may change the index by displacing ions with respect to their valence electrons. In this case the lowest non-vanishing coefficients are of the quadratic form, i.e. the refractive index changes proportionally to the square of the applied field: "Kerr effect".

4.2.1 Ferroelectrics have anisotropic electronic bonds: Birefringence

In anisotropic materials, the electronic bonds may have different polarizabilities for different directions (you may think of different, orientation-dependent spring constants for the electronic harmonic oscillator). Remembering that only the $\vec{E}$-vector of the light interacts with the electrons, we may use polarized light to test the polarizability of the material in different directions. LNO is one of the most important electro-optic materials and we use it as an example. The common notations are shown in Figure 4.7. If the $\vec{E}$-vector is in plane with the surface of the crystal, the wave is called a TE wave. In this example, the TE wave would experience the ordinary index n_o of LiNbO$_3$ ($n_o \approx 2.20$). If we rotate the polarization by 90°, the $\vec{E}$-vector will be vertical to the surface and the wave is called TM. In LNO, it will experience the extraordinary index $n_e \approx 2.29$. Therefore these two differently polarized waves will propagate with different phase velocities $v = c/n$. In the example of Figure 4.7, the TE mode is faster than the TM mode.

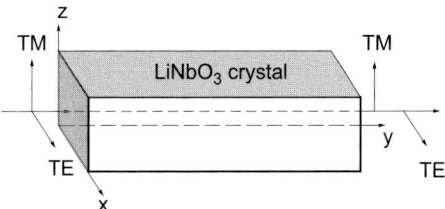

Figure 4.7: Explanation of the notations: The light beam propagates to the right. The LiNbO$_3$ z-axis is vertical to the main surface of the crystal. This is a "z-cut". If the $\vec{E}$- vector of the light beam lies parallel to the surface, this is called TE polarization. If the $\vec{E}$-vector is oriented normal to the surface plane, this is a TM polarization

How do we describe the polarization states of light, which are not polarized parallel to one of the main crystalline axes? We have to look at the different components of $\vec{E}$ parallel to the main axes, evaluate their propagation and at the end of the optical path add them up to the resulting beam.

With respect to the crystal, this is visualized by the index ellipsoid or indicatrix. (In a homogeneous medium this would be a sphere.) Without external field, the indicatrix is oriented along the main crystallographic directions.

$$\frac{x^2}{n_x^2} + \frac{y^2}{n_y^2} + \frac{z^2}{n_z^2} = 1 \tag{4.3}$$

This relation is shown in Figure 4.8. Using a simple language, we may say that linearly polarized light impinging on the indicatrix with the polarization parallel to a main axis will propagate with the corresponding phase velocity and will remain linearly polarized. Otherwise different phase velocities apply for the orthogonal polarizations and an elliptically polarized beam will result.

4.2 The electro-optic effect

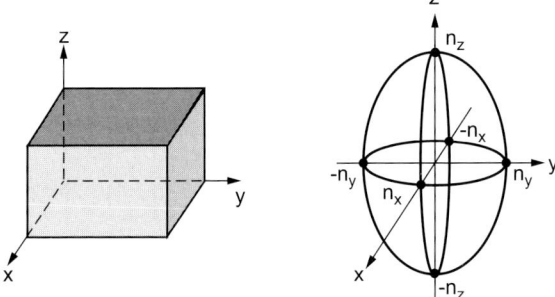

Figure 4.8: Relation between the crystal and the index ellipsoid (right). The application of an electric field rotates and deforms the ellipsoid

The subject may become a little involved, if an external electrical field is applied: Ions move, bonds are deformed and even the first-order equations for the change of the indicatrix must include the cross-terms, which describe the rotation of the ellipsoid. We follow the notation of [2] and write for the indicatrix in the presence of an electric field

$$B_{11}x^2 + B_{22}y^2 + B_{33}z^2 + 2B_{23}yz + 2B_{31}zx + 2B_{12}xy = 1 \quad (4.4)$$

This ellipsoid is tilted with respect to the x, y, z-axes and the parameters B_{ij} are a function of the electric field $\vec{E}$. In order to couple the six constants B_{ij} to the three components of $\vec{E}$, 18 coefficients are needed. They are arranged in the form of a 3×6 matrix, which is sometimes called the "electro-optic tensor":

$$\begin{pmatrix} B_{11} - \dfrac{1}{n_x^2} \\ B_{22} - \dfrac{1}{n_y^2} \\ B_{33} - \dfrac{1}{n_z^2} \\ B_{23} \\ B_{31} \\ B_{12} \end{pmatrix} = \begin{pmatrix} r_{11} & r_{12} & r_{13} \\ r_{21} & r_{22} & r_{23} \\ r_{31} & r_{32} & r_{33} \\ r_{41} & r_{42} & r_{43} \\ r_{51} & r_{52} & r_{53} \\ r_{61} & r_{62} & r_{63} \end{pmatrix} \cdot \begin{pmatrix} E_x \\ E_y \\ E_z \end{pmatrix} \quad (4.5)$$

The 18 r_{ij} are called "electro-optic coefficients", and the problem starts to look a little complex and confusing. Fortunately, the symmetry and the physics of the crystal frequently reduce the complexity nicely. Look at four important examples:

$$\begin{pmatrix} r_{11} & r_{12} & r_{13} \\ r_{21} & r_{22} & r_{23} \\ r_{31} & r_{32} & r_{33} \\ r_{41} & r_{42} & r_{43} \\ r_{51} & r_{52} & r_{53} \\ r_{61} & r_{62} & r_{63} \end{pmatrix} : \begin{pmatrix} 0 & -3.4 & 8.6 \\ 0 & 3.4 & 8.6 \\ 0 & 0 & 30.8 \\ 0 & 28 & 0 \\ 28 & 0 & 0 \\ -3.4 & 0 & 8.6 \end{pmatrix} \begin{pmatrix} 0 & 0 & 8 \\ 0 & 0 & 8 \\ 0 & 0 & 30.8 \\ 0 & 820 & 0 \\ 820 & 0 & 0 \\ 0 & 0 & 8.6 \end{pmatrix} \qquad (4.6)$$

Electro − optic tensor | LiNbO$_3$ 3m | BaTiO$_3$ 4mm

$$\begin{pmatrix} r_{11} & r_{12} & r_{13} \\ r_{21} & r_{22} & r_{23} \\ r_{31} & r_{32} & r_{33} \\ r_{41} & r_{42} & r_{43} \\ r_{51} & r_{52} & r_{53} \\ r_{61} & r_{62} & r_{63} \end{pmatrix} : \begin{pmatrix} 0 & 0 & 0 \\ 0 & 0 & 0 \\ 0 & 0 & 0 \\ 1.6 & 0 & 0 \\ 0 & 1.6 & 0 \\ 0 & 0 & 1.6 \end{pmatrix} \begin{pmatrix} 0 & 0 & 0 \\ 0 & 0 & 0 \\ 0 & 0 & 0 \\ 8.6 & 0 & 0 \\ 0 & 8.6 & 0 \\ 0 & 0 & 10.6 \end{pmatrix} \qquad (4.7)$$

Electro − optic tensor | GaAs 43m | KDP 42m

All coefficients are in units of 10^{-12} m/V and the symmetry classes are denoted under the tensors.

Frequently it is possible to avoid the complications of the cross-terms by applying the external field parallel to a main orientation of the crystal and by choosing the corresponding polarization of the light beam.

Important examples are the trigonal crystals LiNbO$_3$ and LiTaO$_3$. For $\vec{E} = (0, 0, E)$ the ellipsoid equation changes to the following form:

$$\left(\frac{1}{n_o^2} + r_{13}E\right)(x^2 + y^2) + \left(\frac{1}{n_e^2} + r_{33}E\right)z^2 = 1 \qquad (4.8)$$

The principal axes have changed length, but they are not rotated. This ellipsoid gives for $n_o(E)$ and $n_e(E)$

$$\frac{1}{n_o^2(E)} = \frac{1}{n_o^2} + r_{33}E \qquad (4.9)$$

$$\frac{1}{n_e^2(E)} = \frac{1}{n_e^2} + r_{13}E \qquad (4.10)$$

Using the approximation $\sqrt{1+\alpha} = 1 - 1/2\alpha$, we find

$$n_o(E) = n_o - \frac{1}{2}n_o^3 r_{13}E, \qquad \vec{E} = (0, 0, E) \qquad (4.11)$$

$$n_e(E) = n_e - \frac{1}{2}n_e^3 r_{33}E \qquad (4.12)$$

4.2 The electro-optic effect

In the following, we will use equations of this rather simple form, but we keep in mind that the general case is not always easy to analyze. The sign of the index change depends on the sign of the applied electric field. The changes are generally very small. For LNO, even the largest coefficient $r_{33} = 31 \cdot 10^{-12}$ m/V = 0.31 Å/V. That tells us that it takes external fields of volts on an atomic length scale for strong changes of n. Of course, such high values for the applied electric field are not realistic. As the electrical breakdown of LNO limits the usable fields to approximately 10 V/μm, a maximum index change of $1.65 \cdot 10^{-3}$ is possible.

4.2.2 Applied fields change the optical pathlength: Phase modulators

We have seen that a bulk transparent electro-optic crystal changes its index by a small amount, if an electric field is applied. Figure 4.9 shows a phase modulator for transverse geometry. Let us look at some numbers for LiNbO$_3$: $d = 1$ mm, $L = 10$ mm, $r_{33} = 30.9$ pm/V, $n_e = 2.2$, $U = 180$ V, and $\lambda_o = 0.6$ μm. We find
$\lambda = \lambda_o/n_e = 0.273$ μm (wavelength inside the crystal),
$m = L/\lambda = 36630$ (number of waves within 10 mm of crystal),

$$\begin{aligned}\Delta n_e/n_e &= (1/2) \cdot n_e^2 \cdot r_{33} \cdot E_z \\ &= (1/2) \cdot 4.84 \cdot 30.9 \cdot 10^{-12} \cdot 180 \cdot 10^3 \\ &= 1.35 \cdot 10^{-5} \text{ (relative change of } n_e)\end{aligned} \quad (4.13)$$

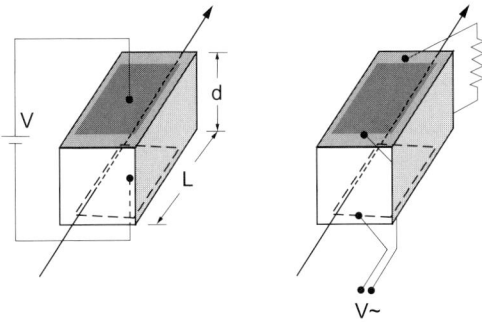

Figure 4.9: Schematic of bulk electro-optic phase modulator in transverse geometry: (a) standard electrode arrangement for static measurements, (b) travelling wave configuration

This change of the optical pathlength corresponds to $\Delta m = 1.35 \cdot 10^{-5} \times 36630 = 0.5$ (half a wavelength).

This is an important message: Even the best electro-optical materials provide only small index changes. Therefore it takes a long optical path (corresponding to $4 \cdot 10^4$ wavelengths in this example) to provide a total phase delay of π. Taking a closer look at the proceeding calculation, we find the same half-wave voltage $V_\pi = 180$ V for all geometries with a constant ratio $L/d = 10$. The smaller we can design the electrode spacing d, the lower the voltage or the shorter the modulator length may become. Other geometries are also possible. In most cases the transverse geometry has the lowest half-wave voltage V_π amongst the electro-optic modulators.

If we think about it, we come to the conclusion that a phase modulator alone probably is not yet a very useful device. Only if we can create interference patterns between the phase-modulated signal and a reference beam we will be able to detect a phase modulation. This is realized in the Mach-Zehnder geometry, see Figure 4.10. Another possible setup uses the phase modulator as a controlled optical retarder, see Figure 4.11. If the "slow" beam is retarded by π, the vector sum of the total beam is again linearly polarized, but rotated by 90°. Therefore a retarder between two crossed polarizers (45°/-45°) makes an on-off-amplitude modulator. This is shown for the case of a KDP Pockels cell in Figure 4.6. In this case, a longitudinal E-field geometry has been used.

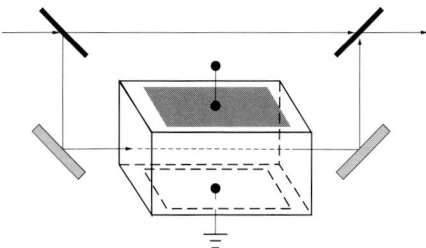

Figure 4.10: Free beam interferometric Mach-Zehnder setup. Two beam splitters permit the interference between phase-modulated and primary beam. This is an intensity modulator

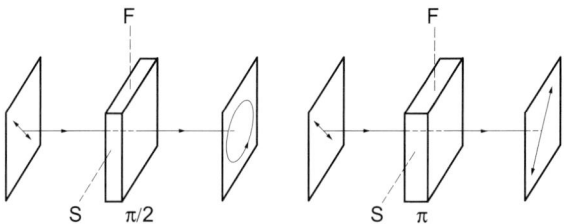

Figure 4.11: Two optical retarders. F: fast axis, S: slow axis. In both cases the incident beam is polarized under 45°. Left: a quarter wave plate retards the slow component of the polarization by $\pi/2$ and elliptical polarization is achieved. Right: a half-wave plate rotates the plane of polarization from 45° to -45°.

4.2 The electro-optic effect 89

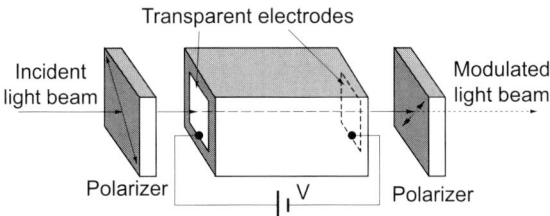

Figure 4.12: Longitudinal Pockels cell. The KDP electro-optic crystal between two crossed polarizers makes an intensity modulator.

4.2.3 Optical waveguides improve the device efficiency

Take a look at Figure 4.13. It demonstrates that a focussed Gaussian beam may have a narrow waist at a certain point, but then will diverge again. In contrast, an optical waveguide maintains a better optical confinement over its entire length. A waveguiding device maintains a high optical intensity within a small cross-section. This results in a narrow spacing for the modulator electrodes and correspondingly low drive voltages of the modulator. (In addition, waveguiding devices are advantageous for non-linear functions for most functions of non-linear optics, because always a high electrical field strength of the interacting beams is needed and will be provided most efficiently by a waveguiding device.) In addition, waveguides are well matched and easily coupled to diode lasers, optical fibers and all devices of integrated optics.

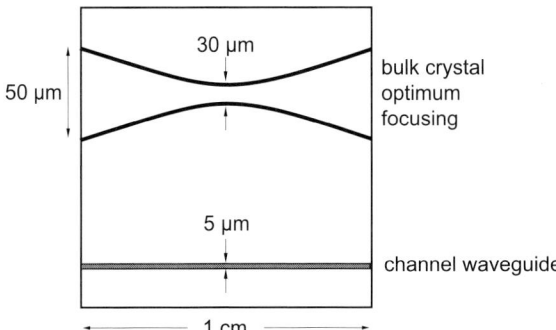

Figure 4.13: Schematic of the possible light beam confinement over a length of 1 cm. Top: free laser beam with optimum focusing; bottom: optical channel waveguide

But not all waveguides are suitable for a modulator application. In (large crossection) multimode guides, every mode propagates with its own phase velocity. Therefore an optimized single-mode waveguide sustaining one TE or TM mode (depending on the correct polarization state) is needed for a waveguiding modulator device.

A Mach-Zehnder waveguide modulator is shown in Figure 4.14 (a). It has the same sinusoidal on-off characteristics as the device shown in Figure 4.12. Actually its off-state is realized by the mode-coupling geometry seen in Figure 4.15. The two modes with a phase difference of π couple to an antisymmetric higher mode, which is not sustained by the single-mode waveguides, but instead is radiated into the substrate. The phase mismatch of π corresponds to the off-state. The on-off ratio can reach values exceeding 30 dB.

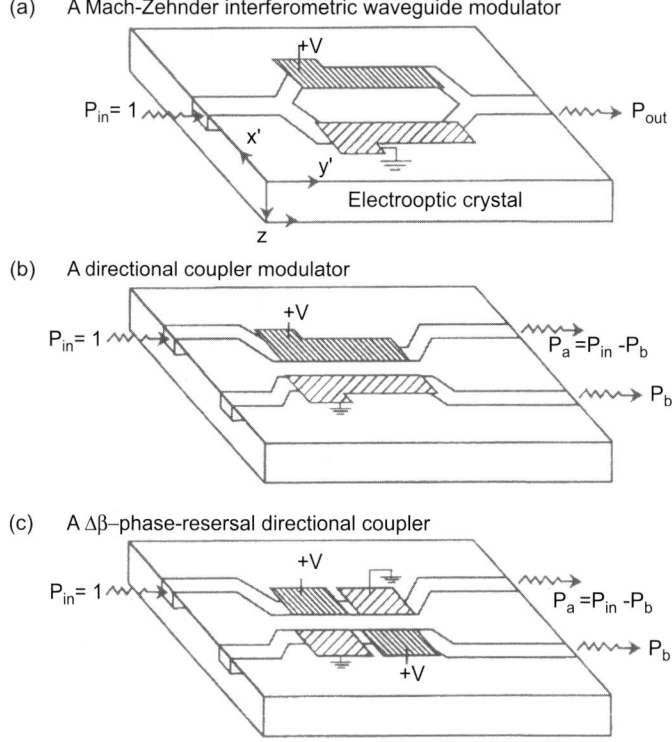

Figure 4.14: Three different electrooptic waveguide devices [1]; (a) Mach-Zehnder modulator with push-pull arrangement of the electrodes, (b) electrically controlled directional coupler, (c) coupler with phase-reversal electrodes.

Nevertheless, the device is not really a linear device, as can be seen from the sinusoidal characteristics of Figure 4.16. The balancing of the two arms has to be (electrically) adjusted to achieve the high extinction ratio. On the other hand, the linear part of the characteristics may be used for analog applications, see Figure 4.11. For instance, this type of linear phase modulation is used in the so-called "electro-optical needle", which measures rapidly varying electrical fields in integrated semiconductor circuits [8]. An LNO phase modulator is a very fast device, which converts electrical fields into optical signals. In the case of the electro-optical needle, electrical signals of up to 1000 GHz can be detected.

4.2 The electro-optic effect

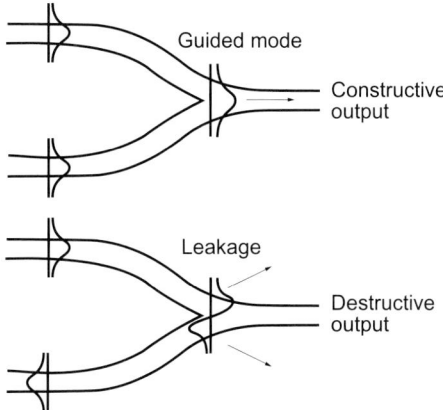

Figure 4.15: Illustration of the constructive and destructive interference pattern in a Mach-Zehnder waveguide modulator. Only if the outgoing waveguide is single mode, the resulting antisymmetric mode will be completely radiated into the substrate. A multimode waveguide cannot be used for an efficient Mach-Zehnder modulator

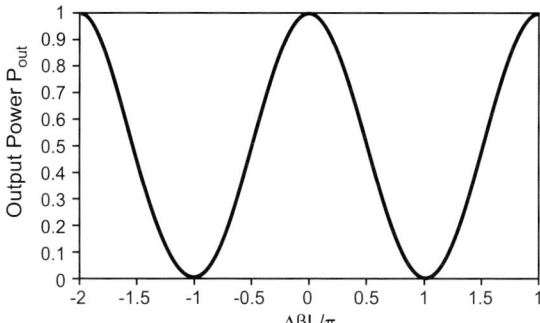

Figure 4.16: The observed $\cos^2(\nu)$ signal from a Mach-Zehnder modulator. A phase mismatch of π gives one full modulation amplitude. The pattern is completely periodic for higher values of the phase mismatch

For waveguide phase modulators and especially the Mach-Zehnder modulator, there are two main electrode configurations, which are explained in Figure 4.18.

For high data rate modulators, the dynamic behavior becomes an important issue. Today, a modulation rate of 100 GHz has become a realistic objective. For a better understanding of the modulator dynamics, we have to consider the phase velocity of the two interacting waves:
a) the optical wave moves with its phase velocity of $v = c/n$ ($\nu = 500$ THz),
b) the microwave propagates with its phase velocity of $v = \sqrt{\epsilon(\nu)}$ ($\nu = 100$ GHz).

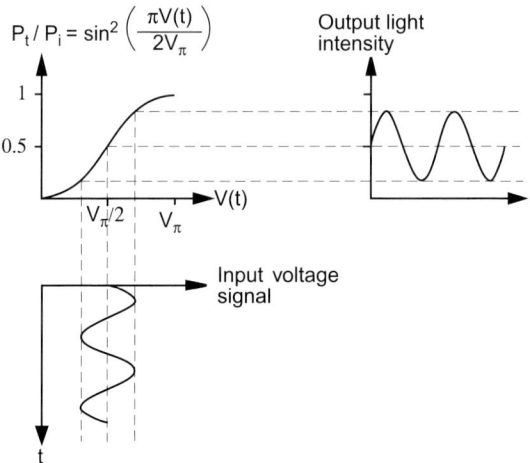

Figure 4.17: Demonstration of the limited linear part of electro-optical modulators. In digital optocommunication applications, the full modulation depth (on/off) is used

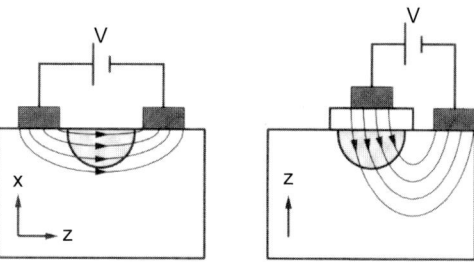

Figure 4.18: Two main electrode configurations; left: x-cut crystal, two electrodes parallel to the optical waveguide, right: z-cut crystal, one electrode lies on top of a separating buffer layer directly over the waveguide

For the optical signal, actually the effective index n_{eff} of the guided mode determines the phase velocity, but the value of n_{eff} is very close to the bulk index n of the waveguiding material. The microwave is more difficult to analyze. Parts of its electrical field propagate in the LNO. Depending on orientation, the microwave dielectric constant ϵ of LNO is between 25 and 45. Assuming for the moment a complete propagation of the microwave within the LNO and an average value of $\epsilon = 30$, a 100 GHz signal with $\lambda_{vac} = 3$ mm has a reduced wavelength of 0.55 mm. As we have seen, typical modulator electrodes are 10 mm long. For the modulating microwave, this length corresponds to 18 full or 36 half waves. In this exaggerated example, the modulator electrical signal would change its polarity 36 times along the optical modulator electrode.

4.3 Non-linear optics

How is it possible to design 100 GHz modulators?

- The electrical signal propagates partly in air. This reduces the effective microwave index and brings it closer to the optical index.

- Very thick metallic electrodes (up to 30 μm Au have been used) provide low ohmic resistance and lower RC constants.

- A favorable design uses a coplanar ridge structure waveguide for the microwave, which travels parallel to the optical signal. As has been shown, this structure in fact allows a good match between microwave and light signal. In this case, microwave and light travel with the same phase velocity and strong modulation is achieved.

- Some authors have designed electrode configurations, which reverse the sign of the modulating field along the path of interaction: "Phase-reversal electrodes". This is similar to the electrode configuration shown in Figure 4.14 (c). In this design the polarity reversal of the electrodes compensates the sign change of the modulating signal due to the phase velocity mismatch.

4.3 Non-linear optics

The previous paragraphs have been concerned with linear optical properties. Throughout the long history of optics, and indeed until relatively recently, it was thought that all optical media were linear. The assumption of linearity of the optical medium has far-reaching consequences:

- The optical properties, such as the refractive index and the absorption coefficient, are independent of light intensity

- The principle of superposition, a fundamental of classical optics holds.

- The frequency of light cannot be altered by its passage through the medium.

- Light cannot interact with light; two beams of light in the same region of a linear optical medium can have no effect on each other. Thus light cannot control light.

The invention of the laser in 1960 allowed to examine the behavior of light in optical materials at higher intensities than previously possible. Many of the experiments carried out made it clear that optical media do in fact exhibit nonlinear behavior, as exemplified by the following observations:

- The refractive index, and consequently the speed of light in an optical medium, does change with the light intensity.

- The principle of superposition is violated.

- Light can alter its frequency as it passes through a nonlinear optical material (e.g., from red to blue!).

- Light can control light; photons do interact.

Especially the optical ferroelectrics are famous for their nonlinear properties and the application of nonlinear optics opens many fascinating possiblities. Linearity or nonlinearity is a property of the *medium* through which light travels, rather than a property of the light itself. Nonlinear behavior is not exhibited when light travels in free space. *Light interacts with light only via the medium.* The presence of an optical field modifies the properties of the medium which, in turn, modify another optical field or even the original field itself.

The properties of a dielectric medium through which an electromagnetic (optical) wave propagates are completely described by the relation between the polarization density vector $\vec{P}(r,t)$ and the electric-field vector $\vec{E}(r,t)$. It was suggested that $\vec{P}(r,t)$ could be regarded as the output of a system whose input was $\vec{E}(r,t)$. The mathematical relation between the vector functions $\vec{P}(r,t)$ and $\vec{E}(r,t)$ defines the system and is governed by the characteristics of the medium. The medium is said to be nonlinear if this relation is nonlinear.

4.3.1 Nonlinear optical media

A linear dielectric medium is characterized by a linear relation between the polarization density and the electric field, $P = \epsilon_0 \chi E$, where ϵ_0 is the permittivity of free space and χ is the electric susceptibility of the medium. A nonlinear dielectric medium, on the other hand, is characterized by a nonlinear relation between P and E, as illustrated in Figure 4.19.

The nonlinearity may be of microscopic or macroscopic origin. The polarization density $P = Np$ is a product of the individual dipole moment p, which is induced by the applied electric field E, and the density of dipole moments N. The nonlinear behavior may have its origin in either p or N.

The relation between p and E is linear when E is small, but becomes nonlinear as E acquires values comparable with interatomic electric fields (typically, 10^5 to 10^8 V/m). This may be explained in terms of the simple Lorentz model in which the dipole moment is $p = -ex$, where x is the displacement of a mass with charge $-e$ to which an electric force $-eE$ is applied. If the restraining elastic force is proportional to the displacement (i.e., if Hooke's law is satisfied), the equilibrium displacement x is proportional to E; P is then proportional to E, and the medium is linear. However, if the restraining force is a nonlinear function of the displacement, the equilibrium displacement x and the polarization density P are nonlinear functions of E and, consequently, the medium is nonlinear.

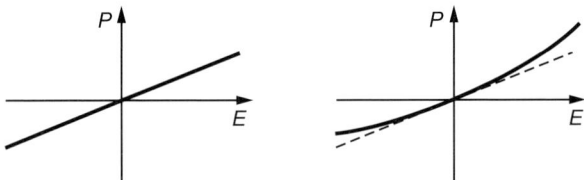

Figure 4.19: The P-E relation for (a) a linear dielectric medium, and (b) a nonlinear medium

Since externally applied optical electric fields are typically small in comparison with characteristic interatomic or crystalline fields, even when focused laser light is used, the nonlinearity is usually weak. The relation between P and E is then approximately linear for small E, deviating only slightly from linearity as E increases. Under these circumstances, it is possible to expand the function that relates P to E in a Taylor's series about $E = 0$,

$$P = \alpha_1 E + \frac{1}{2}\alpha_2 E^2 + \frac{1}{6}\alpha_3 E^3 .... \tag{4.14}$$

and to use only few terms. The coefficients α_1, α_2, and α_3 are the first, second, and third derivatives of P with respect to E at $E = 0$. These coefficients are characteristic constants of the medium. The first term, which is linear, dominates at small E. Clearly, $\alpha_1 = \epsilon_0/\chi$, where χ is the linear susceptibility, which is related to the dielectric constant and the refractive index by $n_2 = \epsilon/\epsilon_0 = 1 + \chi$. The second term represents a quadratic or second-order nonlinearity, the third term represents a third-order nonlinearity, and so on. It is customary to write

$$P = \epsilon_0 \chi E + 2dE^2 + 4\chi^{(3)} E^3 .... \tag{4.15}$$

where $d = 1/4\alpha_2$ and $\chi^{(3)} = 1/24\alpha_3$ are coefficients describing the second- and third-order nonlinear effects, respectively. Equation (4.15) provides the basic description for a nonlinear optical medium. Anisotropy, dispersion, and inhomogeneity have been ignored.

In centrosymmetric media the P-E function must have odd symmetry, so that the reversal of E results in the reversal of P without any other change. The second-order nonlinear coefficient d must then vanish, and the lowest order nonlinearity is of third order.

Typical values of the second-order nonlinear coefficient d for dielectric crystals, semiconductors, and organic materials used in photonics applications lie in the range $d = 10^{-24}$ to 10^{-21} (MKS units, As/V^2). Typical values of the third-order nonlinear coefficient $\chi^{(3)}$ for glasses, crystals, semiconductors, semiconductor-doped glasses, and organic materials of interest in photonics are $\chi^{(3)} = 10^{-34}$ to 10^{-29} (MKS units).

4.3.2 The nonlinear wave equation

The propagation of light in a nonlinear medium is governed by the wave equation, which was derived from Maxwell's equations for an arbitrary homogeneous dielectric medium,

$$\nabla^2 E = -\frac{1}{c_0^2}\frac{\partial^2 E}{\partial t^2} = \mu_0 + \frac{\partial^2 P}{\partial t^2} \tag{4.16}$$

It is convenient to write P as a sum of linear and nonlinear parts,

$$P = \epsilon_0 \chi E + P_{NL} \tag{4.17}$$

$$P_{NL} = 2dE^2 + 4\chi^{(3)} E^3 .... \tag{4.18}$$

Using (4.17) and the relations $n^2 = 1 + \chi$, $c_0 = (\mu_0 \epsilon_0)^{-1/2}$, and $c = c_0/n$, Equation (4.16) may be written as

$$\nabla^2 E = -\frac{1}{c_0^2}\frac{\partial^2 E}{\partial t^2} = -S$$

$$S = -\mu_0 + \frac{\partial^2 P_{NL}}{\partial t^2} \tag{4.19}$$

It is useful to regard (4.19) as a wave equation in which the term $S = -\mu_0 \partial^2 P_{NL}/\partial t^2$ acts as a source radiating in a linear medium of refractive index n. Because P_{NL} (and therefore S) is a nonlinear function of E, Equation (4.19) is a nonlinear partial differential equation in E. This is the basic equation that underlies the theory of nonlinear optics.

Since $S(E_0)$ is a nonlinear function, new frequencies are created. The source therefore emits an optical field E_1, with frequencies not present in the original wave E_0. This leads to numerous interesting phenomena that have been utilized to make useful nonlinear-optics devices.

4.3.3 Second order nonlinear optics

We examine the optical properties of a nonlinear medium in which nonlinearities of order higher than the second are negligible, so that

$$P_{NL} = 2dE^2 \tag{4.20}$$

We consider an electric field E comprising one or two harmonic components and determine the spectral components of P_{NL}. In accordance with the first Born approximation, the radiation source S contains the same spectral components as P_{NL}, and so, therefore, does the emitted (scattered) field.

Consider the response of this nonlinear medium to a harmonic electric field of angular frequency ω (wavelength $\lambda_0 = 2\pi c_0/\omega$) and complex amplitude $E(\omega)$:

$$E(t) = Re\left\{E(\omega)\exp(j\omega t)\right\} \tag{4.21}$$

The corresponding nonlinear polarization density P_{NL} is obtained by substituting (4.20) into Equation (4.19),

$$P_{NL}(t) = P_{NL}(0) + Re\left\{P_{NL}(2\omega)\exp(j2\omega t)\right\}$$

where $P_{NL}(0) = dEE^*$ (4.22)

and $P_{NL}(2\omega) = dE(\omega)E(\omega)$

This process is illustrated graphically in Figure 4.20.

Second-harmonic generation

The source $S(t) = -\mu_0 \partial^2 P_{NL}(t)/\partial t^2$ corresponding to (4.22) has a component at frequency 2ω and complex amplitude $S(2\omega) = 4\mu_0\omega^2 dE(\omega)E(\omega)$, which radiates an optical field at frequency 2ω (wavelength $\lambda_0/2$). Thus the scattered optical field has a component at the second harmonic of the incident optical field. Since the amplitude of the emitted second-harmonic light is proportional to $S(2\omega)$, its intensity is proportional to $|S(2\omega)|^2 \alpha \omega^4 d^2 I^2$, where $I = |E(\omega)|^2/2\eta$ is the intensity of the incident wave. The intensity of the second-harmonic wave is therefore proportional to d^2, to $1/\lambda_0^4$, and to I^2. Consequently, the efficiency of second-harmonic generation is proportional to $I = P/A$, where P is the incident power and A is the cross-sectional area. It is therefore essential that the incident wave have the

4.3 Non-linear optics

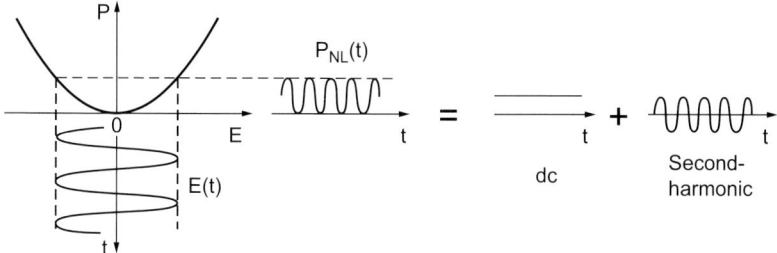

Figure 4.20: A sinusoidal electric field of angular frequency ω in a second-order nonlinear optical medium creates a polarization with component at 2ω(second-harmonic) and a steady (dc) component

largest possible power and be focused to the smallest possible area to produce strong second-harmonic radiation. Pulsed lasers are convenient in this respect since they deliver large peak powers.

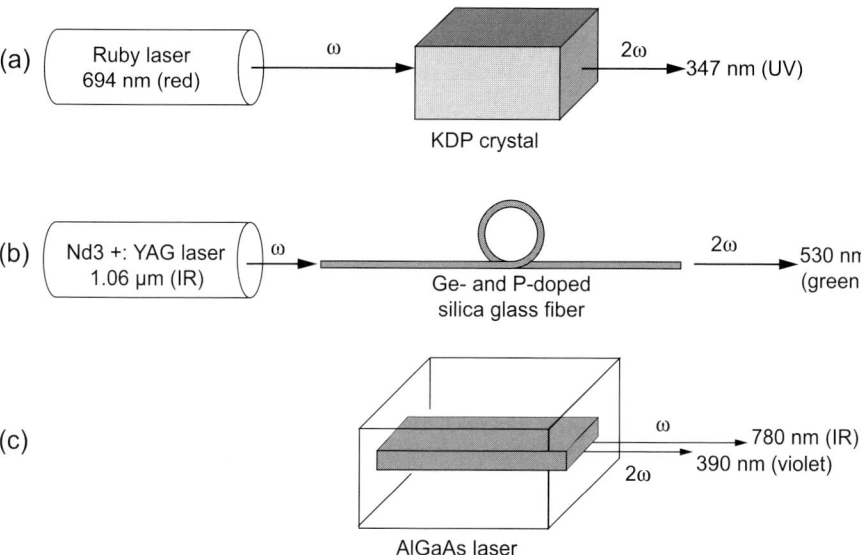

Figure 4.21: Optical second-harmonic generation in (a) a bulk crystal; (b) a glass fiber; (c) within the cavity of a semiconductor laser

Figure 4.21 illustrates several optical second-harmonic-generation configurations in bulk crystals and in waveguides, in which infrared light is converted to visible light and visible light is converted to the ultraviolet. Efficient second harmonic generation is also provided by

$KNbO_3$, and to a lesser extend by $LiNbO_3$ or $BaTiO_3$. A very important issue is "phase matching", which describes the fact, that for efficient conversion the phase velocities of primary wave and second harmonic has to be matched.

Optical rectification

Optical rectification is rarely discussed or used. Nevertheless it is quite interesting. The component $P_{NL}(0)$ in (4.22) corresponds to a steady (non-time-varying) polarization density that creates a dc potential difference across the plates of a capacitor within which the nonlinear material is placed (Figure 4.22). The generation of a dc voltage as a result of an intense optical field represents optical rectification (in analogy with the conversion of a sinusoidal ac voltage into a dc voltage in an ordinary electronic rectifier). An optical pulse of several MW peak power, for example, may generate a voltage of several hundred μV.

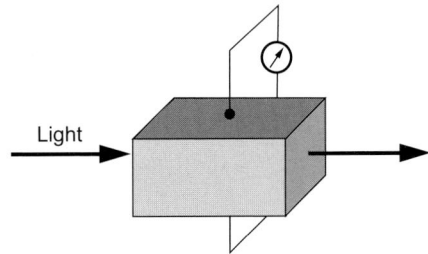

Figure 4.22: The transmission of an intense beam of light through a nonlinear crystal generates a dc voltage across it.

Bibliography

[1] B. Saleh and M. Teich, *Fundamentals of Photonics*, Wiley, New York, 1991.
[2] H. Nishihara, M. Haruna, T. Suhara, *Optical Integrated Circuits*, McGraw-Hill, New York, 1989.
[3] A. Yariv, *Optical Electronics in Modern Communications*, Oxford University Press, 1997.
[4] A. Yariv, *Quantum Electronics*, Wiley, New York, 1989.
[5] A. Yariv, P. Yeh, *Optical Waves in Crystals*, Wiley, New York, 1983.
[6] C. C. Davis, *Lasers and Electro-Optics*, Cambridge University Press, 1996.
[7] P. Gay, *Introduction to Crystal Optics*, Longman, London, 1982.
[8] A. Prokhorov, Yu Kuzminov, O. Khachaturyan, *Ferroelectric Thin Film Waveguides in Integrated Optics*, Cambridge International Science Publ., 1996.

5 Microwave Properties and Measurement Techniques

Norbert Klein

Research Center Juelich, Germany

5.1 Introduction

Functional oxide materials play an important role for applications in microwave communication and sensor systems. Whereas silicon and GaAs represent the basic materials for the digital part of communication and sensor systems, the analogue parts require high quality factors and low losses, which cannot be fulfilled by semiconductors. Oxide insulators provide extremely low microwave losses expressed by the value of its loss tangent $\tan \delta = \mathrm{Im}(\epsilon_r)/\mathrm{Re}(\epsilon_r)$. The functionality of oxides in microwave devices or circuits can be classified as follows:

- high and temperature stable dielectric constant for use as substrates and dielectric resonators,

- voltage dependent dielectric constant to achieve tuneability,

- piezoelectric actuation for mechanical tuning, switching and acoustic resonators,

- ferrimagnetic properties for tuning, switching and nonreciprocal propagation *and*

- low surface resistance of superconducting oxides for low losses and high selectivity.

However, the key components to provide the power generating functionality are based on semiconductors. Novel materials and transistor structures based on InP, SiGe and GaN represent an area of extensive R&D activities with emphasis on high power, low noise and high operation frequencies. The performance of a real device such as the phase noise of a microwave oscillator depends *both* on the noise properties of the transistor *and* on the loss tangent of the oxide material which forms the stabilising resonator. Therefore, material and device related R&D activities both on semiconductors and on oxides are essential to open new horizons for microwave communication and sensor applications.

Beside the issue of performance and integration capability costs represent an even a more urgent driving force for material development. Integration of oxide films and devices in silicon technology as well as integration of low-loss metals like silver with low-loss oxide insulators play in important role. Apparently, the ultimate low-loss metals for the fabrication of planar integrated microwave circuits are functional oxides, namely oxide high-temperature superconductors. High-temperature superconductor films have been used to build ultra-high performance microwave filters for highly sensitive and selective receiver front ends with some practical use for base stations of cellular networks.

Polar Oxides: Properties, Characterization, and Imaging
Edited by R. Waser, U. Böttger, and S. Tiedke
Copyright © 2005 WILEY-VCH Verlag GmbH & Co. KGaA, Weinheim
ISBN: 3-527-40532-1

This article give an overview about the microwave properties of insulating and superconducting oxide dielectrics. In addition, microwave measurement techniques for bulk and thin film oxides will be reviewed.

5.2 Basic relations defining microwave properties of dielectrics and normal/superconducting metals

For the physical description of the microwave properties of dielectric materials and metals we consider Maxwell's equations in the absence of localized charges:

$$\nabla \times E = -\frac{\partial B}{\partial t}; \quad \nabla \times H = \frac{\partial D}{\partial t} + j; \quad \nabla B = 0; \quad \nabla E = 0 \tag{5.1}$$

The universal treatment of dielectrics and metals including superconductors relies on the fact that for a harmonic time dependence according to $\exp(i\omega t)$ one can describe the displacement current $\partial \mathbf{D}/\partial t$ and the Ohmic current density $\mathbf{j}$ by a generalized complex dielectric function ϵ or a generalized complex conductivity σ (assuming that the current density obeys a generalized Ohm's law with complex conductivity, i.e. nonlocal effects can be neglected):

$$\sigma = i\omega\epsilon; \quad \sigma = \sigma_1 - \sigma_2; \quad \epsilon = \epsilon_0 \epsilon_r (1 + i \tan \delta) \tag{5.2}$$

The quantity ϵ_r is the relative dielectric constant or permittivity of a dielectric medium, $\epsilon_0 = 8.85 \times 10^{-12}$ As/Vm. The quantity $\tan \delta$ represents the loss tangent of a dielectric medium. Metals in the microwave range are usually described by a complex conductivity with dominant real part for normal metals and dominant imaginary part for superconductors.

It is useful to consider the solution of Maxwell's Equations (5.1) for plane electromagnetic waves in the absence of boundary conditions, which can be written as $\exp[i(\beta z - \omega t)]$ assuming propagation in z-direction of cartesian coordinates. The quantity β is the complex propagation constant of the medium with dominant real part for dielectrics and dominant imaginary part for metals. The impedance of the medium, Z, defined as ratio of electric to magnetic field is related to β by $Z = \omega \mu_0 / \beta$ with $\mu_0 = 1.256 \times 10^{-6}$ Vs/Am. As it can be derived from Maxwell's equations, the impedance is related to the conductivity/dielectric function by the following expression:

$$Z = \sqrt{\frac{i\omega\mu}{\sigma}} = \sqrt{\frac{\mu}{\epsilon}} \tag{5.3}$$

In case of vacuum ($\epsilon = \epsilon_0$) the impedance is real and its absolute value is 377 Ω (wave impedance of vacuum). In case of a metal an incident wave decays rapidly from the surface, thus the impedance of a metal is called "surface impedance" $Z_s = R_s + iX_s$. The real part of Z_s is called surface resistance R_s, the imaginary part surface reactance X_s.

5.3 Surface impedance of normal metals

For a normal metal at microwave frequencies the imaginary part of the conductivity can be neglected and the real part is equal to the dc conductivity σ_{DC}. In this case a simple expression for the surface impedance follows from Maxwell's equations and Ohm's law:

$$Z = \sqrt{\frac{\omega\mu_0}{2\sigma_{DC}}}(1+i) = \frac{1}{\delta\sigma_{DC}}(1+i); \quad \delta = \sqrt{\frac{2}{\omega\mu_0\sigma_{DC}}} \quad \text{(skin depth)} \quad (5.4)$$

Equation (5.4) is valid as long as the skin depth is large in comparison to the mean free path of the electrons in the metal. This holds true in the microwave range at room temperature, for cryogenic temperature the surface resistance lies above the values predicted by Equation (5.4) and exhibits a $f^{2/3}$ rather than a $f^{1/2}$ frequency dependence (anomalous skin effect [7]).

As an example, for copper with a room temperature conductivity of $5.8 \times 10^7 (\Omega\,\text{m})^{-1}$ the surface resistance at 10 GHz is 26 mΩ, the skin depth is 0.66 μm. Therefore, the Q of a cavity resonator with a geometric factor of several hundred is in the 10^4 range. However, for planar resonators like the ones shown in Figure 5.8 the G values are only a few Ohms leading to Q values of only a few hundred. This is too small for many filter and oscillator applications.

5.4 Surface impedance of high-temperature superconductor films

The microwave response of a superconductor can be understood by two physical phenomena: First, by the Meissner effect, which causes any magnetic field applied to the surface of a superconductor drops exponentially to zero inside the superconductor on the length–scale of the London penetration depth λ_L, which is about 160 nm for the most relevant high-temperature superconductor (HTS) compound YBa$_2$Cu$_3$O$_7$ (called "YBCO") at $T \to 0$ K and increases with temperature according to $(\lambda_L(0)/\lambda_L(T))^{-2} \approx 1 - (T/T_c)^2$ (transition temperature T_c = 92 K for YBCO). This shielding behavior is almost frequency independent up to the THz range for YBCO leading to a frequency independent skin depth equal to λ_L. The second fact is that at finite temperatures below T_c Cooper pairs (corresponding to "ballistic" charge carriers without dissipation) and quasiparticles (corresponding to normal conducting charge carriers with a temperature dependent density) coexist. Therefore, the complex conductivity of a superconductor is composed of two parts:

$$\sigma = \sigma_1(T) - i\sigma_2(T) = \sigma_1 - \frac{i}{\omega\mu_0\lambda_L^2(T)} \quad (5.5)$$

The real part describes the conductivity of the quasiparticles, which are assumed to behave like conventional electrons, i.e. obey Ohm's law. The imaginary part of the conductivity corresponds to the inductive response of the Cooper pairs. The explicit relation between σ_2 and λ_L follows from the London equations (see any textbook about superconductivity, e.g. [20]). At microwave frequencies, $\sigma_1 \ll \sigma_2$ holds true resulting in a simple expression for the surface resistance and reactance of a superconductor, by a Taylor expansion of Equation (5.3) using Equation (5.5):

$$R_s = \frac{1}{2}\omega^2 \mu_0^2 \sigma_1(T) \lambda_1^3(T) \quad \text{and} \quad X_s(T) = \omega \mu_0 \lambda_L(T) \tag{5.6}$$

In contrast to normal metals, the surface resistance exhibits a quadratic frequency dependence. The absolute values for thin films of YBCO at a temperature of 77 K are shown in Figure 5.1. According to these data, there is a clear advantage by orders of magnitude upon using HTS for the whole range of microwave communications bands. Above 77 K, R_S increases strongly towards the critical temperature due to the strong increase of the London penetration depth. Below 77 K, R_S decreases gradually towards zero temperature by less than one order of magnitude [12].

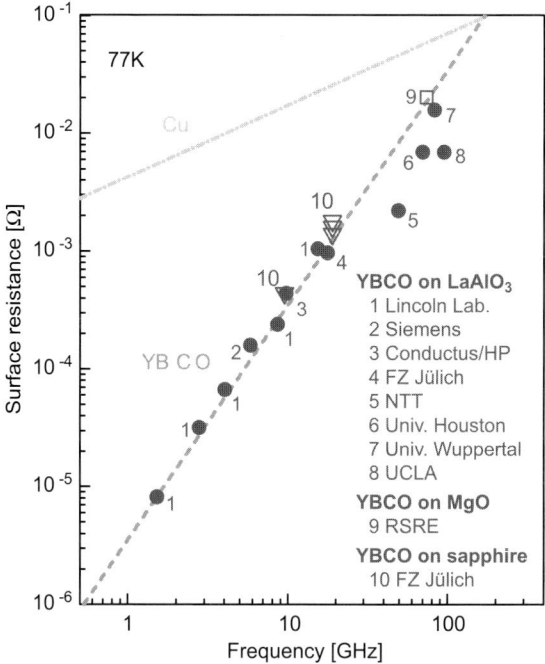

Figure 5.1: Measured surface resistance of epitaxially grown YBCO films grown at different laboratories at $T = 77$ K in comparison to that of copper.

The surface resistance data discussed so far correspond to the linear regime corresponding to low levels of microwave power. For higher power levels corresponding to higher values of the high frequency magnetic fields nonlinearities occurs resulting either in an increase of R_S or/and intermodulation distortion [7]. This is of particular importance, because - as a result of the Meissner effect - the current density across a superconducting microstrip line is strongly enhanced by about one order of magnitude at the edges on a lengthscale of the London penetration depth. Typically, intermodulation distortion as well as a field dependent surface resistance occurs at power levels which depend strongly on the film quality. As an example, misoriented grains in the strongly anisotropic materials and grain boundaries give raise to strong nonlinearities. In general, one can claim that planar resonators made from high quality HTS can handle power levels in the milliwatt range, which is sufficient for most of the receiver applications. In contrast, applications in the transmit circuits of communications systems, i.e. power levels above several watts, can only be handled in "edge current free" geometries (see Figure 5.8 (e)).

The nonlinear effects in HTS films are strongly temperature dependent, in particular close to T_c. Therefore, operation temperature of 50 to 65 K are more favourable than 77 K. Temperatures in this range can be attained by low-power closed-cycle cryocoolers.

Apart from YBCO, thin films with reasonable microwave properties have been prepared from the thallium-based compounds $Tl_2Ba_2CaCu_2O_8$ ($T_c \approx 105$ K) and $Tl_2Ba_2Ca_2Cu_3O_{10}$ ($T_c \approx 115$ K) [10]. HTS films with reasonable and qualified microwave properties nowadays can be grown on wafers up to more than 4" in diameter, the most common size are 2" and 3" for microwave applications. A very important step was the preparation of double-sided coating, which have turned out to be essential for planar microwave devices, where the metal ground plane needs to be superconducting in order to achieve high quality factors.

5.5 Microwave properties of dielectric single crystals, ceramics and thin films

Until now, the physical understanding of the origin of microwave losses of dielectrics is very rare. So far, only for single crystals a complete microscopic understanding has been achieved. This is because single crystals of very high purity are at their theoretical limit of absorption by the phonon system. In contrast, for dielectric bulk ceramics and thin films the losses are mostly determined by defects. There is a large variety of defects and possible loss mechanisms associated with defects like acoustic phonon excitation by point defects, point defect Debye relaxation, unwanted doping and losses initiated by unwanted, possibly conducting phases. In this section, such mechanisms are discussed briefly and the state-of-the art of existing materials will be presented. Apart from linear microwave dielectrics, nonlinear dielectrics will be discussed with respect to frequency agility.

The most simple model of dielectrics in the microwave and near infrared regime is based on classical harmonic oscillators with damping [2]:

$$\epsilon(\omega) = \epsilon_\infty + \sum_{j=1}^{N} \frac{S_j \omega_j}{\omega_j^2 - \omega^2 - i\omega\Gamma_j} \qquad (5.7)$$

The quantity ϵ_∞ represents the optical permittivity, which is determined by the electronic polarizability. The second term represents the ionic crystal lattice as a sum of N classical harmonic oscillators with eigenfrequencies ω_j, damping constants Γ_j and oscillators strengths S_j. In order to fit Equation (5.7) to the observed far infrared spectra, these parameters are used as fit parameters. For the microwave regime, the frequency ω is small in comparison to the phonon frequencies and simplified expressions for ϵ_r and $\tan \delta$ can be derived:

$$\epsilon_r = \frac{1}{\epsilon_0}[\epsilon_\infty + \sum_{j=1}^{N} S_j]; \quad \tan \delta = \frac{\omega}{\epsilon_0 \epsilon_r} \sum_{j=1}^{N} \frac{S_j \Gamma_j}{\omega_j^2} \quad (5.8)$$

In agreement with experimental observations, most ionic crystals exhibit no significant frequency dependence of the permittivity and a linear frequency dependence of the loss tangent in the microwave regime. However, measured absolute values of the loss tangent and its temperature dependence differ significantly from Equation (5.8) using parameters determined by infrared measurements. The theoretical problem relies on a proper quantum mechanical calculation of the relaxation rate $\Gamma_{TO}(\mathbf{q} = 0, \mathbf{j}, \omega, T)$ of transverse optical phonons of branch $\mathbf{j}$ at a wave vector $\mathbf{q}$. Since the microwave frequencies are much smaller than ω_{TO}, multiphonon absorption and emission processes will determine the loss tangent because of energy conservation. In any case, one expects a monotonously increase of the loss tangent with temperature due to the fact that the occupation of phonon states is determined by the Bose statistics. Therefore, for $T = 0\,\text{K}$ the intrinsic phonon losses disappear.

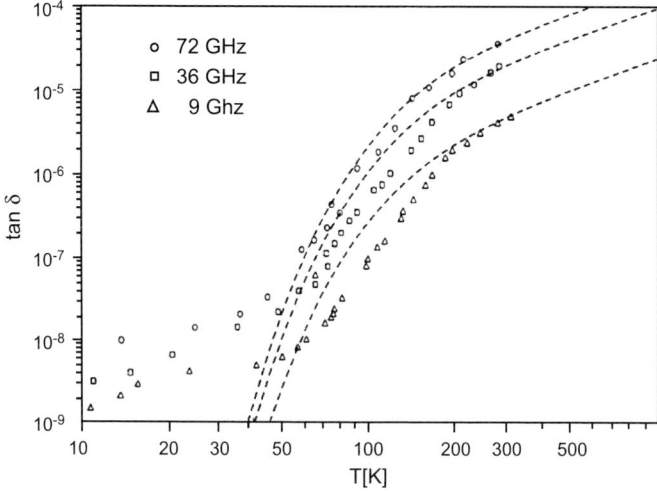

Figure 5.2: Measured temperature dependence of the loss tangent of sapphire and theoretical calculations based on the SKM model (from [4]).

Sparks, King and Mills [17] and Gurevich and Tagantsev [6] investigated theoretically two- and three phonon difference absorption processes and calculated the temperature dependence of the loss tangent of various single crystals. Figure 5.2 shows experimental results on the temperature dependence of the loss tangent of high-purity single crystals of sapphire ($\epsilon_r = 9.2-11.4$) [4] and calculations based on the SKM model [21]. Apart from some deviations below 50 K, which are due to extrinsic losses, there is a good agreement between theory

5.5 Microwave properties of dielectric single crystals, ceramics and thin films

and experiment. It is worth to mention that sapphire is the microwave dielectric with the lowest losses at all. In particular, at cryogenic temperatures, $\tan \delta$ values in the 10^{-8} range allow for Q values as high a several 10^7. A similar good agreement between theory and experiment has been achieved for single crystals of rutile with a permittivity of about 100.

However, the situation becomes already more complicated for ternary single crystals like lanthanum-aluminate (LaAlO$_3$, $\epsilon_r = 23.4$). The temperature dependence of the loss tangent depicted in Figure 5.3 exhibits a pronounced peak at about 70 K, which cannot be explained by phonon absorption. Typically, such peaks, which have also been observed at lower frequencies for quartz, can be explained by defect dipole relaxation. The most important relaxation processes with relevance for microwave absorption are local motion of ions on interstitial lattice positions giving rise to double well potentials with activation energies in the 50 to 100 meV range and color-center dipole relaxation with activation energies of about 5 meV.

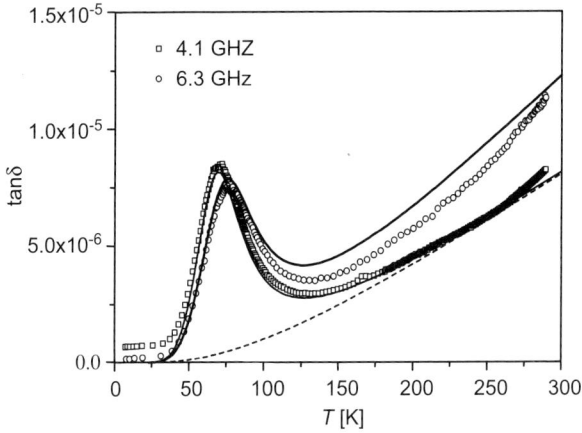

Figure 5.3: Measured temperature dependence of the loss tangent of LaAlO$_3$ single crystals and a theoretical fit employing the SKM model and defect dipole relaxation (from [22]).

Figure 5.3 shows the measured temperature dependence of the loss tangent of LaAlO$_3$ single crystals and a fit employing the SKM model plus a Debye relaxation term. The best fit was achieved for an activation energy of 31 meV, the estimated defect concentration is only 10^{16}/cm^3. It was suggested that aluminium atoms on interstitial lattice positions are responsible for the observed relaxation phenomena. This indicates, that the dielectric losses are extremely sensitive to very small concentrations of point defects [21].

Charged point defects on regular lattice positions can also contribute to additional losses: the translation invariance, which forbids the interaction of electromagnetic waves with acoustic phonons, is perturbed due to charged defects at random positions. Such single-phonon processes are much more effective than the two- or three phonon processes discussed before, because the energy of the acoustic branches goes to zero at the Γ point of the Brillouin zone. Until now, only a classical approach to account for these losses exists, which has been successfully used to describe the losses in garnets, glasses and in doped sapphire at low temperatures [22]. At $T \to 0\,\mathrm{K}$ this mechanism may become dominant, because intrinsic phonon losses and thermally activated Debye losses disappear.

Finally, any doping leading to a finite conductivity σ will lead to a loss contribution according to $\tan\delta = \sigma/(\omega\epsilon_0\epsilon_r)$, see Equation (5.3), which is the dominant mechanism for semiconductors. However, different conduction mechanism like hopping conductivity may give rise to complicated temperature and frequency dependences, which cannot be discussed in detail here.

Most of the materials discussed so far (e.g. sapphire, lanthanum aluminate, magnesium oxide ...) are common substrate materials for HTS films, thus their low dielectric losses are essential to utilize the low conduction losses of HTS films in planar microwave circuits. For dielectric resonators operating at room temperature they are not very attractive, first, because single crystals are quite expensive, and secondly, because they exhibit a relatively strong temperature dependence of the dielectric constant leading to a temperature coefficient of the resonant frequency of a dielectric resonator of about 100 ppm/K (typically for sapphire) and up to more than 1000 ppm/K for rutile. For HTS or cryogenically cooled dielectric resonator devices this is not very important, because the temperature has to be controlled anyway (because of the temperature dependent London penetration depth of a superconductor), but for room temperature devices, e.g. in mobile phones, base stations or satellites the temperature can vary between typically -40 and +50°C. In addition, devices operating at high power levels might heat up due to power dissipation in the dielectric material.

Fortunately, they are several species of low-loss dielectric ceramics with tailored temperature coefficient of dielectric constant, which can be made lower than 1 ppm/K for a certain temperature window around room temperature. Physically, this can be accomplished either by intrinsic compensation of the temperature dependence of thermal volume expansion $V(T)$ and lattice polarizability $\alpha(T)$ via the Clausius-Mossotti relation:

$$\frac{\epsilon_r(T) - 1}{\epsilon_r(T) + 1} = \frac{4\pi\alpha(T)}{3V(T)} \tag{5.9}$$

For a suitable doping, both effects compensate each other for some materials and lead to a broad turning point $TP(\epsilon_r^{-1}d\epsilon_r/dT\,|_{T=TP}= 0)$ in the temperature dependence of $\epsilon_r(T)$. As a second approach, polycrystalline mixtures of grains of different materials with opposite temperature coefficient of ϵ_r lead to an effective medium with zero temperature coefficient.

Table 5.1 summarizes the most relevant commercially available microwave dielectrics with compensated temperature coefficients. Usually, the quantity $Q \times f$ is quoted for the microwave losses: Assuming that $Q = 1/\tan\delta$ for a dielectric resonator and $\tan\delta \propto f$ according to Equation (5.8) $Q \times f$ is a measure for the loss tangent. However, for most dielectrics $Q \times f$ drops below a certain frequency gradually to lower values indicating that the frequency dependence of the loss tangent is weaker than linear. As an example, for high-quality BMT-ceramic $Q \times f$ =350 THz at 10 GHz and only about 200 THz at 4 GHz [11].

This behavior is not yet fully understood. So far, most of the material development has been performed in companies, indicating that the market for microwave dielectrics has been grown tremendously over the last years (as an example, each commercial satellite receiver contains a small microwave ceramic disk as frequency stabilising element for the local microwave

5.5 Microwave properties of dielectric single crystals, ceramics and thin films

oscillator). In fact, the optimization of microwave ceramics with respect to low losses has been pursued by empirically improving the preparation conditions (with particular emphasis on the sintering procedure) rather than due to a systematic study of physical loss mechanisms (which now is appreciated as a challenge within the microwave ceramic community). As a tendency, the general requirements for low losses are:

- dense materials with large grains,
- small amount of impurity phases,
- high structural order,
- low level of free charge carriers.

For sure, this list is incomplete and may have neglected important details.

MATERIAL	ε_r	τ_f	$\tan \delta^{-1}$	f[GHz]	
Al_2O_3	10	-60	50,000/ 100,000	10	BC SC
$LaAlO_3$	24	-60	40,000	10	SC
$Ba(Mg_{1/3}Ta_{2/3})O_3$	24	0	26,000	10	BC
Ba-Zn-Ta-O	29	-3...3	50,000	2	BC
Zr-Sn-Ti-O	38	-3...+3	8000	7	BC
Ca-Ti-Nd-Al	47	20	6000	6.5	BC
Ba-Nd-Ti-O	80	90	2500	5.5	BC
TiO_2	100	450	16,000	3	BC, SC
$CaTiO_3$	155	800	5000	2.3	TF
$SrTiO_3$	270	1200	1500	2	TF*
$Ba_xSr_{1-x}TiO_3$	420	?	1500	1.5	TF*

Table 5.1: Microwave properties of the most important microwave dielectrics (SC= "bulk single crystals", BC = "bulk ceramics", TF = "thin films", τ_f = "temperature coefficient of resonant frequency". The materials marked with * are tuneable dielectrics.

The materials listed in Table 5.1 are also not complete but the selection represents the most relevant composites which are of commercial interest. In addition to the performance issue, there is a strong tendency in research and development to reduce the costs. For that reason, a lot of research is devoted to the niobates as possible replacement for the tantalates, because niobium is cheaper than tantalum. In addition, compensated materials with very high values of the permittivity are currently under development. Recently, for the compound $Ag(Nb_{1-x}Ta_x)O_3$ with $0.35 < x < 0.65$, ϵ_r values of 450 were achieved for potential use as filters (to replace the surface-acoustic-wave devices) and planar antennas in mobile phones [19].

The microwave ceramics discussed so far are prepared by sintering pellets of pressed power. Usually, the sintering temperatures are very high (typically around 1200 to 1500°C) in order to obtain large and highly ordered grains, which is essential for low values of the loss tangent.

For planar single- and multilayer microwave circuits dielectric ceramics are important as low-cost substrates. One very cost effective method of manufacturing integrated microwave circuits is the so-called LTCC process. LTCC stands for "Low Temperature Co-fired Ceramics" and is based on unsintered ceramic foils (in most cases composed of aluminium oxide with organic additives), where metal circuits and vias connections through a foil are prepared by printing, stamping and metal thick film deposition processes. Several (typically up to 30) of such foils are stacked together and subsequently the entire stack is sintered at relatively low temperatures (typically around 850°C). The LTTC technology allows for a very cheap and versatile manufacturing of multilayer circuits with complex wiring. The permittivity values of LTCC ceramics are between 4 and 10. Due to the low temperatures sintering temperatures loss tangent value are in the range of about 5×10^{-4} at 10 GHz, which is higher than for high-temperature ceramics.

In summary, the field of microwave ceramics has been continuously progressing over the last ten years. More recently, a strong interest has come up to utilize the non-linear dielectric properties of thin or thick films of ferroelectrics (or incipient ferroelectrics) above the Curie-temperature to build tuneable microwave devices. Figure 5.4 shows the temperature dependence of the permittivity at different applied voltages for thin films and bulk single crystals of $SrTiO_3$. Whereas $SrTiO_3$ exhibit a maximum tuneability at about 80 K and therefore ideally suited for planar tuneable HTS devices, $Ba_xSr_{1-x}TiO_3$ with $x = 0.5$–0.7 is ideal for room temperature applications. However, due to the high values of the static electric field required for a significant change of ϵ_r only thin-film based electrode designs with micrometer-sized capacitive gap is able to accomplish significant tuning ranges at practicable voltages of several ten volts [14]. This fact limits ferroelectric tuning mostly to planar integrated microwave circuits.

Currently, the most important aspect of research on these materials is the improvement of the loss tangent of about 10^{-3} to 10^{-2} [3] towards the theoretical values of several 10^{-4} (achieved in single crystals of $SrTiO_3$) [13]. The best results obtained until now are already competitive (or even better) with commercial semiconducting varactor diodes. As a clear advantage, ferroelectric thin films varactors are easy to integrate in MMICs or LTCC devices. Finally, the question of intermodulation distortion caused by the nonlinear dielectric constant has just been started to be investigated [15]. Moreover, new materials with tailored properties and low losses are highly desired, but the number appears to be quite limited.

5.6 General remarks about microwave material measurements

In general, the quantities being determined by microwave measurements are complex reflection and transmission coefficients or complex impedances normalized to the impedances of the transmission lines connecting a network analyser and the device-under-test (DUT). In addition to linear frequency domain measurements by means of a network analyser the determination of possible non-linear device (and thus material) properties requires more advanced measure-

5.7 Non resonant microwave measurement techniques

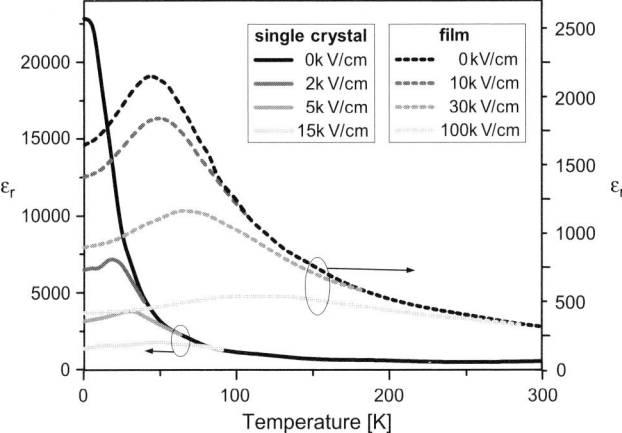

Figure 5.4: Temperature dependence of the dielectric constant of thin films and bulk single crystals of SrTiO$_3$ thin films and bulk single crystals for different values of the applied dc electric fields.

ment schemes like two-tone intermodulation and pulsed response. In any case, the DUT needs to be a well defined microwave device with the sample under test being a well defined part of it. "Well defined" means that the electromagnetic field distribution in the device and its alteration by the sample needs to be carefully investigated.

The main difficulty arises due to the low losses of many microwave oxide materials including insulator and superconducting metals. Therefore, approaches based on a sample being just a small perturbation of the DUT, such as a cavity resonator with a small piece of material inside, are of limited use, because the effect of the sample absorption on the resonator Q is too small. Therefore, in most cases the geometry of the sample has to be selected in a way that the sample itself represents a dominant part of the DUT.

For bulk dielectrics, a dielectric resonator can be formed by a cylindrically shaped piece of dielectric material. Dielectric thin films are more difficult to investigate, in particular when the loss tangent is very small. Planar resonator techniques as well as specially designed dielectric resonators can be used to examine their properties. For high-temperature superconductors both dielectric resonators and planar resonators represent an ideal tool to examine their surface impedance values.

5.7 Non resonant microwave measurement techniques

Non resonant techniques are only of limited use to determine microwave losses with high precision, in particular when the losses are very small. However, for the investigation of nonlinear absorption phenomena (i.e. rf power dependent on surface impedance or loss tangent) by intermodulation distortion measurements broad-band test devices are more common. Typically, a planar transmission line with an impedance of 50 Ohms can be employed for intermodulation studies on thin films materials. In case of high-temperature superconductors the metalization forming the planar transmission line is fabricated by patterning the superconducting film. In case of nonlinear dielectric films a planar capacitor composed of a dielectric-metal bilayer with a micron-sized disruption of the metal sublayer, as shown in Figure 5.5, can be used. Such a gap provides a high concentration of rf electric field in the portion of dielectric film within the gap.

Figure 5.5: SEM picture of a planar STO capacitor (gap width 2 μm) (from [15])

A typical intermodulation measurement setup is shown in Figure 5.6. The stripline is excited with a two tone rf signal of frequencies f_1 and f_2. In case of a nonlinear capacitance caused by a nonlinear response of the dielectric film (i.e. ϵ_r and/or $\tan \delta$ changes with the amplitude of the rf voltage) intermodulation signals at $2f_1 - f_2$ and at $2f_2 - f_1$ emerge. The experimental determination of the dependences of the intermodulation signal amplitude on power level, applied dc bias electric field, frequency separation between f_1 and f_2 and operation temperature allows (a) for an analysis of the physical phenomena of nonlinear microwave properties and (b) to determine the power handling capability of microwave devices and circuits employing tuning by nonlinear dielectric films.

5.8 Resonator measurement techniques

In general, electromagnetic resonators are of common use for material characterization at microwave frequencies. In addition, a resonator represents a basic element of a multipole filter or an oscillator circuit. Any type of electromagnetic resonator is characterized by the resonant frequency f_0 and the unloaded quality factor Q_0 of the selected resonant mode and its spectrum of spurious modes. In order to measure the resonator properties or to use a resonator as part of a filter structure, the resonator needs to be equipped with one or two

5.8 Resonator measurement techniques

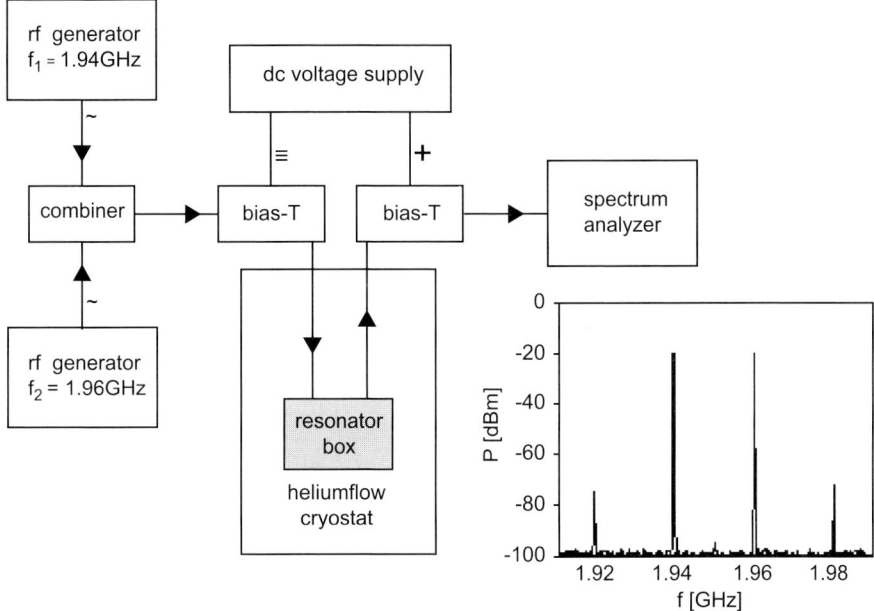

Figure 5.6: Scheme of the experimental setup for IMD measurements (from [15]).

coupling ports, which are characterized by external quality factors Q_{ext1} and Q_{ext2}. The loaded quality factor Q_L, which determines the 3 dB bandwidth or resonant halfwidth Δf_{3dB} by $Q_L = f_0/\Delta f_{3dB}$ depends on Q_0 and on the coupling strength:

$$\frac{1}{Q_L} = \frac{1}{Q_0} + \frac{1}{Q_{ext1}} + \frac{1}{Q_{ext2}} \qquad (5.10)$$

The information about the losses in the material under investigation lies within the unloaded quality factor. In case of measurements at low levels of rf power weak coupling ($Q_{ext1}, Q_{ext2} \gg Q_L$) resulting in $Q_0 \approx Q_L$ is of common use. For measurements of the power dependence matched input coupling ($Q_{ext1} \approx Q_0$) and weak output coupling conditions should be adjusted (e.g. by a mechanical adjustment of coupling antennae position inside a cavity resonator).

From the value of the resonant frequency and its change with temperature or other external parameters the permittivity of a dielectric sample and its temperature or field dependence can be determined. In case of superconductors, the temperature dependence of the magnetic field penetration depth can be determined [8]. Since the mode spectrum of a resonator is controlled both its physical dimensions and by the material properties, the physical dimensions of all resonator components have to be known with tight tolerances. Relative changes of permittivity or penetration depth can be determine with much higher accuracy than absolute values.

Dielectric resonators - In general, a dielectric resonator consists of one (sometimes more than one) piece(s) of dielectric material characterized by its relative permittivity $\epsilon_r \equiv Re\{\epsilon\}$

and its loss tangent $\tan\delta \equiv Im\{\epsilon\}/Re\{\epsilon\}$, with $\epsilon(\omega, T)$ representing the complex dielectric function of the dielectric medium. For many applications, the dielectric body is of cylindrical shape with circular cross section. The dielectric resonator is arranged inside a metallic shielding cavity machined from a highly conducting metal like copper or lightweight aluminium based alloys with silver coating, but wall segments based on high-temperature superconducting thin films can also be used. The unloaded quality factor of such a shielded dielectric resonator is given by

$$\frac{1}{Q_0} = \kappa \tan\delta + \frac{R_S}{G}; \quad \kappa = \frac{\epsilon_r \int_{DR} \mathbf{E}^2 dV}{\int_{DR+C} \mathbf{E}^2 dV}; \quad G = \frac{\omega\mu_0 \int_{DR+C} \mathbf{H}^2 dV}{\int_C \mathbf{H}^2 dA} \quad (5.11)$$

with R_s representing the surface resistance of the shielding cavity material, κ the filling factor indicating the fraction of electric field energy stored in the dielectric resonator (DR = "dielectric resonator", C = "shielding cavity filled with air/vacuum") and G a geometric factor (in units of "Ohms") representing the ratio of cavity volume to surface area which increases with increasing cavity size for a fixed frequency. For a given resonator geometry, the quantities k and G can be calculated for each eigenmode by solving the integrals in Equation (5.11), if the distribution of the electric field $\mathbf{E}(\mathbf{x})$ and the magnetic fields $\mathbf{H}(\mathbf{x})$ have been calculated either analytically or - nowadays very convenient with commercial state-of-the art eigenmode solver programs - numerically based on finite element methods.

The first term in Equation (5.11) represents the dielectric losses, the second one the losses associated with the metallic walls of the shielding cavity. In general, the quantities G and κ depend on the geometry of the cavity and the DR, on the permittivity of the DR, and on the employed resonant mode. Among all possible geometries the cylindrical geometry is the most commonly used one. This is because of the simplicity of fabrication by pressing pellets from ceramic powder and subsequent sintering. The other reason is that the resonant modes of a cylindrically shaped DR can be solved either analytically (in case of an infinitely extended cylinder or a piece of cylinder clamped between two metallic walls) or by mode matching techniques [9]. However, due to recent achievements in numerical simulation techniques there are novel approaches based on different geometries.

Table 5.2 summarizes the relevant properties of the most commonly used modes of cylindrically shaped dielectric resonators. Depending on the values of G, which is a measure for the loss contribution of the metallic shielding cavity (see Equation (5.11)), a reduction of the loss tangent leads to an increase of the resonator Q value. In particular, for the whispering-gallery modes the loss contribution of the metal walls becomes negligible, which results in extremely high Q values if a high-purity single crystalline dielectric like sapphire is employed. Whispering gallery modes (named after the "whispering gallery" at St. Pauls Cathedral in London) are modes with high azimuthal mode number (typically $n \geq 7$) corresponding to a standing wave propagating along the circumference of the dielectric cylinder with $2n$ field maxima/minima per circumference. According to geometrical optics, total reflection occurs at the circumference upon transition from a dense medium (dielectric puck) to the surrounding air, leading to a strong confinement of electromagnetic field energy in the DR.

In Table 5.2 the possible applications of the different modes are given. For most filters, only modes of low order (in many cases only the fundamental mode can be used) because of

5.8 Resonator measurement techniques

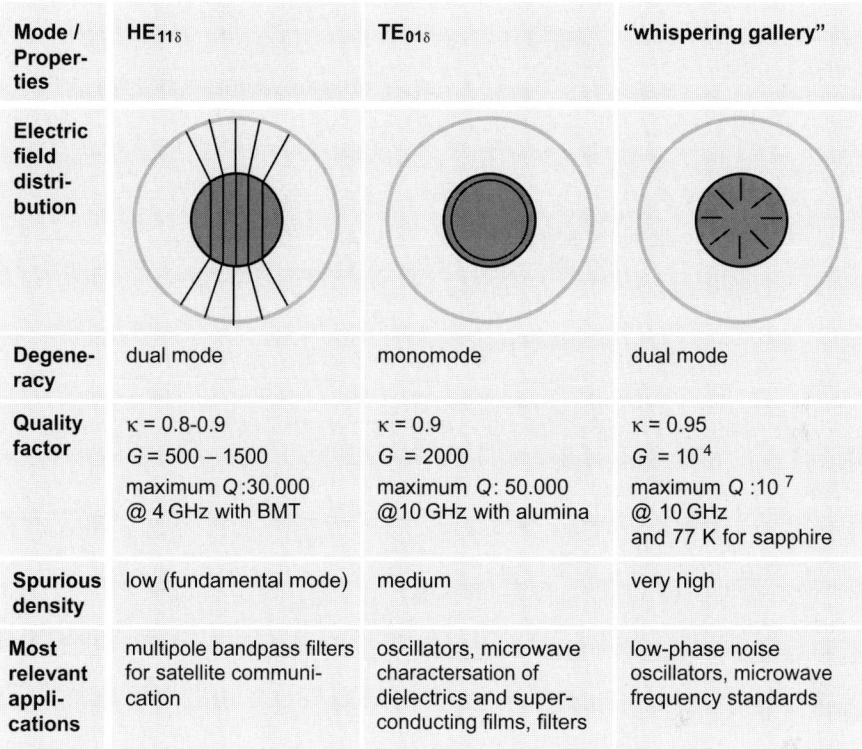

Table 5.2: Modes in cylindrically shaped dielectric resonators.

specifications with regard to the off-band rejection properties of the filter. The $TE_{01\delta}$ mode and higher order $TE0_{0np}$ - modes are very convenient for measurements on bulk dielectrics and on superconducting films. Figure 5.7 shows different configurations being used for various measurement tasks. The Courtney holder (a) [9] allows a very precise determination of the permittivity and its temperature dependence, but is of limited sensitivity for loss tangent measurements because of losses in the metal endplates. In addition, a semi-analytical formula for its resonant frequency exists. For high-sensitive loss tangent measurements configuration (d) is most appropriate, in particular, when whispering-gallery type modes (see Table 5.2) are being used [21]. In this case, even loss tangent values as low as 10^{-8}, as for sapphire at cryogenic temperatures, can be determined by this technique [18]. For both configurations (d) and (e) the mechanical support should be made form a low-permittivity and low-loss material like quartz. Configurations (b) and (c) have been developed for measurements on high-temperature superconductor films. In order to determine their low surface resistance values the dielectric resonator material needs to be sapphire or rutile [8].

Figure 5.7 (e) shows a so-called split post resonator, which allows for a highly precise determination permittivity and loss tangent of dielectric sheet materials [5]. In addition, it is also considered to be useful for measurements of nonlinear dielectric films.

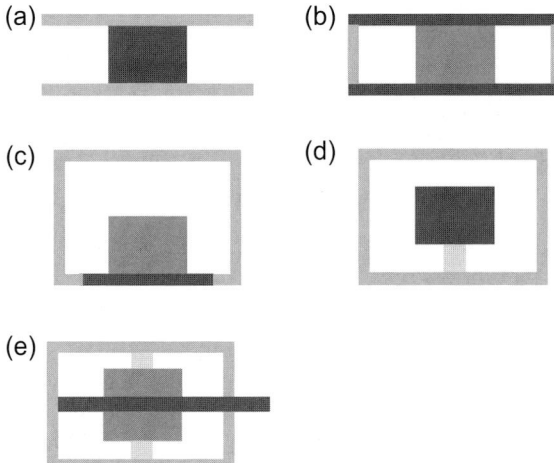

Figure 5.7: Material measurement configuration employing TE$_{0np}$ dielectric resonator modes. The sample under investigation is depicted in red.
a) Courtney holder for ϵ_r and $\tan\delta$ determination of bulk dielectrics
b) sapphire resonator for superconductor high power surface resistance measurements
c) sapphire resonator for superconductor surface impedance measurements
d) resonator for high-sensitivity determination tan $\tan\delta$ determination of bulk dielectrics
e) split-post resonator for ϵ_r and $\tan\delta$ determination of dielectrics sheets and layers

Planar resonators - Equation (5.11) is not only valid for dielectric resonators. Any other type of electromagnetic resonator employing dielectric parts, like metal ceramic coaxial-type resonators (e.g. used as filters in mobile phones) and microstrip or coplanar resonators (used in microwave integrated circuits) have a Q-contribution due to dielectric losses. For the latter type of resonator the dielectric losses are negligible in comparison to metallic losses, unless high temperature superconducting metallization layers are applied.

Figure 5.8 shows typical examples of planar resonators which have been used as building blocks of multipole filter structures or can be us as frequency stabilising element of an integrated oscillator circuit. Planar resonators are either segments of TEM (transverse electromagnetic) transmission lines with length L of half a wavelength corresponding the resonances to be at frequencies $f_n \approx n/2L (n = 1, 2, ...)$ ((b), (d)) or lumped element resonators each composed of a discrete inductor L and capacitor C with its resonance at a frequency $f = 1/[2\pi(LC)^{1/2}]$. Pure lumped element resonators are mostly based on interdigital capacitors and small segments of metallization defining the circuit inductance (a). In many cases hybrids of lumped elements microstriplines are used, e.g. folded microstriplines disrupted by a capacitive gap (c).

Planar resonators and microstrip and coplanar transmission lines represent the passive elements in almost any integrated circuit technology: at first, on chip integration in socalled MMICs (monolithic microwave integrated circuits) with SiGe or III/V semiconducting active

5.8 Resonator measurement techniques

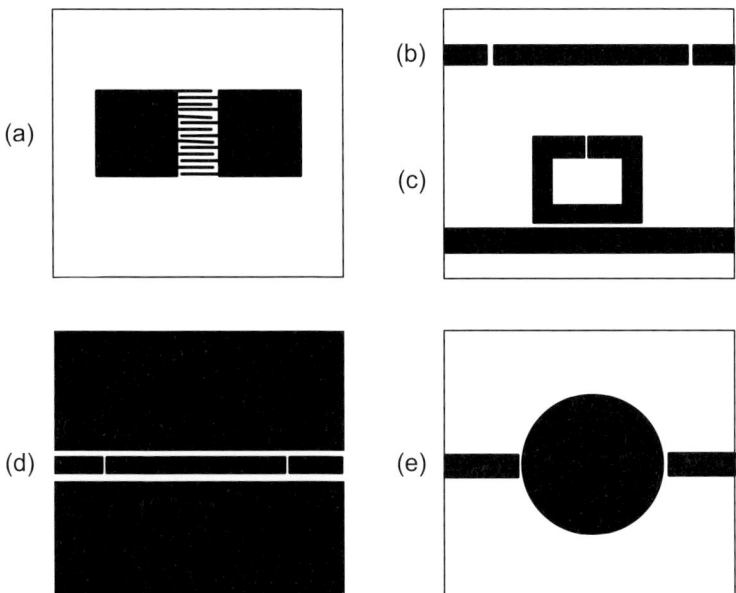

Figure 5.8: Typical planar resonators being used as building blocks for filters: lumped element (a), microstrip (b), folded microstrip with integrated capacitors (c), coplanar (d), and 2-D microstrip resonator. Omitting the capacitive gap in the folded microstrip design (c) leads to a ring resonator (square if circular shaped), which also represents a quite commonly used microstrip resonator design.

devices has been the most important technological effort with regard to the commercial exploitation of the microwave frequency range. In addition, so-called LTCC (Low-Temperature Co-fired Ceramics) technology based on stacked ceramic plates with integrated metallization layers has become extremely important to achieve reasonable Q_s in mobile communication units. On the high-performance end, planar resonators based on high-temperature superconducting (HTS) thin films to be operated at cryogenic temperatures between about 50 and 80 K exhibit Q values even higher than those for cavity or dielectric resonators. According to Equation 5.11 the unloaded Q of a microstrip resonator can be approximated by $Q_0 \approx \omega\mu_0 t/R_s$ (neglecting substrate losses), if the stripline width is large in comparison to the substrate thickness t. In the case of a lumped element resonator with feature size small in comparison to the wavelength the unloaded Q can be approximated by $Q_0 \approx \omega L/R$ with L being the inductance of the resonator and R its resistance. In the case of ring of diameter d, $R \approx R_s \pi d/w$ and $L \approx \mu_0 D/2$ resulting in $Q_0 \approx f\mu_0 w/R_s$. In general, the Q of a planar resonator needs to be calculated by numerical simulation tools.

As mentioned before, the highest Q values can be achieved with HTS thin film technology. In this case also ground plane metalization needs to be formed by a HTS film, e.g. in general double sided films are required for ultimate performance. As an alternative, coplanar resonators have been used (d) with some compromize on Q in comparison to microstrip lines of the same size but with the advantage of single sided coating. For measurements on nonlinear dielectric films at cryogenic temperatures coplanar resonators are very attractive if the gaps between the strip and the groundplanes are filled with the dielectric film [12]. For the highest Q and ultimate power handling capability circular 2-D microstrip resonators excited in the TM_{010}-mode have been used (see [1]), also requiring double sided coating.

Apart from their use as devices, planar resonators are of common use for microwave characterization of high-temperature superconducting films in a "device-specific" configuration.

As already discussed above, nonlinear dielectric films can be characterized by planar structures. Resonant planar structures allow - to a certain extend - the determination of the loss tangent. Figure 5.9 shows a resonator assembly which has been used for loss tangent measurements on $SrTiO_3$ thin film varactors [16].

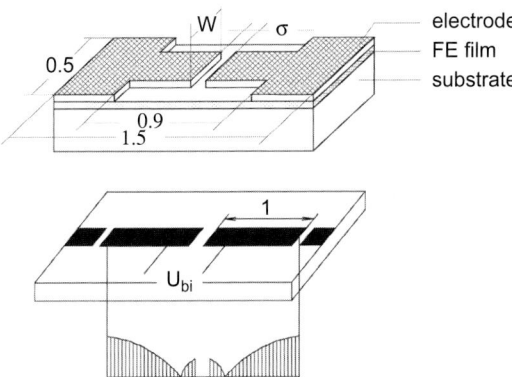

Figure 5.9: Planar nonlinear dielectric based thin film varactor structure (top) and planar microstrip resonator with capacitive gap. The microwave dielectric properties of the varactor structure are measured in a flipchip configuration of both substrates with the narrow (few microns) capacitive gap of the varactor structure being placed in the center of the large (500 microns) capacitive gap of the resonator structure (from [16]).

This measurements assembly is based on a planar thin film varactor structure and a planar microstrip resonator with capacitive gap. The microwave dielectric properties of the varactor structure are measured in a flipchip configuration of both substrates with the narrow (few microns) capacitive gap of the varactor structure being placed face-to-face in the center of the large (500 microns) capacitive gap of the resonator structure. Due to the low Q of the normal conducting planar resonator of only a few hundred the measurement resolution for $\tan\delta$ is limited to about 10^{-3} at a frequency of 2 GHz. As an alternative, a superconducting planar resonator can be used. In that case, measurements can only be performed at cryogenic temperatures.

5.9 Conclusions

The microwave properties of oxide based dielectric bulk material, thin film nonlinear dielectric materials and oxide high temperature superconducting materials were reviewed in this article. In addition, the most important microwave measurement techniques have been discussed. Important future directions of related material research aiming towards further integration both on chip and subsystem level, increase of performance and cost reduction are:

- tantalum free bulk dielectrics with low losses and low temperature coefficients,
- LTCC compatible dielectrics with low losses,
- nonlinear dielectric films with high tuneability and low losses both on silicon and LTCC substrates,
- on-chip integrated low-loss piezoelectric thin film based switches and acoustic resonators and
- fabrication processes for periodic metallodielectric electromagnetic bandgap and metamaterial structures.

The last two items of this listing were not discussed in this review because of being stronger related to design and fabrication rather than to material issues. However, since microwave material properties are strongly related to fabrication technology, size effects due to downscaling of physical dimensions towards the micro- and submicrometer range will become highly important in the future.

Bibliography

[1] B. A. Aminov *et al.*, IEEE *Transactions on Applied Superconductivity* **9**, 4185 (1999).
[2] see e.g. N. W. Ashcroft and N. D. Mermin, *Solid State Physics*, Sunders College, Philadelphia 1976.
[3] K. Bouzehouane *et al.*, *Appl. Phys. Lett.* **80**, 109 (2001).
[4] V. B. Braginsky, V. S. Ilchenko, and Kh. S. Bagdassarov, *Phys. Lett.* A **120**, 300 (1987).
[5] B. L. Givot *et al.*, *14th International Conference on Microwaves, Radar and Wireless Communications*, MIKON-2002, 401 (2002).
[6] V. L. Gurevich and A. K. Tagantsev, *Adv. Phys.* **40**, 719 (1991).
[7] M. Hein, *High-temperature superconductor thin films at microwave frequencies*, Springer Tracts in Modern Physics, **155**, Springer 1999.
[8] B. B. Jin *et al.*, *Phys. Rev.* B. **66**, 104521 (2002).
[9] D. Kajfez and P. Guillon, *Dielectric resonators*, Artech House Inc., 1986.
[10] N. Klein, *Electrodynamic properties of oxide superconductors*, Juelich report Jül-3773, Habilitationschrift 1997.
[11] N. Klein, M. Winter, and H. R. Yi, *Superc. Science and Technology* **13**, 527 (2000).
[12] N. Klein, *Rep. Prog. Phys.* **65**, 1387-1425 (2002).

[13] J. Krupka et al., IEEE Trans. on Microwave Theory and Techniques **42**, 1886 (1994).
[14] R. Ott and R. Wördenweber, *Appl. Phys. Lett.* **80**, 2150 (2002).
[15] R. Ott and R. Wördenweber, submitted for publication.
[16] S. T. Popov, University of St. Petersburg, Internal Report, 2000.
[17] M. Sparks et al., *Phys. Rev. B.* **26**, 6987 (1982).
[18] M. E. Tobar et al., *Joint Meeting of the European Frequency and Time Forum and the* IEEE *International Frequency Control Symposium* 573, 1999
[19] M. Valant et al., *Journal of European Ceramic Society* **21**, 2647 (2001).
[20] J. R. Waldram, *Superconductivity of metals and cuprates*, Institute of Physics Publishing, Bristol and Philadelphia, 1996.
[21] C. Zuccaro et al., *Journal of Applied Physics* **82**, 5695 (1997).
[22] C. Zuccaro, *Mikrowellenabsorption in Dielektrika und Hochtemperatursupraleitern für Resonatoren hoher Güte*, Juelich report Jül-3631, Dissertation 1998.

6 Advanced X-ray Analysis of Ferroelectrics

Keisuke Saito[1], Toshiyuki Kurosawa[1], Takao Akai[1], Shintaro Yokoyama[2], Hitoshi Morioka[2] and Hiroshi Funakubo[2]

1. PANalytical, Tokyo, Japan
2. Tokyo Institute of Technology, Japan

Abstract

Examples of applications of a high-resolution X-ray diffractometer are described. Major advantages of X-ray diffraction reciprocal space mapping and grazing incidence X-ray diffraction are discussed. X-ray diffraction reciprocal space mapping reveals that epitaxial thin films of Pb(Zr,Ti)O_3 with composition near the morphotropic phase boundary have a mixed structure of tetragonal and pseudocubic phases. Moreover, XRD-RSM is applied to distinguish phases in bismuth-layer-structured $Sr_{0.88}Bi_{2.2}(Ta_{0.7}Nb_{0.3})_2O_9$ and fluorite-type SBTN (fluorite-SBTN) prepared on a conventional platinum-coated silicon substrate, which is difficult to achieve by conventional XRD measurement. It is shown that XRD-RSM is also effective for such an application, since it collects diffraction spots lying in the off-axis direction from the surface normal in reciprocal space. Grazing incidence X-ray diffraction is introduced for depth profile analysis of a 'twist' of crystals. It is advantageous for depth profile analysis, since the penetration depth can be controlled by controlling the incident angle of X-ray to the film surface at around the critical angle for total reflection. This is applied to MPB-PZT thin films. We found that the twist of the crystal is large at the film/substrate interface but decreases with increasing film thickness.

6.1 Introduction

The recent modern multi-axes high-resolution X-ray diffractometer is becoming brighter in intensity and higher in resolution. Although it is still darker than synchrotron radiation, X-ray flux from a laboratory scale X-ray tube is becoming similar to that from a rotating anode generator. The reason lying behind this is the development of X-ray optics which enhance the X-ray flux, such as multilayer X-ray reflection mirror [1], multicrystal monochromators [2, 3] and polycapillaries [4]. The refractive index of media for X-ray is almost 1 and is only smaller than unity by about the order of 10^{-5} for Cu kα radiation. Therefore, X-ray cannot be monochromatized or focused using a lens or a prism type of bulk crystal optic; the only way to achieve this is to use multidiffractions using perfect single crystals, such as germanium or silicon (four or two bounce monochromator), or the total reflection of the media (capillary). In the last decade, single-crystal multibounce monochromator and polycapillaries were developed. As a result, the characterization using a laboratory-scale X-ray diffractometer was largely changed and advanced.

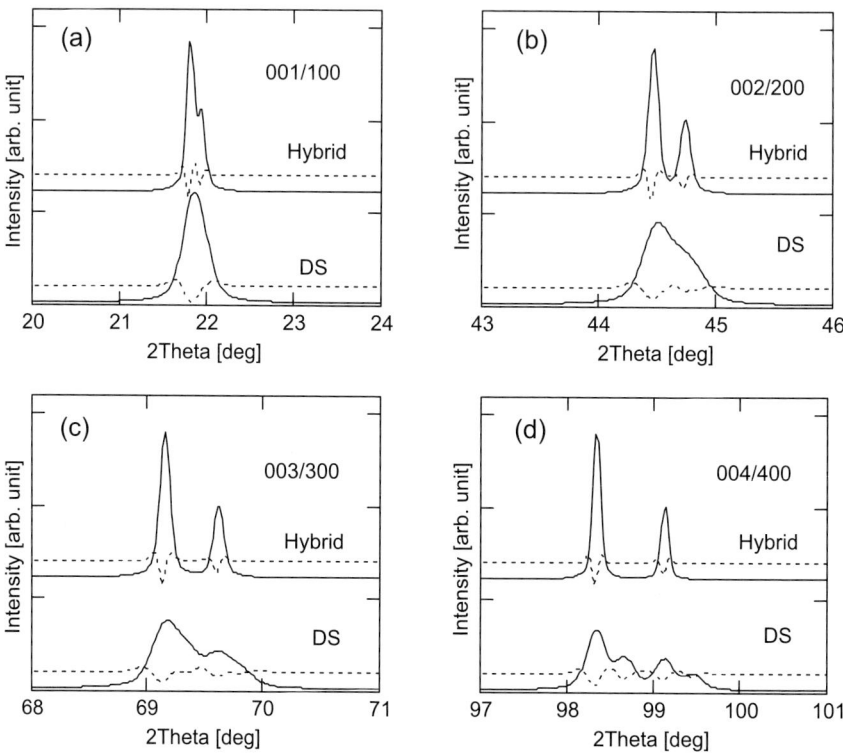

Figure 6.1: Comparison of $2\theta - \theta$ scan profiles obtained by a monochromatized (pure Cu kα1) parallel beam configuration (hybrid x-ray mirror) and a conventional parallel beam configuration achieved by divergence slit (DS) module measured at *001/100* (a), *002/200* (b), *003/300* (c), *004/400* (d) of 500 nm-thick Pb(Zr$_{0.54}$Ti$_{0.46}$)O$_3$ thin film. Dotted lines represent the second derivative of the profiles, indicating the peak positions. Note that the profiles are simulated fitted profiles for obtained spectrum using pseudo-Voight function (mixed Lorentz and Gauss function).

Among the developments, particularly, the single-crystal multibounce monochromator combined with the multilayer X-ray reflection mirror (hybrid X-ray mirror) significantly improves the quality of the data obtained by the laboratory-scale X-ray diffractometer. The monochromator reduces the X-ray beam divergence to 1.7×10^{-4} rad and reduces the wavelength spread to $dl/l \approx 3 \times 10^{-4}$. The X-ray mirror reflects and converges X-ray from the source and, as a result, it enhances X-ray flux by a factor of 10 compared to the same beam dimension from the source with a conventional divergence slit module. Highly monochromatized X-ray is particularly useful and important for characterizing the epitaxial layer grown

6.1 Introduction

on a single-crystal substrate or material that shows a complicated diffraction pattern. Figures 6.1 (a) to 6.1 (d) show conventional $2\theta - \theta$ scans measured at *001/100*, *002/200*, *003/300* and *004/400* diffractions of Pb(Zr$_{0.5}$Ti$_{0.5}$)O$_3$ thin film measured by hybrid X-ray mirror and conventional divergence slit module, respectively. Although all of the diffractions originate on the same lattice plane, it is clear that the peak width is narrower and peak separation is better in the profiles measured by the hybrid X-ray mirror compared to those measured by the conventional divergence slit module. Furthermore, since the conventional parallel beam achieved by the divergence slit module contains a kα1/kα2 doublet, it is difficult to analyze the profile when it is complicated.

X-ray diffraction reciprocal space mapping (XRD-RSM) is a technique that collects diffracted X-ray intensities two dimensionally in reciprocal space [5,6]. This technique enables the determination of the state of residual strain in epitaxial thin film [5,6], in-plane and out-of-plane lattice parameters [7–11] and the constituent crystallographic phase in strongly oriented thin film [12–14]. XRD-RSM was first employed for the determination of the residual strain state in compound semiconductor thin films, such as (In$_x$Ga$_{1-x}$)(As$_y$P$_{1-y}$) layer grown on a InP substrate or (Al$_x$Ga$_{1-x}$)As layer grown on a GaAs substrate [5,6]. Thereafter, this technique was applied to domain structural characterizations of epitaxially grown Pb(Zr,Ti)O$_3$ thin films [15,16]. However, the problem in analyzing ferroelectric thin films was the intensity from the laboratory-scale X-ray tube with optics configuration applied to the compound semiconductor. In the case of compound semiconductor thin films, the layer is typically grown on a single-crystal substrate perfectly as an epitaxial thin film, resulting in a sufficient diffraction intensity even at a higher diffraction angle and at a thinner film thickness. On the other hand, ferroelectric thin films typically have weaker film orientations than compound semiconductor thin films, resulting in weak diffraction intensities at a higher diffraction angle. Recent novel optics mentioned above eliminate this difficulty and enable the measurement of ferroelectric thin films by laboratory-scale high-resolution X-ray diffractometer.

In thin film applications, the depth profile information is sometimes the matter of interest. In synchrotron radiation, grazing incidence X-ray diffraction (GIXRD) is employed to characterize thin films along the depth direction [17–19]. This technique limits the penetration depth of X-ray by decreasing the incident angle against the wafer surface. Although X-ray has strong penetration power against most of the media, the penetration depth can be controlled and reduced by controlling the incident angle around the critical angle of X-ray. The critical angle is a function of density and absorption of the media and is about $0.2°$ to $0.4°$ for the most of inorganic materials. The penetration depth is largely reduced at an incident angle below the critical angle ($\approx$ nm for Pb(Zr$_{0.5}$Ti$_{0.5}$)O$_3$) and it exponentially increases above the critical angle, then it obeys the well-known penetration depth formula for the higher incident angle case.

In this work, XRD-RSM and grazing incidence X-ray diffraction are applied to ferroelectric Pb(Zr$_{0.52}$Ti$_{0.48}$)O$_3$ (PZT) thin films for structural characterizations. Moreover, XRD-RSM is employed for phase distinguishing of Sr$_{0.88}$Bi$_{2.2}$(Ta$_{0.7}$Nb$_{0.3}$)$_2$O$_9$ (SBTN) from impurity paraelectric phases. It is known that PZT ceramics with a composition near the morphotropic phase boundary (MPB, Zr/(Zr+Ti) $\approx$ 0.54) have the mixed structure of tetragonal and rhombohedral phases. At this composition, PZT ceramics show enormously large piezoelectric and dielectric properties [20, 21]. If this special characteristics of MPB-PZT is originated due to the phase coexistence, it is important to determine the crystallographic structure of MPB-PZT as a form of thin film, since MPB-PZT thin film has been receiving much attention for application to thin-film-based micro-electro-mechanical systems (MEMS). On the other hand, SBTN is a bismuth-layer-structured ferroelectric (BLSF) [22] and is known to be grown as impurity fluorite- or pyrochlore-type paraelectric phases when deposited at a relatively low temperature or a excess-bismuth or bismuth-deficient composition [12, 23, 24]. Since both of the impurity paraelectric phases show their major diffractions at scattering angles similar to those of SBTN, it is fairly difficult to distinguish them by conventional X-ray diffraction measurements [12].

6.2 Experimental

6.2.1 X-ray diffractometer

PANalytical's X'Pert PRO-MRD system (PANalytical, Almelo, Netherlands) was employed for high-resolution X-ray diffraction studies (Figure 6.2 (a)). A hybrid X-ray mirror, the combination of a Ge220 double-crystal monochromator with a multilayer X-ray reflection mirror, and a parallel-plate collimator with $\delta(2\theta) = 1.6 \times 10^{-3}$ rad in resolution were used for X-ray diffraction reciprocal space mapping (PRO-MRD) measurement (Figure 6.2 (b)). The phase identification of the SBTN phase from the impurity paraelectric phase by XRD-RSM and the grazing incidence X-ray diffraction (GIXRD) for PZT thin film were carried out by X-ray polycapillaries ($\delta\theta = 3.5 \times 10^{-3}$ rad) and a parallel-plate collimator with 3.5×10^{-3} rad resolution. As schematically shown in Figure 6.2 (b), the diffractometer has an incident angle of ω, a scattering angle of 2θ, a rotation about the film surface normal of ϕ and an inclination perpendicular to the scattering plane of ψ. Also, two detectors that have two different resolutions are mounted on the diffracted beam side simultaneously and can be switched automatically by operating software (double-diffracted beam arm).

6.2.2 Method of X-ray diffraction

In this study, two methods of XRD-RSMs and GIXRD were applied to ferroelectric thin film. Which XRD-RSM is chosen depends on the orientation of the thin film media. In the following session, details of these two XRD-RSMs and the GIXRD and the comparison of them to the conventional $2\theta - \theta$ scan method are discussed.

X-ray diffraction reciprocal space mapping

In choosing beam optics to measure XRD-RSM, one must consider "resolution function" in the reciprocal space. The resolution function is defined by the incident beam divergence and the acceptance window of the diffracted beam side optic. Figure 6.3 schematically shows the definition of the resolution function in the reciprocal space. The X-ray detector is located at the tip of the scattering vector **H** in the reciprocal space. The incident beam divergence $\delta\omega$ and the acceptance window of the diffracted beam optic $\delta 2\theta$ define the resolution function (gray area in Figure 6.3) in the reciprocal space. The form of the obtained diffracted intensity distribution of the crystal by XRD-RSM is defined by the convolution of the resolution function and the reciprocal lattice point of the crystal. Therefore, a resolution function smaller than

6.2 Experimental

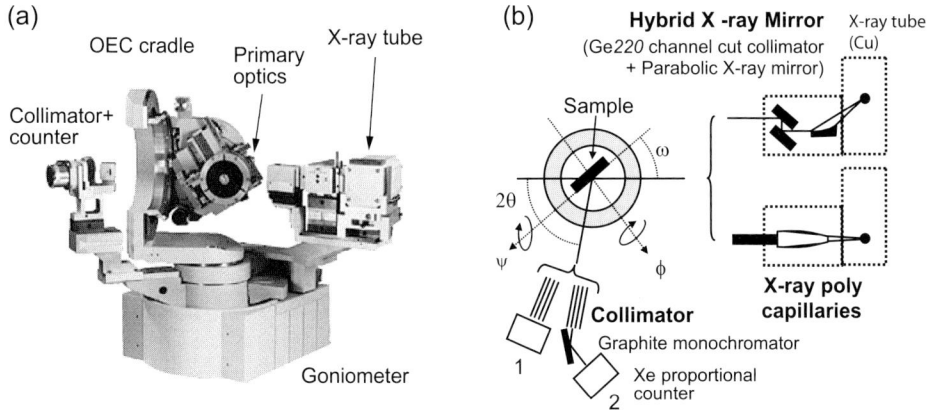

Figure 6.2: (a)Example of overview of high-resolution X-ray diffractometer (PANalytical's X'Pert PRO-MRD system), and (b) schematics of incident and diffracted side optics used in the present study. For incident beam optics, a hybrid X-ray mirror and X-ray polycapillaries are used in combination with line focus and point focus of X-ray tube, respectively.

the reciprocal lattice point of the crystal measured is preferred, otherwise the form of the reciprocal lattice point of the crystal is deformed and that of the resolution function will be obtained.

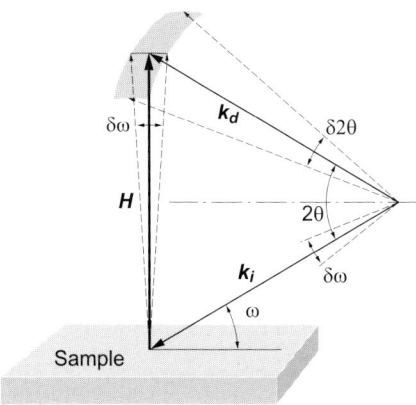

Figure 6.3: Schematic of the resolution function in the reciprocal space. Two vectors k_i and k_d represent the incident and the diffracted wave vectors and **H** represents the scattering vector. The divergence of the incident X-ray and the acceptance window of the diffracted beam side optic are represented as $\delta\omega$ and $\delta(2\theta)$.

Typically, crystals as a form of thin films exists in three different states. Three dimensionally aligned perfect crystals are called "epitaxial" thin films (Figure 6.4 (a)). The epitaxial thin film can be grown on a single-crystal substrate that has lattice parameters very similar to those of the thin film. To resolve the X-ray diffractions from both the layer and the substrate, it is necessary to employ highly monochromatized and paralleled X-ray and measure diffractions at the highest scattering angle possible (see Figure 6.1). Therefore, the hybrid X-ray mirror is the ideal incident beam optic and the fine parallel plate collimator is ideal for diffracted-beam optics from the point of view of the resolution function. The ω scan axis is parallel to the direction of the circumference of the scattering vector $\mathbf{H}$ and the $2\theta - \omega$ scan is parallel to the radial direction from the origin of the reciprocal space. By combining these two scan axes, namely ω and $2\theta - \omega$ scan axes, diffracted intensity distribution in the gray area shown in Figure 6.4 (a) can be mapped.

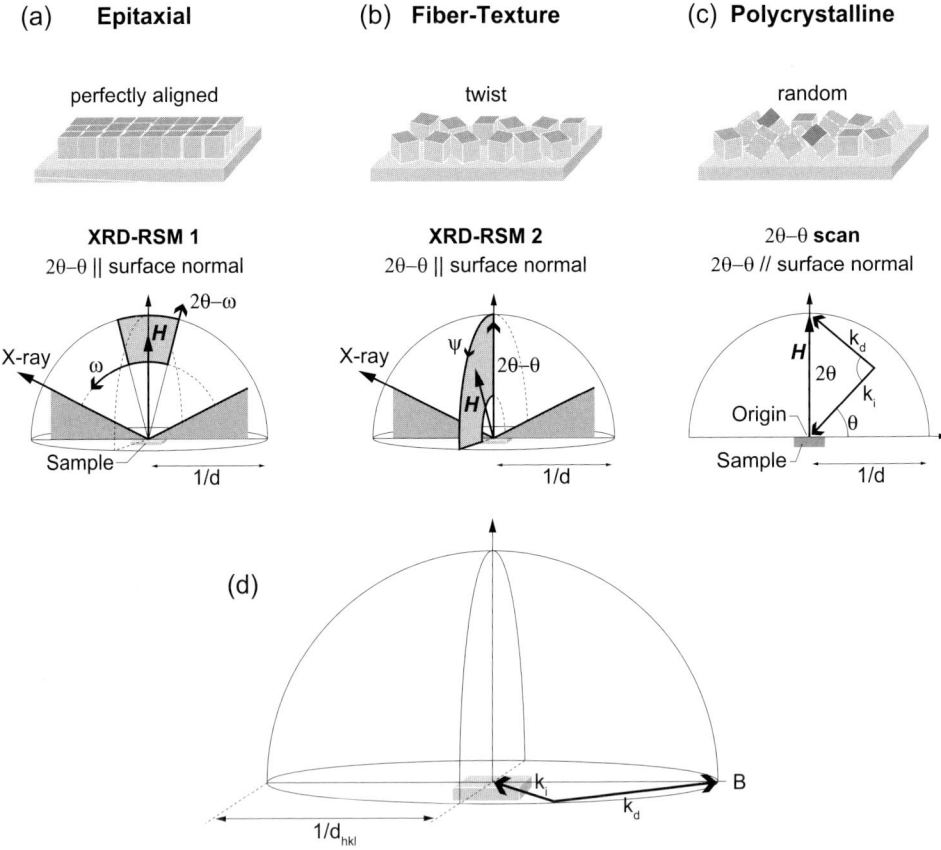

Figure 6.4: Schematics of the relationship between X-ray characterization and the crystal state of the thin films, (a)-(c), and the scattering geometry of the grazing incidence X-ray diffraction in the reciprocal space, (d).

6.2 Experimental

When there is an orientation in the film surface and not in the in-plane, this is called "fiber-texture". XRD-RSM with ω offset mentioned above can also be applied to this film. However, the intensity of diffraction in the inclined direction (diffraction along off-specular direction) is very weak compared to epitaxial thin films, since the diffracted intensity is distributed along the ϕ axis direction at a constant ψ angle. Moreover, the intensity of diffraction is typically very weak at a higher scattering angle (larger **H**). Since XRD-RSM with ω offset is only available at a higher scattering angle due to the problem of limiting sphere (The Ewald sphere in the reciprocal space that the incident wave vector is just parallel to the surface of the film), XRD-RSM with ω offset is not suitable in this case. In such a case, it is advantageous to incline the scattering vector **H** along the ψ axis, since there is no limitation for ψ to be inclined at any scattering angle (Figure 6.4 (b). When applying XRD-RSM with ψ inclination, it is necessary to consider the vertical (axial) divergence of the optic. The vertical divergence is the divergence perpendicular to the scattering plane (the plane includes both the incident and the diffracted wave vectors). This only causes an asymmetry of the diffraction peak profile at a low scattering angle if the measurement is carried out in the scattering plane. However, when the diffraction is measured with ψ offset, the diffraction peak intensity decreases markedly and the peak profile is broadened due to the vertical divergence. It is effective to use polycapillaries to minimize this effect, since a cylindrically paralleled X-ray is obtained from the capillary.

A polycrystalline thin film does not have any preferred orientation (Figure 6.4 (c)). In such a case, the diffraction from the crystal is not a spot but a so-called Debye-Scherrer ring. Therefore, the sample does not have to be inclined to obtain the diffraction pattern. Conventional $2\theta - \theta$ scans move the scattering vector **H** in the radial direction along the film surface normal. Thus, these scans give sufficient information when the film is polycrystalline. The obtained diffracted intensity must be corrected in terms of the absorption and the Lorentz polarization. These two terms and the obtained diffracted intensity have the following relation:

$$I_{hkl} = I_0 \left|F_{hkl}\right|^2 (LP_{hkl}) A_{hkl} \frac{V}{V_c} \tag{6.1}$$

where I_0, F_{hkl}, LP_{hkl}, A_{hkl}, V and V_c are the intensity of the incident X-ray, the structure factor, the Lorentz polarization factor, the absorption factor, the volume of the sample and the volume of the unit cell, respectively. The absorption factor A_{hkl} and the Lorentz polarization factor LP_{hkl} are represented as follows.

$$A_{hkl} = \frac{A \sin\alpha \cdot \sin\beta}{\mu(\sin\alpha + \sin\beta)} \cdot \left(1 - \exp\left\{-\mu \cdot t \left(\frac{1}{\sin\alpha} + \frac{1}{\sin\beta}\right)\right\}\right) \tag{6.2}$$

$$LP = \frac{1 + \cos^2(2\theta_m) \cdot \cos^2(2\theta)}{1 + \cos^2(2\theta_m) \cdot \sin(2\theta)} \tag{6.3}$$

$$\mu = \rho \cdot \sum_i \left(\frac{\mu}{\rho}\right)_i \cdot w_i \tag{6.4}$$

where A, α, β, μ, t, ρ, w, θ_m and θ are the area on the sample surface illuminated by X-ray, the incident angle, the exit angle from the surface, the linear absorption coefficient, the thickness of the film, the density of the film, the weight fraction of the constituent element, the Bragg angle of the primary monochromator and the Bragg angle of the film, respectively. In Equation (6.4), μ/ρ is called the mass absorption coefficient.

Grazing incidence X-ray diffraction

The refractive index n of the media is defined as:

$$n = 1 - \delta - i\beta \tag{6.5}$$

$$\delta = \frac{\lambda^2 e^2}{2\pi m c^2} \sum_a (Z_a + f'_a) N_a \tag{6.6}$$

$$\beta = \frac{\lambda^2 e^2}{2\pi m c^2} \sum_a (f''_a) N_a \tag{6.7}$$

where λ is the X-ray wavelength, e is the charge, m is the mass of the electron, c is the velocity of light, N_a is the number of the atoms of type a in the unit cell, Z_a is the atomic number, and f' and f'' are the anomalous scattering factors. The δ and β for Pb(Zr$_{0.5}$Ti$_{0.5}$)O$_3$ are 2.0×10^{-5} and 1.6×10^{-6} at Cu kα wavelength, respectively, indicating that the refractive index at this wavelength is only about 10^{-5} smaller than unity. Therefore, the critical incident angle $\theta_c = (2\delta)^{1/2}$ for total reflection exists outside of the media and is about the order of 10^{-3} rad. The $1/e$ penetration depth $D(\alpha)$ is defined as

$$D(\alpha) = \frac{\lambda}{4\pi B} \tag{6.8}$$

$$2B^2 = \left(-\alpha^2 + 2\delta^2\right) + \sqrt{(\alpha^2 - 2\delta)^2 + 4\beta^2} \tag{6.9}$$

where α is the incident angle of X-ray to the surface. Figure 6.5 shows the calculated penetration depth dependent on the incident angle for Pb(Zr$_{0.5}$Ti$_{0.5}$)O$_3$ of X-ray with wavelength of 0.154 nm. The critical angle of PZT 50/50 is $\theta_c = (2\delta)^{1/2} = 0.36$ deg. The calculation shows that the reflectivity and the penetration depth are significantly decreased and increased at this incident angle, respectively. Moreover, the calculation shows that the penetration depth can be controlled by controlling the incident angle of X-ray around the critical angle of the media.

The scattering configuration in the reciprocal space is schematically drawn in Figure 6.4 (d). In contrast with the XRD-RSM results mentioned in paragraph "X-ray diffraction reciprocal space mapping", both the incident and diffracted wave vectors lie in the plane parallel to the surface. Thus, the rocking curve scan (a) and $2\theta - \theta$ scan (b) in Figure 6.4 (d) under GIXRD configuration show "twist" and d-spacing of the crystal in the in-plane, respectively.

6.3 Results and discussion

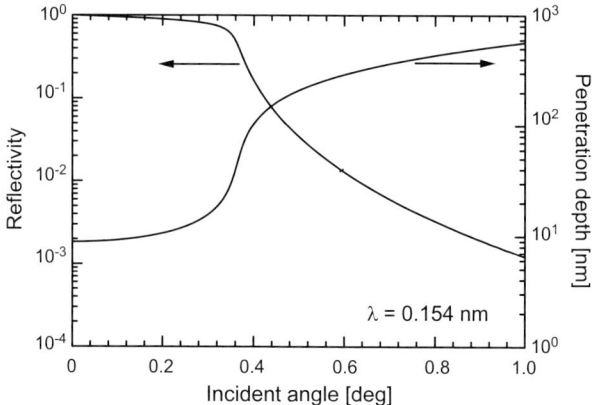

Figure 6.5: The calculated reflectivity and the penetration depth profile for PZT 50/50 of X-ray with 0.154 nm wavelength. The critical angle of PZT 50/50 is 0.36° for this wavelength.

6.2.3 Sample preparation

A thin film of Pb(Zr$_{0.52}$Ti$_{0.48}$)O$_3$ (PZT 52/48) with 5 μm thickness was prepared on a SrRuO$_3$ (100 nm) / (001)–SrTiO$_3$ substrate at 600 °C by pulsed metalorganic chemical vapor deposition (pulsed–MOCVD) using Pb(C$_{11}$H$_{19}$O$_2$), Zr(O·t – C$_4$H$_9$)$_4$, Ti(O·i – C$_3$H$_7$)$_4$ and O$_2$ as starting materials [25]. Temperature-dependent XRD-RSM measurement was performed and the crystal structure of this MPB-PZT thin film was investigated. Moreover, the depth profile of the twist of PZT 52/48 crystal was characterized by employing the GIXRD method.

Thin films of Sr$_{0.88}$Bi$_{2.2}$(Ta$_{0.7}$Nb$_{0.3}$)$_2$O$_9$ and Sr$_{0.90}$Bi$_{4.4}$(Ta$_{0.7}$Nb$_{0.3}$)$_2$O$_9$ ((SBTN) and fluorite–SBTN) with thicknesses of 200 nm were deposited on a platinum-coated (001) silicon substrate at 585 °C by pulsed–MOCVD using Sr(Ta$_{0.7}$Nb$_{0.3}$(OC$_2$H$_5$)$_5$(OC$_2$H$_4$OCH$_3$))$_2$, Bi(CH$_3$)$_3$ and O$_2$ as starting materials [26]. The phase identification of the SBTN phases and fluorite phase using XRD-RSM method is carried out.

6.3 Results and discussion

6.3.1 Structural characterization of PZT 52/48 thin film

Figure 6.6 (a) and Figure 6.6 (b) show XRD-RSM results measured at *004* and *204* diffractions. The results show tetragonal c-domain and a-domain diffractions, indicating that the well-known ...$c/a/c/a$...90° domain structure exists in PZT 52/48 thin film. Moreover, additional diffractions were observed in between c-domain and a-domain diffractions. The *004* and *204* diffractions of this additional phase have the same Q_y coordinate, indicating that the interaxial angle between c and a crystallographic axes is 90°. The lattice parameters of the tetragonal and the additional phase were obtained as $a = 0.4018$ nm and $c = 0.4138$ nm for tetragonal phase and $a = 0.4050$ nm and $c = 0.4080$ nm for additional phase. The additional phase appears to be a tetragonal phase. However, due to its strained nature, we assign this phase as a pseudocubic phase. The interaxial angle between the c and a axes can be determined by the reciprocal space coordinates (Q_x and Q_y) of the surface normal *004* diffraction and the asymmetric *204* diffraction due to the fact that the reciprocal lattice is a Fourier transformed unit cell. Figure 6.7 shows the calculated diffraction spot configuration in the reciprocal space. In the figure, the horizontal and vertical axes are parallel to in-plane [100] STO and out-of-plane [001] STO, respectively. The dotted lines represent the Fourier-transformed unit cell of (001)-oriented tetragonal PZT and rhombohedral PZT. The calculation shows that the *004* and *204* diffractions occupy the corner of the Fourier-transformed unit cell. If the reciprocal space coordinates of *004* and *204* diffractions can be obtained by XRD-RSM, one can determine the interaxial angle δ. Moreover, the figure shows that the relative distance between *004* and *204* equals half of the in-plane lattice parameter. As a result, combining XRD-RSMs measured at both surface normal and off-axis diffractions, the in-plane and the out-of-plane lattice parameters as well as the interaxial angle can be determined.

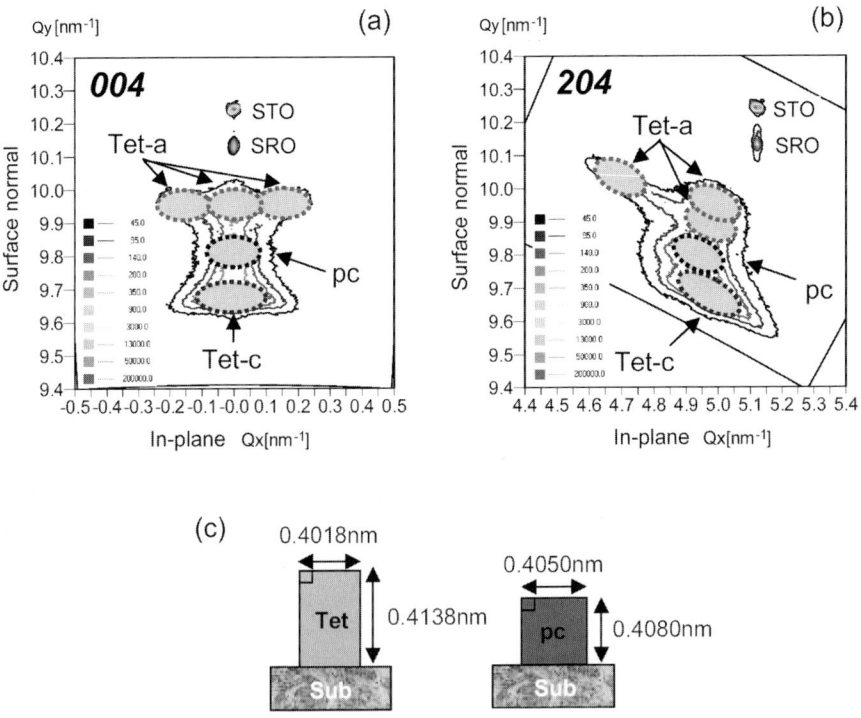

Figure 6.6: XRD-RSM measured at around *004* and *204* diffraction of PZT 52/48 // SrRuO₃ // (001)–SrTiO₃, (a), (b) and obtained lattice parameters for PZT 52/48, (c).

Temperature-dependent XRD-RSM measurement was performed by employing a DHS 900 domed hot stage (Anton Paar) attached to the PANalytical's X'Pert PRO-MRD system. Figure 6.8 shows an overview of the DHS 900 domed hot stage. The sample is mounted on

6.3 Results and discussion

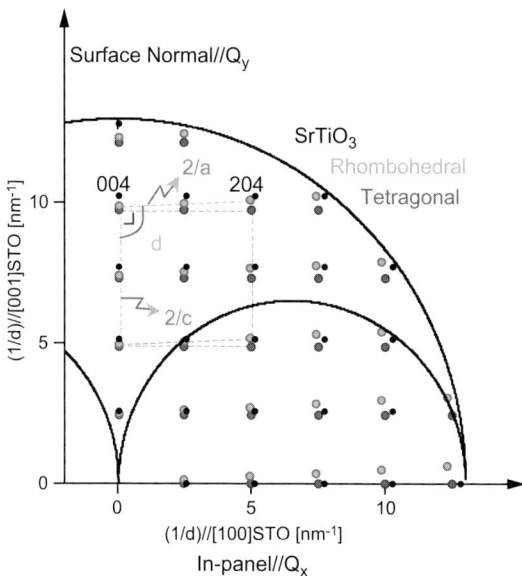

Figure 6.7: Calculated diffraction spot configuration in reciprocal space. δ indicates interaxial angle between c and a crystallographic axes.

a heater made by Inconel (Ni-Cr-Fe based metal) and covered by a dome made by PEEK (Polyetheretherketone). The stage is cooled by compressed air supplied from the air nozzle to prevent a heat conduction to the diffractometer. It is able to heat the sample up to 900° C. The atmosphere inside the dome can be replaced by inert gas or vacuum. In this study, XRD-RSMs were performed under ambient condition and temperatures up to 750° C from room temperature. Figures 6.9 (a) to 6.9 (d) show temperature-dependent XRD-RSMs measured at *204* diffraction. XRD-RSM measured at 750° C (Figure 6.9 (a)) shows that PZT 52/48 is in a high-temperature cubic phase at this composition. At the Curie temperature of about 350° C shown in Figure 6.9 (b), the diffraction spot is divided into two spots and the two diffraction spots change to those originated from tetragonal *c*- and *a*-domain and pseudocubic phases. The results showed that the Curie temperature of PZT 52/48 thin film grown on a SRO/SRO substrate is about 350° C and the pseudocubic phase appeared at the Curie temperature. It is known that the MPB-PZT of ceramics is a mixture of tetragonal and rhombohedral phases. The result of the present study possibly indicates that the rhombohedral phase is strained and changes to the cubic phase due to the applied growth strain from the underlying substrate.

Figure 6.8: Anton Paar's DHS 900 domed hot stage attached to the PANalytical's X'Pert PRO–MRD system.

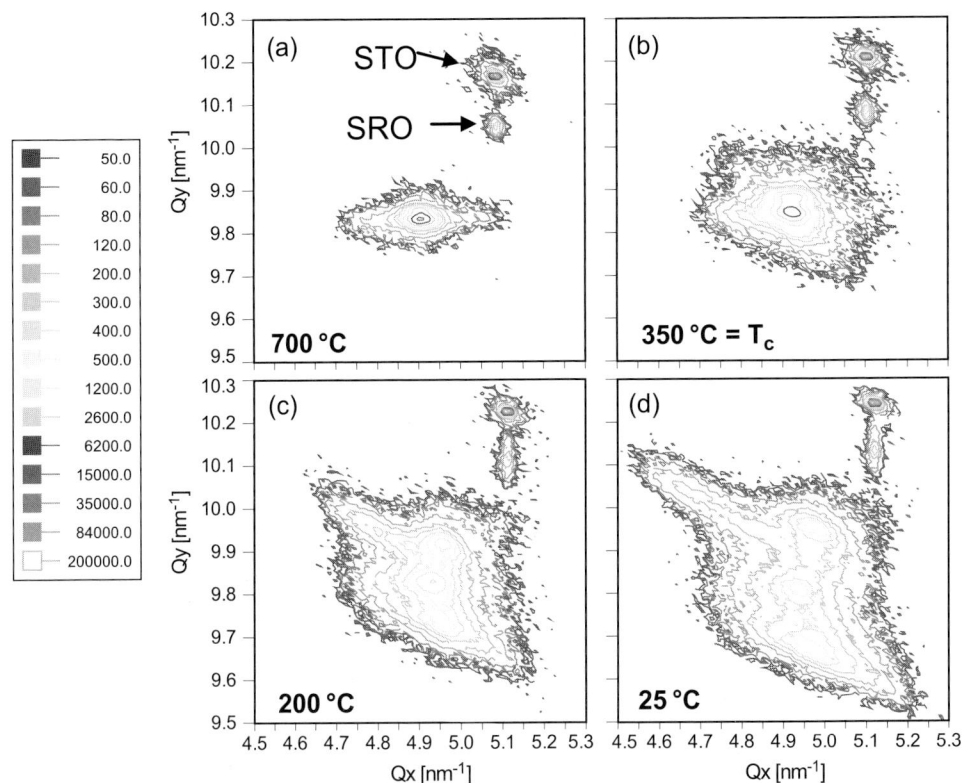

Figure 6.9: Temperature-dependent XRD-RSMs measured at *204* diffraction. Temperature was varied between 25°C and 700°C.

6.3.2 Distinguishing SBTN phase from fluorite–SBTN phase

Figure 6.10 (a) and Figure 6.10 (b) show conventional $2\theta - \theta$ scans of SBTN and fluorite–SBTN thin films, respectively. As a reference, $2\theta - \theta$ scan for $Sr_{1.2}Bi_2Ta_2O_9$ was also measured and shown in Figure 6.10 (c). Conventional $2\theta - \theta$ scans in Figure 6.10 (a) and Figure 6.10 (b) show strong diffraction around 29° in 2θ for both SBTN and fluorite–SBTN. These peaks are indexed as *115* SBTN and *111* fluorite–SBTN, respectively. The results indicate that the films have strong *115* and *111* orientation in SBTN and fluorite–SBTN, respectively.

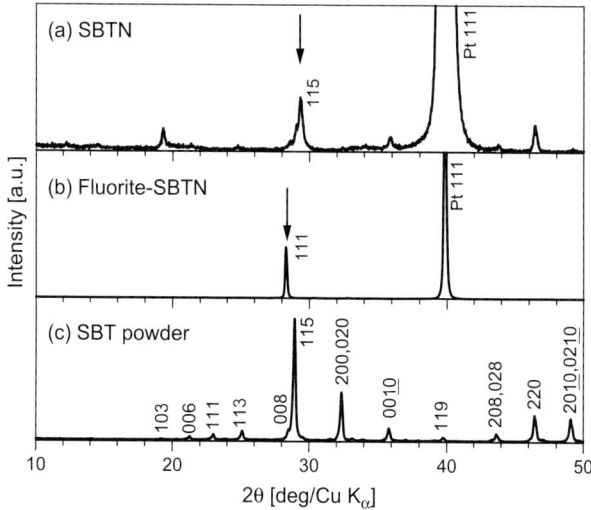

Figure 6.10: Conventional $2\theta - \theta$ scans measured for (a) a SBTN and (b) a fluorite–SBTN thin film and (c) SBT powder.

Figures 6.11 (a) and 6.11 (b) show results of XRD-RSMs with ψ inclination. The horizontal and the vertical axes are the $2\theta - \theta$ scan axis and ψ axis, respectively. In the measurement, the $2\theta - \theta$ and ψ axes were varied from 25° to 50° and for 0° to 90°, respectively. Both films showed diffraction spots, indicating that the films have a strong orientation. To identify the film orientation, the diffraction patterns were simulated and fitted. Figures 6.11 (c) and 6.11 (d) show the calculated diffraction pattern distributions assuming that the SBTN thin film has $\{103\}_T$ (T denotes the pseudotetragonal Mirror index) orientation and the fluorite–SBTN thin film has a $\{111\}$ orientation, respectively. It can be seen that the calculated patterns well match to the XRD-RSM results shown in Figures 6.11 (a) and 6.11 (b), indicating that the SBTN and fluorite–SBTN thin films have $\{103\}_T$ and $\{111\}$ orientations. The powder $2\theta - \theta$ scan shown in Figure 6.10 (c) showes that the relative intensity of the *103*$_T$ diffraction of SBTN is less than 1% compared to that of the *115* diffraction. This suggests that the $\{103\}_T$ orientation is hardly detected by the conventional $2\theta - \theta$ scan due to the intrinsically weak diffraction intensity of the *103*$_T$ diffraction. Moreover, although the *115* diffraction is observed as the most intense diffraction in the conventional $2\theta - \theta$ scan profile shown in Figure 6.10 (a), the

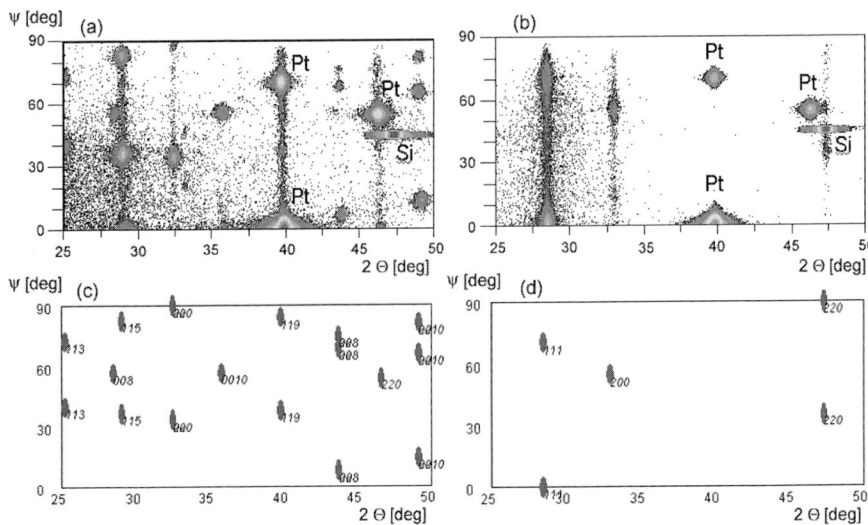

Figure 6.11: XRD-RSMs with y offset measured for (a) SBTN and (b) fluorite–SBTN thin films and fitted calculated diffraction patterns for the obtained results, (c) and (d).

relative volume fraction of the *115* orientation is fairly small, since the *115* diffraction is the intrinsically strongest diffraction in SBTN as shown in Figure 6.10 (c). The results show that the film orientation can be effectively determined by measuring not only the diffraction lying along the surface normal direction (oriented hkl diffraction) but also by those lying in the inclined direction enabled by the XRD-RSM with y offset configuration.

6.3.3 Grazing incidence X-ray diffraction study on PZT 52/48 thin films

The rocking curve full width at half maximum intensity (FWHM) was characterized at the *100/001* diffraction for PZT 52/48 thin films with GIXRD geometry. Figure 6.12 shows the calculated diffraction pattern distribution for PZT 52/48 thin films based on the lattice parameters obtained from XRD-RSMs shown in Figure 6.6. As shown in Figure 6.12, the *100/001* diffraction includes *100* diffraction of c-domains, the *001* diffraction of a-domains and the *100* diffraction of the pseudocubic phase. Therefore, FWHM obtained by this measurement indicates a "twist" of all these phases.

Figure 6.13 summarizes the rocking curve FWHM measured at *100/001* diffraction of a PZT 52/48 thin film with GIXRD geometry. In Figure 6.13, the calculated penetration depth of X-ray with 0.154 nm wavelength for PZT 52/48 is also plotted. The result shows that FWHM measured at low angles of the incidence of X-ray is small and it increases with increasing the incident angle of X-ray. This indicates that the twist of the crystal is rather large in the region close to the interface between the PZT 52/48 and SRO/STO substrate and is rather small in the region close to the surface. Figure 6.14 shows the cross-sectional TEM observation for a tetragonal PZT 39/61 thin film grown on a (001)–STO substrate. Figure 6.14 shows that

6.3 Results and discussion

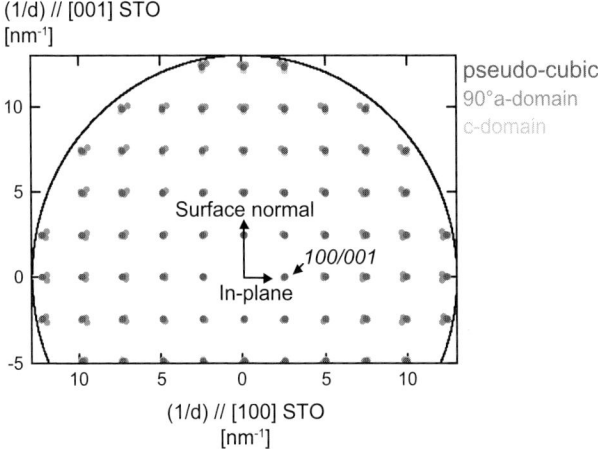

Figure 6.12: Calculated diffraction patterns for Pb($Zr_{0.52}Ti_{0.48}$)O_3 thin films with a ...$c/a/c/a$... domain-structured tetragonal phase and a pseudo-cubic phase. The lattice parameters for the two phases were measured by XRD-RSMs.

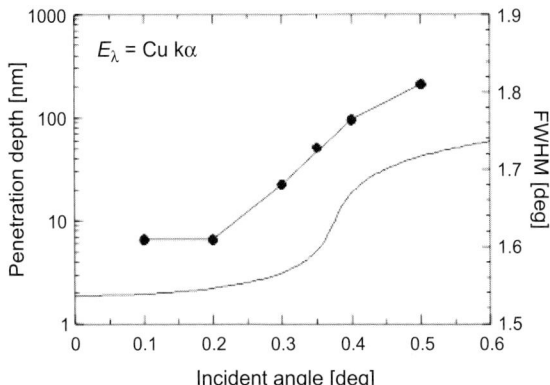

Figure 6.13: Incident angle dependence on the rocking curve FWHM of Pb($Zr_{0.52}Ti_{0.48}$)O_3 thin films and the penetration depth (calculation).

the film has ...$c/a/c/a$... domain structure. The film shows a dense a-domain structure in the region close to the layer/substrate interface and the density of a-domains decreases with increasing film thickness. This is possibly the reason for the depth profile of the rocking curve FWHM summarized in Figure 6.13. Namely, at the film/substrate interface, the high density of a-domains causes a large twist due to the high density of the domain boundary and, at the surface region, the twist decreases with decreasing density of the a-domain.

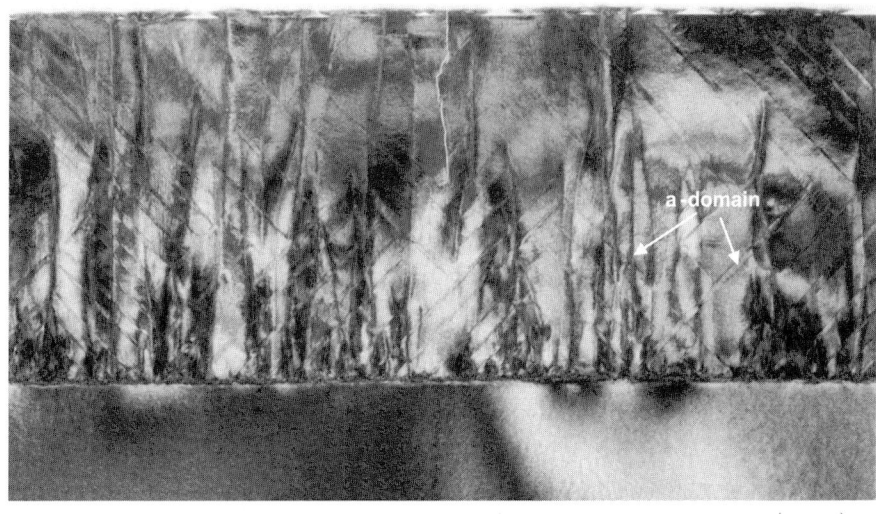

Figure 6.14: Cross-sectional high-resolution transmission electron micrograph of 1 mm-thick $Pb(Zr_{0.39}Ti_{0.61})O_3$ thin film grown on (001) $SrTiO_3$ substrate.

6.4 Conclusions

Examples of applications of X-ray diffraction reciprocal space mapping (XRD-RSM) together with the high-temperature stage and the grazing incidence X-ray diffraction (GIXRD) technique were introduced. Moreover, it was shown that a high-resolution X-ray diffractometer is effective in resolving diffractions from similar d-spacings. XRD-RSM with an ω offset was applied to the structural characterization of PZT 52/48 thin films and revealed that the MPB-PZT thin film consists of both tetragonal and pseudocubic phases. Temperature-dependent XRD-RSM showed that both phases have similar Curie temperatures. Rocking curve FWHM measured by GIXRD geometry showed that the FWHM is rather small in the top surface region and it increases with increasing penetration depth of X-ray, indicating that the twist of the crystal is large at the film/substrate interface and rather small in the top surface region. The cross sectional TEM view suggested that the depth profile of rocking curve FWHM is due to the variation of the density of the a-domain across the thickness of thin film.

XRD-RSM with ψ inclination was demonstrated to distinguish the SBTN phase from the fluorite–SBTN phase. It is known that both phases are difficult to distinguish by conventional $2\theta - \theta$ measurement, since both phases show diffractions at 2θ angle that are close to each other. It was revealed that XRD-RSM with ψ inclination is effective for phase distinguishing even when conventional $2\theta - \theta$ scans are similar and preferred orientation is strong.

Acknowledgement

The authors would like to thank Fujitsu laboratories, Japan for TEM observations. The author (KS) thanks Dr. P. Munk of PANalytical, Netherlands for financial support for the present study. Thanks are also due to K. Seo of PANalytical, Japan for his support.

Bibliography

[1] M. Schuster, and H. Göbel, *J. Phys. D* **28**, A270 (1995).
[2] W. Bartels, *Philips Tech. Rev.* **41**, 183 (1983).
[3] P. van der Sluis, *J. Appl. Cryst.* **27**, 50 (1994).
[4] H. Nozaki, and H. Nakazawa, *J. Appl. Cryst.* **19**, 453 (1986).
[5] P. van der Sluis, *J. Phys. D* **26**, A188 (1993).
[6] P. Fewster, *J. Appl. Cryst.* **24**, 178 (1991).
[7] K. Saito, K. Ishikawa, A. Saiki, I. Yamaji, T. Akai, and H. Funakubo, *Integrated Ferroelectrics* **33**, 59 (2001).
[8] K. Saito, M. Mitsuya, T. Suzuki, Y. Nishi, M. Fujimoto, M. Nukaga, I. Yamaji, T. Akai, and H. Funakubo, *Mat. Res. Soc. Symp. Proc.* **655** CC10. 13. 1 (2001).
[9] K. Saito, K. Nagashima, M. Aratani, I. Yamaji, T. Akai, and H. Funakubo, *Proc. 12th IEEE International Symposium on Applications of Ferroelectrics* (2000).
[10] K. Saito, T. Oikawa, T. Kurosawa, T. Akai, and H. Funakubo, *J. Appl. Phys.* **93**, 545 (2003).
[11] K. Saito, T. Kurosawa, T. Akai, S. Yokoyama, H. Morioka, T. Oikawa, and H. Funakubo, *Mat. Res. Soc. Symp. Proc.* **748** U13. 4. 1 (2003).
[12] K. Saito, M. Mitsuya, N. Nukaga, I. Yamaji, T. Akai, and H. Funakubo, *Jpn. J. Appl. Phys.* **39**, 5489 (2000).
[13] K. Saito, M. Mitsuya, I. Yamaji, T. Akai, and H. Funakubo, *Jpn. J. Appl. Phys.* **42**, 539 (2003).
[14] K. Saito, T. Oikawa, T. Kurosawa, T. Akai, and H. Funakubo, *Jpn. J. Appl. Phys.* **41**, 6730 (2001).
[15] W. -Y. Hsu and R. Raj, *Appl. Phys. Lett.* **67**, 792 (1995).
[16] C. M. Foster, Z. Li, M. Buckett, D. Miller, P. M. Baldo, L. E. Rehn, G. R. Bai, D. Guo, H. You, and K. L. Merkle, *J. Appl. Phys.* **78**, 2607 (1995).
[17] W. Marra, P. Eisenberger, and A. Cho, *J. Appl. Phys.* **50**, 6927 (1979).
[18] M. Doerner, and S. Brennan, *J. Appl. Phys.* **63**, 126 (1988).
[19] M. Toney, and S. Brennan, *Phys. Rev. B* **39**, 7963 (1989).
[20] B. Jaffe, W. Cook, and H. Jaffe, *Piezoelectric Ceramics* (Academic, London, 1971).
[21] J. Kuwata, K. Uchino, and S. Nomura, *Jpn. J. Appl. Phys.* **21**, 1298 (1982).
[22] B. Aurivillius, *Arkiv Kemi.* **54**, 463 (1949).
[23] T. Ami, K. Horonaka, C. Isobe, N. Nagel, M. Sugiyama, Y. Ikeda, K. Watanabe, A. Machida, K. Miura, and M. Tanaka, *Proc. MRS Symp.* **415**, 195 (1996).

[24] M. A. Rodriguez, T. J. Boyle, B. A. Hernandez, C. D. Buchheit, and M. O. Eatough, *J. Mater. Res.* **11**, 2282 (1996).
[25] K. Nagashima, M. Aratani, and H. Funakubo, *J. Appl. Phys.* **89**, 4517 (2001).
[26] M. Mitsuya, N. Nukaga, K. Saito, M. Osada, and H. Funakubo, *Jpn. J. Appl. Phys.* **40**, 3337 (2001).

7 Characterization of PZT-Ceramics by High-Resolution X-Ray Analysis

M. J. Hoffmann, H. Kungl, J.-Th. Reszat and S. Wagner

Universität Karlsruhe, Institut für Keramik im Maschinenbau, Karlsruhe, Germany

Abstract

PZT ceramics show the highest performance for compositions near the morphotropic phase boundary (MPB) in the coexistence region of the tetragonal and rhombohedral phase. X-ray diffraction is therefore used to determine the phase content and to analyze the interaction of both structures. The first part of the paper describes a reproducible method to determine the phase composition. It is shown that X-ray measurements from a surface near region of sintered samples agree well with electrical measurements which provide information of the sample volume. Temperature dependent XRD-measurements are used to study the phase transformation during heating for compositions at the MPB. It will be demonstrated that a phase transition from rhombohedral to tetragonal will have a strong impact on the temperature dependent permittivity and strain behavior.

The second part describes high resolution synchrotron X-ray diffraction experiments to investigate the extrinsic (domain switching) and intrinsic (lattice deformation) contributions to the total strain induced by an external electrical field. Evidence is given for both $180°$ and non-$180°$ domain switching. The direction dependence of the intrinsic effect is examined in rhombohedral PZT, which shows the highest lattice strain in [100] direction. The extrinsic contribution to strain is analyzed for tetragonal as well as rhombohedral PZT. It shows a much higher non-$180°$ domain wall mobility in rhomboedral PZT, but the resulting strain is comparable to the tetragonal structure due to the smaller lattice distortion.

7.1 Introduction

Lead zirconate titanate solid solution ceramics $Pb(Zr_{1-x}Ti_x)O_3$ (PZT 1 - x/x) with compositions near to the morphotropic phase boundary (MPB) reveal excellent electromechanical properties and are of high technological interest. Cooling of a PZT solid solution with a composition near the MPB below its Curie-temperature (T_C) induces a phase transformation from paraelectric cubic ($Pm3m$) to the ferroelectric rhombohedral F_R ($R3m$) and ferroelectric tetragonal F_T ($P4mm$) modifications [1, 2]. A distortion of the unit cells occurs at the transformation. For a tetragonal composition, the lattice distortion occurs in [001] direction and is characterized at room temperature by a c/a ratio of 1.0332 for undoped PZT (45/55). The distortion of the rhombohedral PZT is along the [111] direction and characterized by the ratio

Polar Oxides: Properties, Characterization, and Imaging
Edited by R. Waser, U. Böttger, and S. Tiedke
Copyright © 2005 WILEY-VCH Verlag GmbH & Co. KGaA, Weinheim
ISBN: 3-527-40532-1

of the lattice spacing d_{111}/d_{11-1}. This ratio determined as 1.007 for undoped PZT (60/40) is significantly smaller than the tetragonal one [3]. For an undoped morphotropic composition with a Zr/Ti-ratio of 53/47 the c/a-ratio of the F_T-modification decreases to 1.023 [3] and the typical strong overlapping of the $F_{T(200)}$-, $F_{T(002)}$- and $F_{R(200)}$-peaks in the range of $43° \leq 2\Theta \leq 46°$ can be observed. Due to the coexistence of the two phases at the MPB there are 14 possible directions along which the spontaneous polarization can occur, leading to a highly polarizable ceramic by applying an electrical field. More recent investigations indicate that the high polarization of undoped PZT may be attributed to the formation of a ferroelectric monoclinic phase [4, 5]. However, this phase transformation has not yet been detected in doped PZT ceramics. Beside the Zr/Ti-ratio, doping shows a strong influence on the electromechanical properties due to the creation of Pb-vacancies in donor doped PZT ("soft" PZT) and O-vacancies in acceptor doped PZT ("hard" PZT). In general it is observed that dopants lower the c/a-ratio [1], [2], [6–8]. Furthermore, the grain size has an influence on the lattice distortion and the resulting electromechanical properties. This effect has been intensively studied by Randall et al., who also summarized results from other authors [6]. The comparison reveals inconsistencies which they related to differences in processing and grain boundary phases.

Two types of contributions to dielectric and piezoelectric properties of ferroelectric ceramics are usually distinguished [6], [9–12]. One type is called an intrinsic contribution, and it is due to the distortion of the crystal lattice under an applied electric field or a mechanical stress. The second type is called an extrinsic contribution, and it results from the motion of domain walls or domain switching [8]. To provide an understanding of material properties of PZT, several methods to separate the intrinsic and extrinsic contributions were proposed. These methods are indirect, and are based on measurements of the dielectric and piezoelectric properties of ferroelectric ceramics [8], [10–12]. In the experiments reported in this paper a different approach is adopted, which is based on measurements of high-resolution synchrotron X-ray powder diffraction. The shift in the positions of the diffraction peaks under applied electric field gives the intrinsic lattice deformation, whereas the domain switching can be calculated from the change in peak intensities [13, 14].

The present work summarizes opportunities of using high-resolution synchrotron and standard XRD techniques for structural characterization as well as for investigations of structure-property-relationships. XRD will be used to determine quantitatively the phase content of morphotropic PZT. Temperature dependent measurements provide information about the phase transformation of morphotropic donor doped PZT ceramics and high-resolution synchrotron X-ray diffraction gives information about the extrinsic and intrinsic contributions to the electric field induced strain. XRD results are finally compared with electrical measurements to analyze the interactions among microstructure, phase content and properties.

7.2 Experimental

PZT ceramic samples were prepared by the conventional mixed-oxide route, as described elsewhere [15, 16]. Sintering was done at 1225°C for 2 h, and the density of the sintered bodies was always higher than 99 % of theoretical density. High resolution XRD measurements were performed using synchrotron radiation with the powder diffractometer B2 at HASYLAB Hamburg (Germany) in parallel beam geometry [17]. In-situ measurements ($\Omega - 2\Theta$ stepscans) were performed on single-phase and morphotropic PZT ceramics. The samples were disc-shaped with a diameter of $10 - 15$ mm and a thickness of $0.5 - 1.0$ mm. Texture formation due to cutting was avoided by a heat treatment up to 700°C for 4 h after grinding and polishing. The lattice deformation was determined by calculating the c/a-ratio and the d_{111}/d_{11-1}, respectively. The intrinsic strain was determined under bipolar and unipolar electric loading up to 4 kV/mm. The time to measure one bipolar hysteresis-cycle was about 80 min. The extrinsic effect which represents the motion of the non-180° domain walls was characterized from the ratio of the peak intensities I_{111}/I_{11-1} for rhombohedral and I_{002}/I_{200} for tetragonal PZT, respectively [13, 14]. Macroscopic strain measurements were performed using a LVDT and a computer controlled setup. Dielectric properties before and after poling and as a function of temperature were determined by capacity measurements at a frequency of 1 kHz using an HP 4284A analyzer.

7.3 Results and discussion

7.3.1 Quantitative analysis of the F_T and F_R phase content

A typical peak triplet consisting of the two tetragonal (200) and (002) peaks and the rhombohedral (200) peak is shown in Figure 7.1 for a La-donor doped PZT (54/46). The strong overlapping requires a deconvolution for a quantitative analysis of the phase content. The best fit can be achieved by using a Pseudo-Voigt function with an asymmetry parameter and variable half-width of the peaks. The volume fraction of the rhombohedral phase (V_R) can then be calculated based on the integral intensities of the three peaks:

$$V_R[\%] = \frac{I^R_{(200)}}{I^R_{(200)} + I^T_{(200)} + I^T_{(002)}} \cdot 100 \quad \text{and} \quad V_T = 100 - V_R. \tag{7.1}$$

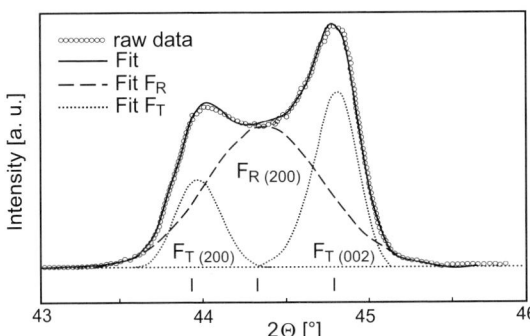

Figure 7.1: Typical XRD spectra of a morphotropic PZT ceramic with the corresponding peaks of the tetragonal (F_T) and rhombohedral (F_R) phase.

Figure 7.2 shows, as an example, the volume fraction of the tetragonal phase (V_T) for compositions across the MPB of La-doped ceramics. The width of the MPB for these ceramics is approximately 2 – 4 mol.% Zr and the MPB (50 % tetragonal and 50 % rhombohedral phase) is located at 53.6 mol.% Zr. However, the data given in Figure 7.2 are only valid for the present material. The width of the MPB and the location depends on the composition, which means type and amount of dopant and also on the grain size. For ceramics with tetragonal volume fractions > 50 % we always found an increase in the tetragonal phase content with increasing grain size, e.g. when the sintering temperature is raised.

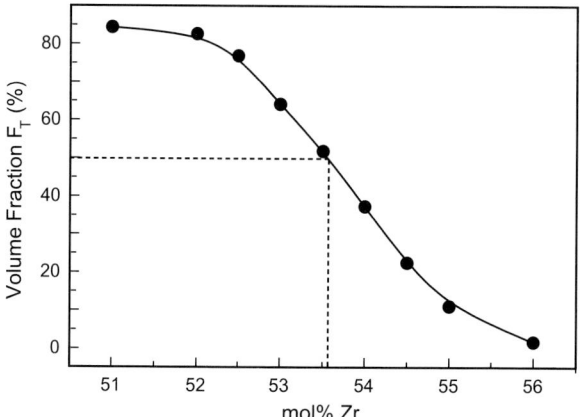

Figure 7.2: Volume fraction of the tetragonal phase as a function of composition for La-donor doped PZT determined by quantitative X-ray analysis.

Quantitative X-ray analysis is not used very often for the determination of the phase contents of PZT ceramics, although this method is a standard technique in materials laboratories. One of the reasons might be the lack of an adequate fit procedure, the other is the concern whether X-ray analysis reflects the composition of the sample, since it gives only information from the surface-near region. In order to proof this argument, we perfomed electrical measurements which give information about the sample volume and compared them to X-ray measurements. The background for the electrical measurements is schematically shown in Figure 7.3 (a). The analysis is based on a change in relative permittivity ε_r during poling. Samples with more than 50 % tetragonal phase indicate an increase in ε_r after poling, while materials on the rhombohedral side show a decrease. The morphotropic phase boundary is defined for a composition which reveals no change after poling.

An analysis of the permittivity change ε_r for La-doped PZTs are given in Figure 7.3 (b). There is a significant increase of ε_r for Ti-rich compositions and a smaller decrease for the Zr-rich ones. The location of the MPB is determined at 53.9 mol.% Zr where ε_r does not change during poling. A comparison of the XRD data (Figure 7.2) and the permittivity measurements (Figure 7.3 (b)) indicate an excellent agreement within 0.3 mol.% Zr. Based on

7.3 Results and discussion

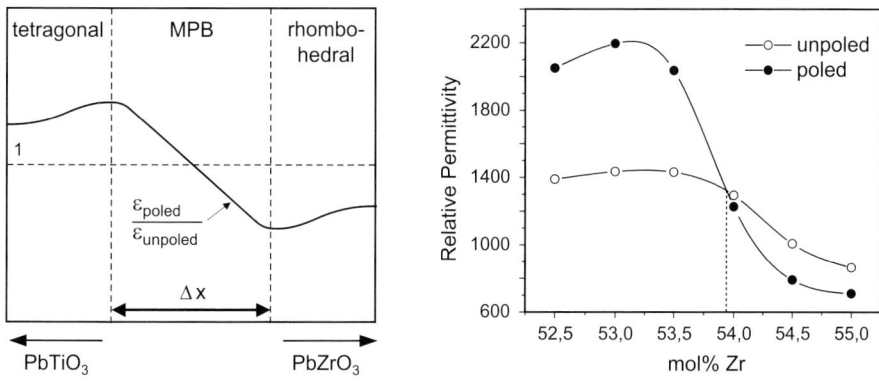

Figure 7.3: Change of relative permittivity (ε_r) as a function of Zr-content before and after poling: (a) Schematic behavior near the MPB after [18] and (b) results for La-doped PZT.

these results we assumed that the XRD data taken from surface near regions will also provide useful information about microscopic processes taking place in the volume.

A further analysis of the XRD spectra shown in Figure 7.1, indicates different half-widths for the tetragonal and rhombohedral phase. This effect has been studied systematically for La-donor doped ceramics with Zr/Ti-ratios between 37.5/62.5 and 70/30. The half-width as a function of composition for the two tetragonal peaks ((200), (002)) and the rhombohedral peak (200) is shown in 7.4. The half-width of the tetragonal peaks are nearly constant in single-phase tetragonal ceramics up to a Zr/Ti-ratio of 50/50 and increase slightly when F_T starts to coexist with F_R. The rhombohedral phase shows a small increase with decreasing Zr-ratio in single-phase ceramics and a pronounced increase near the MPB. The peak broadening at the MPB indicates a wide distribution of lattice parameters which can be attributed to locally different internal stresses within the material. The stresses are created due to the transition from the cubic structure to the ferroelectric ones during cooling after sintering. If we consider now the MPB as a defined concentration for $F_T \leftrightarrow F_R$ phase transition rather than a compositional range, one of the phases should at least be metastable [19]. The metastable phase may form during cooling when it reduces the total free energy of the whole system which includes not only chemical potentials, but also spontaneous polarization und local internal stresses. For further considerations we take into account the lower elastic modulus of F_R (≈ 60 GPa) compared to F_T (≈ 100 GPa) [20]. This means that for F_T-rich compositions (< 53 mol.% Zr) the weaker F_R-phase is embedded in a matrix of the stiffer F_T-phase. Stresses due to a mismatch of different F_T domain orientations can then be accommodated by a reduction of the lattice distortion of the rhombohedral unit cells [19]. A decrease in lattice distortion further leads to a reduction of the spontaneous polarization. Both effects will minimize the Gibb's free energy of the system. The contribution of stress accommodation and reduction of spontaneous polarization of the F_T-phase are less pronounced for F_R-rich compositions due its higher stiffness.

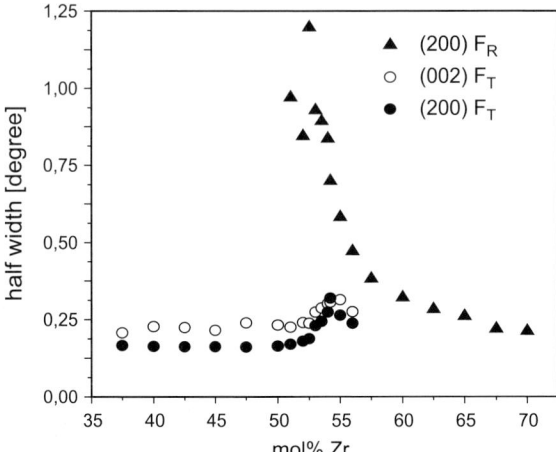

Figure 7.4: Half-widths of the tetragonal (F_T) and rhombohedral (F_R) peaks for different Zr/Ti-ratios.

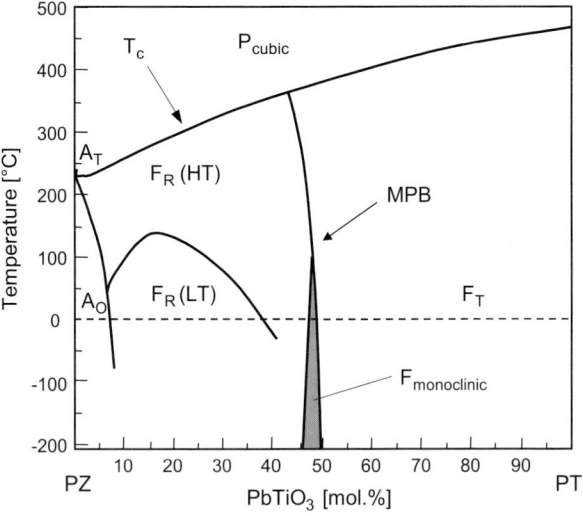

Figure 7.5: Revised sub-solidus phase diagram for undoped PZT solid solution after [5].

7.3.2 Temperature induced $F_R \leftrightarrow F_T$ phase transition

The phase diagram for PZT solid solutions [21] has been revised when Noheda et al. [5] discovered a monoclinic phase in undoped PZT (7.5). Nevertheless, up to now there is no experimental evidence for the existence of a monoclinic structure in doped PZT above room

7.3 Results and discussion

temperature. Therefore, we performed high-temperature XRD investigations with synchrotron radiation to study the phase changes during heating. Figure 7.6 shows two XRD spectra for two La-doped compositions near the MPB at $40°C$ and $160°C$. The tetragonal volume fraction of the ceramic compacts, PZT (53/47) and PZT (54/46) at $20°C$ was alculated as 0.64 and 0.37, respectively. The HT-measurements (Figure 7.6) were taken from crushed ceramic samples, which means that the phase content is slightly different compared to sintered compacts due to the relaxation of internal stresses. The diffraction pattern reveals the expected decrease in lattice distortion for the F_T-rich material, but PZT (54/46) exhibits an increasing F_T-content which is attributed to a phase transition of a certain volume fraction of F_R to F_T. This phase transition could only be observed in a narrow compositional range < 1 mol.% Zr and probably at a well-defined temperature (experiments were performed in steps of $40°C$). From qualitative analysis of the XRD spectra we cannot observe a significant change of the tetragonal volume fraction after the partial transformation (not shown in Figure 7.6 (b)). This would support the existence of a MPB at a defined composition as well as the metastability of one ferroelectric phase in the F_T/F_R-coexistence region. No evidence could be found for the formation or existence of the monoclinic phase.

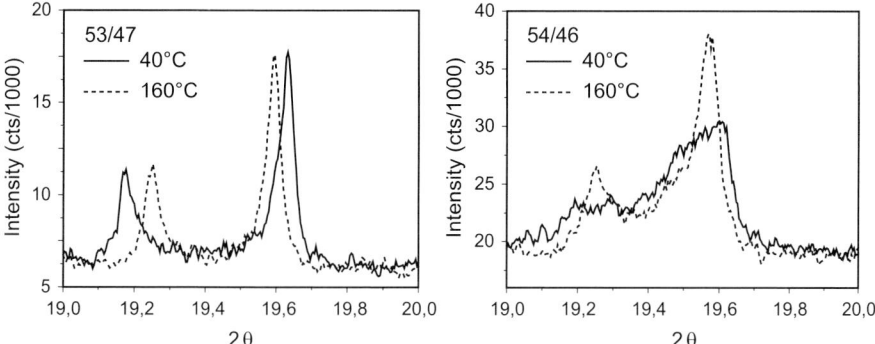

Figure 7.6: X-ray diffraction patterns for two La-doped PZT ceramics at $40°C$ and $160°C$. (a) PZT (53/47) shows a change in lattice distortion and (b) PZT (54/46) a change in phase composition.

The influence of the observed phase transition on the electrical properties are shown in Figure 7.7 and Figure 7.8. PZT (53/47) and (55/45) which do not show a phase transition reveal a moderate increase in relative permittivity with increasing temperature for poled samples. The higher permittivity of PZT (53/47) compared to PZT (55/45) is attributed to the higher tetragonal volume fraction (Figure 7.3 (b)). However, a much stronger temperature dependence is observed for PZT (54/46) which can be attributed to a $F_R \leftrightarrow F_T$ phase transition occurring between $40°C$ and $8°C$.

Figure 7.8 indicates the effect of a phase transition on the large signal strain. Specimens without a detectable phase transformation (PZT (53/47) and (55/45)) reveal a similar slope, whereas a significant change could be observed for PZT (54/46) in the temperature range

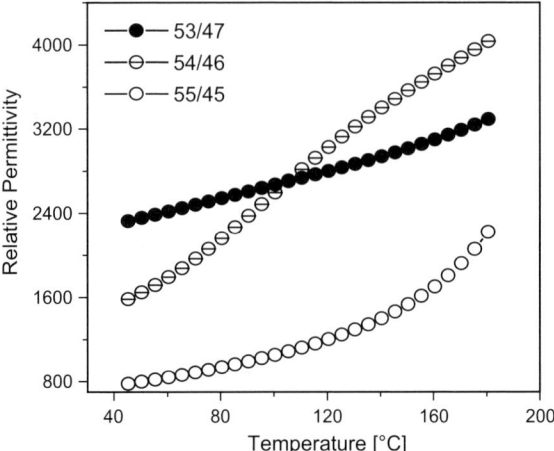

Figure 7.7: Temperature dependence of the relative permittivity for PZT ceramics near the MPB. PZT (54/46) undergoes a partial $F_R \leftrightarrow F_T$ phase transition between 40 and 80°C.

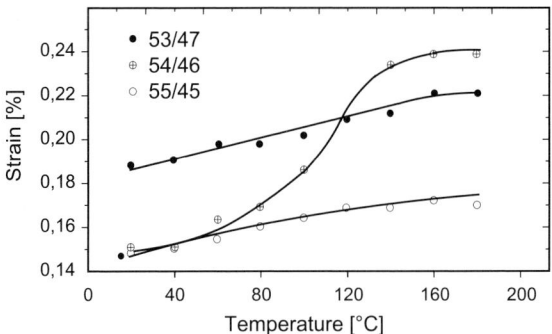

Figure 7.8: Temperature dependence of the unipolar strain at 2.5 kV/mm for PZT ceramics near the MPB. PZT (54/46) undergoes a partial $F_R \leftrightarrow F_T$ phase transition between 40 and 80°C.

between 60°C and 16°C due to the phase transition. The unipolar strain at 160°C is even higher as that of PZT (53/47). The change in slope corresponds well with the observed phase transformation temperature between 40°C and 80°C. The rather broad transition regime over a temperature range of 100°C might be attributed to domain reorientation processes taking place under high electric fields.

The revised phase diagram of Noheda et al. [5] for undoped PZT solid solutions shows a decreasing stability of the monoclinic phase for higher temperatures (Figure 7.5). If the monoclinic phase provides a higher polarizability and piezoelectric coefficients, these properties should decrease above the stability temperature of the monoclinic ferroelectric phase.

7.3 Results and discussion

Our results reveal the opposite effect for compositions next to the MPB which leads us to the conclusion that the monoclinic phase does not play an important role for the performance of doped PZT ceramics.

7.3.3 Analysis of intrinsic and extrinsic contributions to the macroscopic strain

High-resolution synchrotron X-ray powder diffraction is also used to study microscopical changes in a ferroelectrical material, when an electrical field is applied. The analysis is based on the characterization of changes in the XRD diffraction peaks during electrical loading. Details of the experiments are described elsewhere [3, 13, 14]. Figure 7.9 shows the $(200)_R$ diffraction peak for a co-doped PZT ceramic (PZT-SKN, [21]) with rhombohedral structure for three values of the electrical field: at $E = 0$ (poled remanent state), at coercive field E_c (where polarization switching occurs), and at the maximum field $E_{max} = 4\,\text{kV/mm}$, when switching is completed.

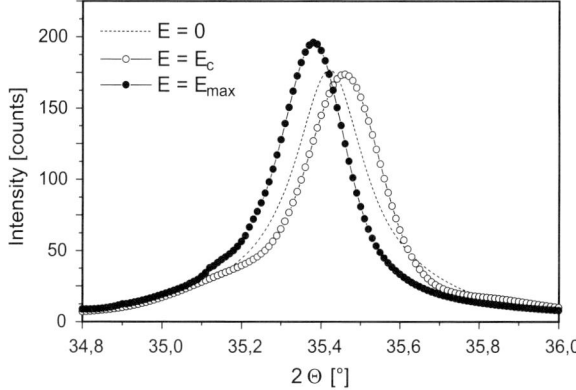

Figure 7.9: Shift of the position of the $(200)_R$ diffraction peak induced by an applied electric field for a PZT with rhombohedral structure.

One can see that as the field changes during bipolar cycling from 0 to E_c and further to E_{max}, the position of the $(200)_R$ peak shifts as follows: first, the peak moves to a higher diffraction angle 2Θ which corresponds to a contraction of the unit cell until the coercive field E_c has been reached. As the field increases beyond E_c, the peak position moves back to lower diffraction angles, which indicates an expansion of the unit cell. This effect can be attributed to 180° domain switching, where contraction and expansion of the unit cell occur due to the converse piezoelectric effect. The unit cell contracts when the spontaneous polarization is directed opposite to the electric field (at $E < E_c$), and expands when the polarization is directed along the field ($E > E_c$) after the switching has occurred.

Non-180° domain switching in rhombohedral PZT is characterized by a change of the ratio of the peak intensity within the (111)/(−111) reflex group, Figure 7.10. At $E = 0$, the sample has already been poled by preliminary cycling. Therefore the magnitude of the (111) peak, $I_{(111)}$, which is proportional to a number of domains with polarization direction along poling field, is higher than that of the (-111) peak, $I_{(-111)}$. From the ratio of the peak intensities $R = I_{(111)}/I_{(-111)}$, the volume fraction of domains, N, aligned along the poling field, can be estimated as follows: $N = R/(1+R)$. The volume fraction obtained from the data at $E = 0$, Figure 7.10, is approximately equal to 75 %. As the applied electric field approaches the coercive field, E_c, the intensity ratio $I_{(111)}/I_{(-111)}$ becomes smaller. This reduction means that the number of domains with spontaneous polarization aligned along the field becomes smaller. At E_c only $N \approx 50\%$ of the total amount of domains have their polarization direction either parallel or anti-parallel to the electric field. In the remaining 25 % of the domains, the spontaneous polarization changes its direction by 71° or 109° (since this material has a rhombohedral structure) and these domains now do not contribute to the (111) peak. As the field increases further from E_c to E_{max}, the intensity ratio changes again to the original value after poling. Taken together, the data shown in Figure 7.10 can be interpreted as a polarization reversal which occurs as two step process with non-180° domain switching: $[111] \rightarrow [-111] \rightarrow [-1-1-1]$.

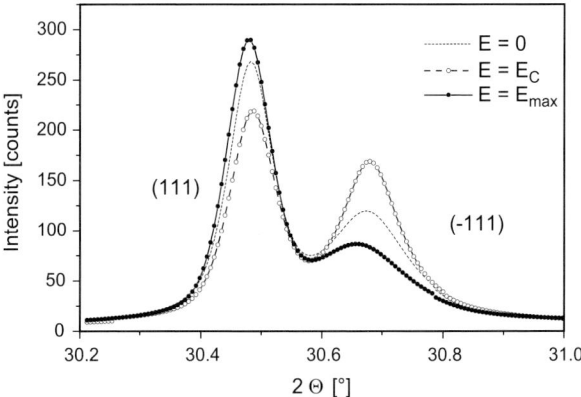

Figure 7.10: Change of the (111)/(−111) peak intensity ratio induced by an applied electric field for a PZT with rhombohedral structure.

The electric field induced intrinsic strain for different crystallographic directions could be calculated from the shift in peak positions (see Figure 7.9). Figure 7.11 shows the result of the measurements for a rhombohedral PZT in [111] and [100] direction. Only one half of each cycle is shown for the sake of a clarity of the plot. The curve for the [100] direction reveals the typical shape of a butterfly loop for the electric field induced strain in ferroelectrics. However in [111] direction, which is parallel to the spontaneous polarization, strain is significantly smaller. From both curves, the so-called unipolar strain can be evaluated as the strain induced at the maximum electric field E_{max} with a reference to the remanent state ($E = 0$). The calculation gives strain values in [111] direction of 0.02 % and for the [100] direction 0.15 %. The observations are in good agreement with theoretical calculations made by Du et al. [22],

7.3 Results and discussion

which showed that the maximum value of the intrinsic component of the piezoelectric coefficient d_{33} should be in [100] direction in case of the rhombohedral structure of the ferroelectric phase.

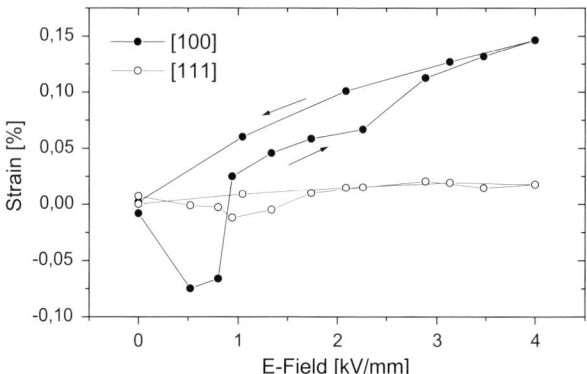

Figure 7.11: Electrical field dependence of the intrinsic strain measured along [111] and [100] direction in PZT with a rhombohedral structure.

Similar experiments had been performed for PZT with a tetragonal structure, where the highest intrinsic strain was observed in [111] and [101] direction, whereas the smallest in [100] direction, which is the direction of spontaneous polarization in tetragonal ferroelectric materials [23]. A comparison of the intrinsic unipolar strain for materials with the same amount and type of dopant, but different crystal structure (Zr/Ti-ratio) indicates that the highest intrinsic strain for tetragonal PZT was approximately half of the value measured for rhombohedral PZT. Measurements of the activity of non-180° domains for the same type of materials are shown in Figure 7.12. Tetragonal PZT reveals also a much smaller domain activity for non-180° switching ($N \approx 7-8\,\%$) compared to rhombohedral PZT ($N \approx 20-25\,\%$). However, macroscopic electric field induced strain measurements indicate comparable values for both types of materials and even a higher unipolar strain for the tetragonal PZT, Figure 7.13. To understand this contradiction, a simple model has been derived to calculate the strain contribution of non-180° domains, taking into account the higher lattice distortion of the tetragonal structure. If we define a length L_{max} and L_c which both correspond to the number of domains with a polarization direction parallel (d_{111}) and perpendicular (d_{-111}) to the poling field at E_{max} and E_c

$$\begin{aligned} L_{\text{max}} &= V_{(111)}^{\text{max}} \cdot d_{(111)} + V_{(-111)}^{\text{max}} \cdot d_{(-111)} \\ L_C &= V_{(111)}^{C} \cdot d_{(111)} + V_{(-111)}^{C} \cdot d_{(-111)} \end{aligned} \quad (7.2)$$

we could calculate the strain contribution of non-180° domains of rhombohedral PZT during a bipolar electrical loading by using the following relationship:

$$\triangle S = 100 \cdot \left(\frac{L_{max}}{L_C} - 1 \right) \tag{7.3}$$

V_{hkl} denotes the volume fraction of the contributing domains and is equal to the corresponding peak intensity. The contribution of tetragonal non-180° domains can be determined based on the (200)/(002) reflex group.

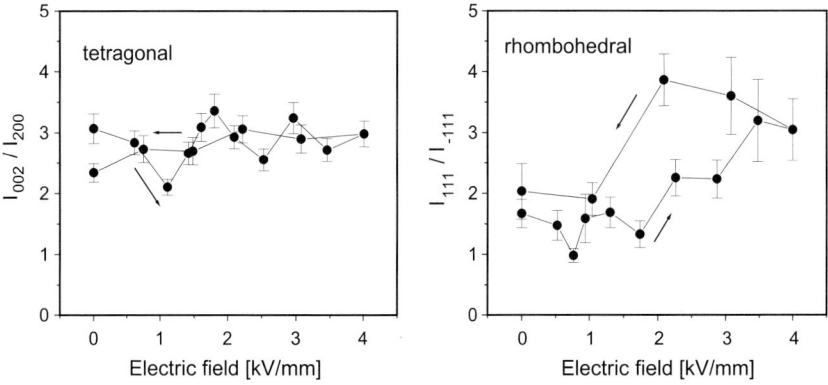

Figure 7.12: Change of the peak intensity ratio for tetragonal (a) and rhombohedral PZT (b) with an applied electric field.

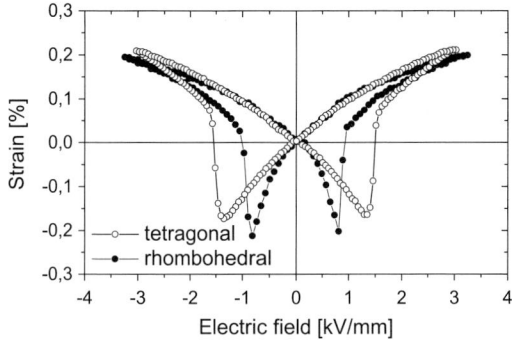

Figure 7.13: Electric field induced macroscopic strain measurements of tetragonal and rhombohedral co-doped PZT using lvdt

Using the change in peak intensities given in (Figure 7.12) and the Equation (7.2) and (7.3), we obtain a non-180° domain contribution during bipolar cycling for tetragonal PZT (c_T/a_T − ratio = 1,0225) of 0.155 % and for rhombohedral PZT ($d_{111}/d_{-111} = 1.0061$) of 0.153 % along the direction of the highest domain activity for each system. The calculation clearly shows that the smaller domain activity in PZT with a tetragonal structure is compensated by the higher lattice distortion which then leads to comparable strain for both structures. A increase of the tetragonal domain switching activity is therefore favorable for an increase of macroscopic strain. The domain activity depends also on the crystallographic direction, similar to the intrinsic deformation. It has been found that the direction of the highest intrinsic strain exhibits the lowest extrinsic contribution and vice versa. This effect could be explained under the assumption that significant lattice distortion (intrinsic strain) will always occur when the corresponding non-180° domain switching will not reduce the total free energy, as it the case for the [100] direction of the rhombohedral structure.

The results have demonstrated that high-resolution synchrotron experiments are useful to understand the ongoing microstructural changes when an electric field is applied. However, we would like to point out that these measurements reflect only the behavior of unit cells and domains which fulfill the Bragg condition. This means in a ceramic with a random distribution of crystallographic orientations, diffraction occurs only for a small volume fraction of the whole sample. The observed behavior is not only determined by the unit cell or domain itself, but also by the surrounding region which could have, i.e. an impact by mechanical stresses. It will be even more complicated in the case of two-phase materials at the morphotropic phase boundary where an additional interaction between rhombohedral und tetragonal domains occur, as discussed above.

7.4 Summary

High resolution synchrotron and standard X-ray diffraction techniques have been used to obtain a better understanding of the correlation among crystal structure, phase composition, and macroscopic properties of donor-doped and co-doped ceramics with compositions near the morphotropic phase boundary. The results on phase composition from X-ray and electrical measurements show a good agreement which leads to the conclusion that the surface inspection by X-ray techniques reflect the interior of the material. Investigations of the half-width of the peaks and temperature dependent X-ray measurements indicate the existence of a morphotropic phase boundary at a defined composition rather than in a compositional range. This means that one of the two phases has to be metastable within the coexistence region near the MPB. The observed phase transition from rhombohedral to tetragonal for La-doped PZT with a Zr/Ti-ratio of 54/46 results in a strong temperature induced increase in strain at 2.5 kV/mm as well as in an increase in permittivity.

The contribution of domain switching and lattice deformation to the electric field induced strain has been analyzed in co-doped rhombohedral and tetragonal PZT. The lattice deformation depends on the crystallographic orientation of the unit cell with respect to the applied electric field and is the highest in [001] direction for rhombohedral structure and in [111] or [110] direction for the tetragonal structure. Non-180° domain activities were investigated by measuring the change of the peak intensities for the [111] peak duplet of the rhombohedral structure and the [200] peak duplet of the tetragonal one. Rhombohedral PZT indicates a much higher domain wall activity with a volume fraction of switchable domains of up to 25 %, whereas tetragonal PZT revealed only $7-8\%$. Nevertheless, the resulting strain was similar for both structures due to the higher lattice distortion of the tetragonal phase.

Acknowledgement

This work was supported by the DFG (German Research Council) under contract No. Ho 1165/9-1 and SFB 595 (DFG center of excellence on electric fatigue of functional materials, Technical University of Darmstadt).

Bibliography

[1] B. Jaffe, R. S. Roth, and S. arzullo, *J. Appl. Phys.* **25**, 809 (1954).
[2] Y. Xu, *Ferroelectric Materials and Their Applications*, Elsevier Science, Amsterdam, 1991.
[3] M. J. Hoffmann, M. Hammer, A. Endriss, and D. Lupascu, *Acta mater.* **49**, 1301 (2001).
[4] B. Noheda et al., *Phys. Rev.* B **61**, 8687 (2000).
[5] B. Noheda et al., *Phys. Rev.* B **63**, 014103 (2000).
[6] A. Randall, N. Kim, J.-P. Kucera, W. Cao, and T. R. Shrout, *J. Am. Ceram. Soc.* **81**, 677 (1998).
[7] M. Hammer, C. Monty, A. Endriss, and M. J. Hoffmann, *J. Am. Ceram. Soc.* **81**, 721 (1998).
[8] M. Hammer and M. J. Hoffmann, *J. Electroceram.* **2**, 75 (1998).
[9] D. Damjanovic and M. Demartin, *J. Phys. Condens. Matter* **9**, 4943 (1997).
[10] Q. M. Zhang, H. Wang, N. Kim, and L. E. Cross, *J. Appl. Phys.* **75**, 454 (1994).
[11] R. Herbiet et al., *Ferroelectrics* **98**, 107 (1989).
[12] V. Müller and Q. M. Zhang, *Appl. Phys. Lett.* **72**, 2692 (1998).
[13] A. Endriss, M. Hammer, M. J. Hoffmann, A. Kolleck, G. A. Schneider, *J. Euro. Ceram. Soc.* **19**, 1229 (1999).
[14] J.-Th. Reszat, A. E. Glazounov, and M. J. Hoffmann, *J. Euro. Ceram. Soc.* **21**, 1349 (2001).
[15] M. Hammer and M. J. Hoffmann, *J. Am. Ceram. Soc.* **81**, 3277 (1998).
[16] M. Laurent, U. Schreiner, P. A. Langjahr, A. E. Glazounov, and M. J. Hoffmann, *J. Euro. Ceram. Soc.* **21**, 1495 (2001).
[17] H. Arnold, H. Bartl, J. Fuess, J. Ihringer, K. Kosten, U. Löchner, P. U. Pennartz, W. Prandl, and T. Wroblewski, *Rev. Sci. Instrum.* **60**, 2380 (1989).
[18] G. Helke, *The Diffuse Phase Transition*, unpublished manuscript, 1992.
[19] W. Wersing, W. Rossner, G. Eckstein, and G. Tomandl, *Silicates Indusriels* **3/4**, 41 (1985).
[20] A. Glazounov et al., *J. Am. Ceram. Soc.* **84**, 2921 (2001).
[21] G. Helke, S. Seifert, and S.-J. Cho, *J. Euro. Seram. Soc.* **19**, 1265 (1999).
[22] X.-H- Du, J. Zheng, U. Belegundu, and K. Uchino, *Appl. Phys. Lett.* **72**, 2421 (1998).
[23] J.-Th. Reszat, *Dissertation*, University of Karlsruhe, 2003

8 In-Situ Synchrotron X-ray Studies of Processing and Physics of Ferroelectric Thin Films

G. B. Stephenson, S. K. Streiffer, D. D. Fong, M. V. Ramana Murty, O. Auciello, P. H. Fuoss, J. A. Eastman[1], A. Munkholm[2], and Carol Thompson[3]

1. Argonne National Laboratory, Argonne,
2. Lumileds Lighting, San Jose,
3. Northern Illinois University, DeKalb, USA

Abstract

To understand the influence of epitaxial strain, depolarizing field, and intrinsic surface effects on the ferroelectric transition, we have investigated ferroelectric thin films using synchrotron x-ray scattering. Here we summarize real-time studies of epitaxial $PbTiO_3$ film growth on $SrTiO_3$ by metalorganic chemical vapor deposition (MOCVD), and in situ studies of the ferroelectric transition as a function of temperature and film thickness. The ability to monitor growth in real time using x-ray scattering allows us to produce films with smooth interfaces, well-defined strain states, and thicknesses controlled to sub-unit-cell accuracy. We find that the ferroelectric phase forms as 180° stripe domains in these films, with a thickness-dependent T_c. The dependence of the stripe period on film thickness is in agreement with theory. For thicker films (e.g. 40 nm), we find that T_c is elevated above the unstressed bulk value by an amount that agrees with theory for the effect of epitaxial strain. However, the observed decrease of T_c for smaller film thickness is significantly larger than that expected solely due to 180° stripe domains, indicating that intrinsic surface or interface effects may also be important.

8.1 Introduction

Ferroelectricity can be either suppressed or enhanced in thin films by the interplay of epitaxial strain, depolarizing field, and intrinsic surface effects. In order to unambiguously understand each of these effects, we have been carrying out a research program to investigate ferroelectric films with well-defined strain states and clean, atomically smooth interfaces. Our approach is to use synchrotron x-ray scattering for real-time control of film growth and for *in situ* studies of the ferroelectric transition as a function of temperature and film thickness. In these studies, coherently-strained epitaxial films of $PbTiO_3$ were grown on $SrTiO_3$ using metalorganic chemical vapor deposition (MOCVD) in a dedicated growth system mounted on an x-ray goniometer [1]. The ability to perform grazing-incidence x-ray scattering in the MOCVD chamber allows us to determine optimum growth conditions [2], to control the thickness of the films to sub-unit-cell accuracy, and to preserve film stoichiometry during high temperature x-ray study [3].

In this article we summarize our results to date on growth and ferroelectricity in PbTiO$_3$ films on SrTiO$_3$. In particular, we discuss the observation of periodic 180° stripe domains below the ferroelectric transition [4]. Such 180° stripe domains have been predicted to occur to minimize the depolarizing field, but have not previously been observed in thin films. We have characterized the stripe period as a function of film thickness and temperature, and find that two stripe phases exist having different stripe periods. The resulting phase diagram is presented. The dependence of stripe period on thickness agrees well with theory, while the suppression of T_c for thinner films is significantly larger than that expected solely from 180° stripe domains, indicating that intrinsic surface or interface effects may also be important. The phase diagram indicates that the ferroelectric phase transition occurs above room temperature in coherently strained epitaxial films of PbTiO$_3$ on SrTiO$_3$ with thicknesses as small as three unit cells [5].

8.2 Growth of ultrathin ferroelectric films

For these studies we have developed an apparatus for in-situ x-ray studies of vapor phase epitaxy located at BESSRC beamline 12-ID-D of the Advanced Photon Source. It consists of a vertical flow MOCVD growth chamber mounted on a horizontal-diffraction-plane z-axis goniometer optimized for grazing-incidence scattering [1]. A normal-incidence laser can be used to monitor film thickness on the 100 nm scale during growth for preparatory studies without x-rays. The use of a standard fused- quartz chamber wall allows studies with sample temperatures up to 1200° C and in oxidizing environments. In order to penetrate the 2-mm-thick chamber walls, a moderately high x-ray energy (24 keV) is employed. The ability to perform x-ray studies in the growth chamber is critical for real-time studies of the growth process, and allows us to precisely control film thickness to sub-unit-cell accuracy for systematic studies of thickness effects. Phase transitions in a film can be studied at high temperature immediately after its growth, avoiding any irreversible relaxation that may occur upon cooling. In addition, for studies of PbTiO$_3$ at temperatures above 500° C, where the equilibrium vapor pressure of PbO is significant, the growth system can be used to continuously supply Pb and O precursors to control the PbO vapor pressure over the film for studies of equilibrium surface structure and to maintain film stoichiometry.

The PbTiO$_3$ films were grown epitaxially on SrTiO$_3$ (001) substrates as described previously [2]. Cation precursors were either titanium isopropoxide (TIP) or titanium tertbutoxide (TTB), and tetraethyl lead (TEL). The oxidant was O$_2$, with N$_2$ carrier gas. Films were typically deposited at 750°C at a total pressure of 10 Torr (P_{O_2} = 2.5 Torr). Under appropriate growth conditions, the PbTiO$_3$ films replicated the high crystalline quality of the substrates (0.01° typical mosaic). Films thinner than 40 nm remained lattice matched with the SrTiO$_3$, while the epitaxial strain was mostly relaxed in thicker films.

Because the volatility of PbO is high and that of TiO$_2$ is low, a stable synthesis window for PbTiO$_3$ can be defined [6] as a function of PbO partial pressure P_{PbO} and temperature T based on the equilibria PbTiO$_3$ ↔ TiO$_2$ + PbO(v) and PbO(v) ↔ PbO(s). These are shown by the dotted lines in Figure 8.1, using literature values [7]. Between these boundaries, single-

8.2 Growth of ultrathin ferroelectric films

phase PbTiO$_3$ is thermodynamically stable and excess PbO volatilizes; the PbTiO$_3$ film can remain exposed to the TEL flow at all times, and deposition can be controlled by the TIP or TTB precursor flow. After growth the film can be maintained at high temperature indefinitely for *in situ* x-ray characterization by owing TEL and O$_2$ to provide a sufficient overpressure of PbO. Because O$_2$ is always present in excess, the value of P_{PbO} in equilibrium with the surface at high temperature is expected to be approximately equal to the partial pressure of TEL in the input flow.

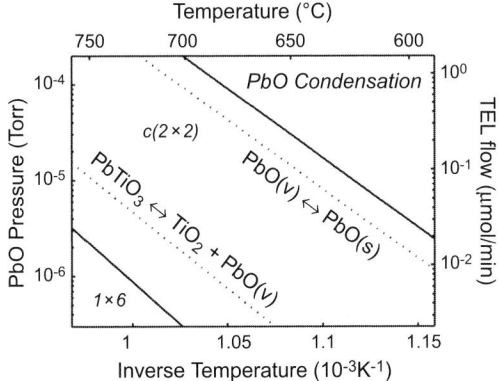

Figure 8.1: Equilibrium phase diagram of the PbTiO$_3$ (001) surface as a function of temperature and PbO partial pressure. Solid lines separate observed phase fields corresponding to PbO condensation, $c(2 \times 2)$, and 1×6 reconstructions. Dotted lines are literature values [7] for the PbO condensation and PbTiO$_3$ decomposition boundaries.

We mapped the equilibrium structure of the PbTiO$_3$ (001) surface [3] by varying T and P_{PbO} while observing the in-plane scattering pattern. Figure 8.1 shows the resulting surface phase diagram. A $c(2 \times 2)$ reconstruction is present under most conditions, while a 1×6 reconstruction occurs under PbO-poor conditions. The surface structure could be reversibly and reproducibly switched between these two reconstructions by changing T or P_{PbO} to cross the phase boundary shown (lower solid line) [8]. At high input ows of TEL and lower temperatures, diffraction peaks from the condensation of polycrystalline PbO began to appear. The observed TEL input pressure at which PbO condensed (upper solid line) is approximately a factor of 2 larger than the literature value for the equilibrium PbO condensation pressure (upper dotted line), validating the equality of P_{TEL} in the input ow and P_{PbO} at the sample surface, within this factor. Since the observed phase field of the 1×6 reconstruction occurs at values of P_{PbO} below the equilibrium stability limit for bulk PbTiO$_3$, we interpret the 1×6 reconstruction to be the first step in the decomposition of PbTiO$_3$. The equilibrium surface phase is the $c(2 \times 2)$ structure over the complete equilibrium phase field of bulk PbTiO$_3$. Using x-ray scattering we determined the structure of the $c(2 \times 2)$ reconstruction [3]. It consists of a single-unit-cell-thick layer of an antiferrodistortive structure with TiO$_4$ octahedra alternately counter-rotated about [001], as in the low-temperature phase of SrTiO$_3$.

One of the primary strengths of the in situ x-ray MOCVD apparatus is the ability to monitor the growth process in real time, which allows understanding of growth mechanisms and determination of optimal growth conditions [2]. In the present study, we used x-ray monitoring to produce ultrathin films of specific thicknesses. For a heteroepitaxial thin film, the scattering intensity along a crystal truncation rod (CTR) exhibits thickness fringes from the interference between the upper and lower interfaces of the film [8]. Figure 8.2 shows the evolution of the $20L$ CTR at $L = 0.5$ during the addition of 9 unit cells to a film with an initial thickness of 1 unit cell. As the thickness fringes form and sweep through a particular value of L, the CTR intensity there oscillates in time with a period proportional to the growth rate. At $L = 0.5$, an oscillation period corresponds to the growth of 2 unit cells. Using in situ x-ray monitoring, the duration of the growth pulse can be adjusted to give film thickness control with sub-unit-cell accuracy.

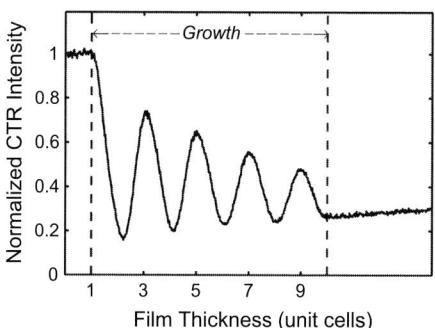

Figure 8.2: Oscillations in the $20L$ CTR intensity at $L = 0.5$ from evolution of thickness fringes during addition of 9 unit cells to 1-unit-cell-thick PbTiO$_3$ film. Vertical dashed lines indicate beginning and end of growth.

8.3 Observation of nanoscale 180° stripe domains

A primary focus of our work has been to understand the ferroelectric phase transition in thin epitaxial films of PbTiO$_3$. It is expected that epitaxial strain effects are important in such films because of the large, anisotropic strain associated with the phase transition. Figure 8.3 shows the phase diagram for PbTiO$_3$ as a function of epitaxial strain and temperature calculated using Landau-Ginzburg-Devonshire (LGD) theory [9]. Here epitaxial strain is defined as the in-plane strain imposed by the substrate, experienced by the cubic (paraelectric) phase of PbTiO$_3$. The dashed line shows that a coherent PbTiO$_3$ film on a SrTiO$_3$ substrate experiences somewhat more than 1 % compressive epitaxial strain. Such compressive strain favors the ferroelectric PbTiO$_3$ phase having the "c domain" orientation, i.e. with the c (polar) axis normal to the film. From Figure 8.3 one can see that the paraelectric-ferroelectric transition temperature T_C for coherently-strained PbTiO$_3$ films on SrTiO$_3$ is predicted to be elevated by 260°C above that of

8.3 Observation of nanoscale 180° stripe domains

bulk, unstressed PbTiO$_3$. The c domain orientation is the equilibrium state of such films below T_C down to low temperature, where a small fraction of a-oriented domains should occur.

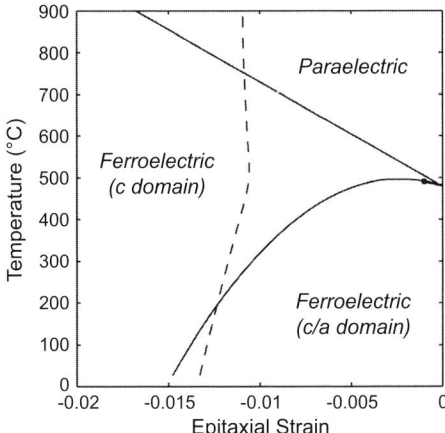

Figure 8.3: Phase diagram for epitaxially-strained PbTiO$_3$ calculated using Landau-Ginzburg-Devonshire theory [9]. The dashed line shows the epitaxial strain corresponding to a SrTiO$_3$ substrate.

We investigated the effect of epitaxial strain on the ferroelectric transition by determining T_C in coherently-strained films of PbTiO$_3$ on SrTiO$_3$. Unexpectedly, we found that the ferroelectric phase forms as highly ordered 180° stripe domains in these films [4]. These stripe domains consist of c-domain lamella with polarization of alternating sign, arranged to minimize the electric field that arises due to polarization (the "depolarizing field"). The presence of 180° stripe domains indicates that the depolarizing field has not been eliminated by alternative mechanisms, such as neutralization by surface charge [10]. Stripe domains have been observed in bulk ferroelectrics, but there had been no previous reports of their detection in thin films. Recently there has been increased interest in the possibility that 180° stripe domains may often occur in and significantly influence the properties of very thin ferroelectric films [11–14]. Their presence is expected to depend sensitively on the conductivity of the film and the electrical boundary conditions [15–17], such as the existence of extremely thin surface dielectric layers [14].

The presence of 180° stripe domains in these films was discovered because of the strong satellite peaks that appear in the diffuse x-ray scattering around PbTiO$_3$ Bragg peaks as the film is cooled through T_C. Figure 8.4 shows in-plane x-ray scattering profiles through a PbTiO$_3$ Bragg peak at various temperatures for a typical film. As T is lowered through T_C, first-order satellite peaks develop around the Bragg peak at positions $\Delta K = \pm K_0$, indicating the presence of an in-plane modulation with well-defined period $\Lambda = a_0 = K_0$. (Note that, for simplicity, H, K, and L values given in this paper are expressed in reciprocal lattice units referred to the room temperature SrTiO$_3$ lattice constant, $a_0 = 3.905$ Å). At about

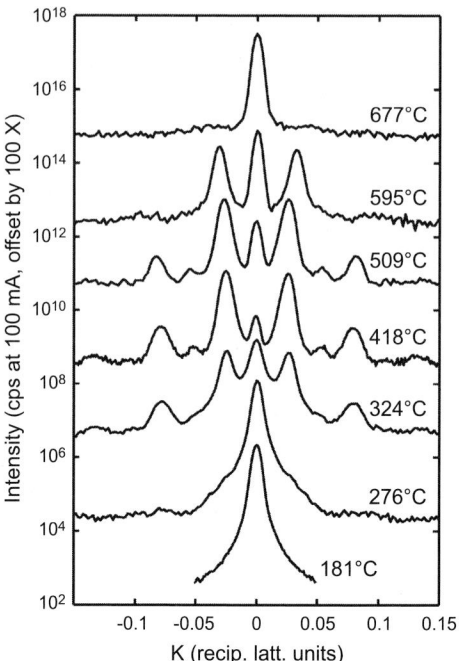

Figure 8.4: Sequence of in-plane scans through the PbTiO$_3$ 304 peak at various temperatures, showing the development and disappearance of satellites in the diffuse x-ray scattering as temperature is lowered (19.2 nm thick film).

100 K below T_C, the first-order satellites shift to smaller K_0, and peaks appear at the higher, odd-order satellite positions $\Delta K = \pm n K_0 (n = 3, 5, ...)$. Finally, at about 400 K below T_C, the satellites disappear. This indicates that some alternative mechanism for neutralizing the depolarizing field is operative at low temperature. At a given temperature the positions and intensities of the satellites do not change with time, and are very reproducible provided that T is lowered monotonically from above T_C.

Several characteristics of these satellites conclusively indicate that they arise from 180° stripe domains [8]. For a given T and film thickness d, similar satellite patterns with identical spacings K_0 are observed around all Bragg peaks HKL except those with no out-of-plane component ($L = 0$). The lack of satellites around $L = 0$ peaks implies that the polarization is solely in the out-of-plane direction (positive or negative c-domains). The lack of even-order satellites is consistent with the 1:1 ratio of positive and negative domains expected for field cancellation. The T dependences of both the Bragg and satellite intensities are consistent with those expected from increasing polarization below T_C. Both the central Bragg peak and the satellites show thickness fringes in the out-of-plane direction with a spacing $\Delta L = a_0/d$, indicating that the stripe structure extends through the film.

Typical in-plane arrangements of the satellite pattern are shown in Figure 8.5. On various samples, the satellites either form rings (a), are aligned in crystallographic directions (b,c),

8.3 Observation of nanoscale 180° stripe domains

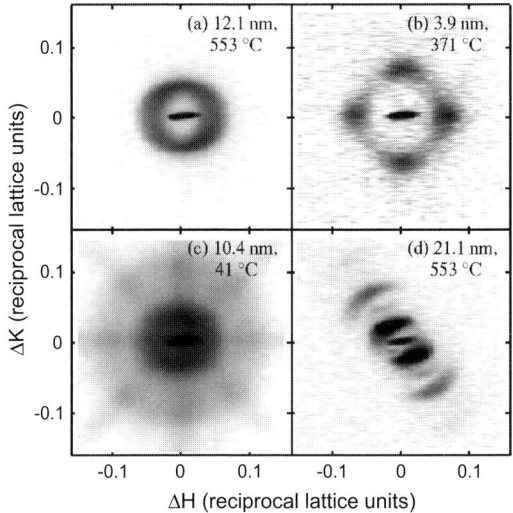

Figure 8.5: Typical in-plane distributions of diffuse x-ray intensity around the PbTiO$_3$ 304 peak, for various film thicknesses and temperatures. Darker shade indicates higher intensity (log scale). Elongation of central Bragg peak is due to asymmetric resolution function.

or are aligned at a particular azimuth (*d*). These are consistent with linear stripe domains oriented either randomly or in a specific direction within the illuminated area (1 mm^2). We confirmed on several samples showing alignment at a non-crystallographic azimuth that the stripe orientation corresponds to that of the terraces on the surface that occur due to substrate miscut (typical 0.25° vicinal angle from (001)).

Figure 8.6 shows the temperature dependence of the stripe period Λ for films of various thickness. As T is lowered, Λ changes fairly abruptly from a lower to a higher value. We interpret this to be a transition between two stripe domain phases, F_α and F_β. We observe that intense higher-order satellites typically occur only in the F_β phase, suggesting that the 180° domain walls have a narrower width in F_β than F_α. The stripe periods vary as the square root of film thickness in both phases, as shown in Figure 8.7 (a). This thickness dependence is expected from the LGD theory of ferroelectric 180° stripe domains [12, 18], although it is remarkable that the parabolic relationship is maintained down to atomic-scale dimensions. Furthermore, a calculation of $\Lambda(d)$ for F_β agrees within a factor of two on an absolute scale [4], which is strikingly good agreement considering that there are no adjustable parameters (i.e. all were determined from the literature).

We determined T_C for films of various thickness by analyzing the T dependence of both the out-of-plane lattice parameter [4] and the satellite intensity [5]. The latter method was more accurate for ultrathin films, where the finite-size broadening of the Bragg peaks becomes severe. The values of $T_C(d)$ obtained are shown in Figure 8.7 (b). For thicker films we observe that T_C approaches the predicted value of 752°C for coherently strained PbTiO$_3$ on SrTiO$_3$ as $d \to \infty$. As d decreases, T_C is gradually suppressed by hundreds of degrees below

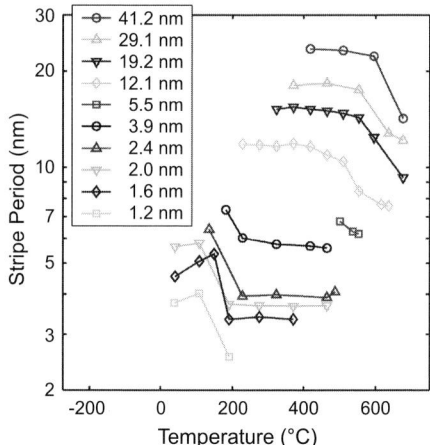

Figure 8.6: Stripe period as a function of temperature for various thickness PbTiO$_3$ films, showing transition between F_α and F_β stripe phases.

this value. At a thickness of 1.2 nm (three unit cells), T_C decreases abruptly. We were unable to find a ferroelectric transition in thinner films (two or one unit cells thick) at temperatures down to -153°C. LGD theory predicts that the stripe domains become ineffective at neutralizing the depolarizing field in sufficiently thin films, so that T_C depends on d. However, the observed suppression of T_C for thin films is much larger than this prediction. One explanation for this discrepancy is the suppression of the polarization at interfaces due to intrinsic effects, which have typically been described using additional terms in the mean-field free energy functional [19]. The combination of epitaxial strain and intrinsic surface effects are a subject of continuing investigation [20]. However, the suppression of T_C we observe does not obey a power law in thickness, as predicted by the simplest analyses.

Figure 8.8 shows the observed phase diagram as a function of d and T. The symbol type indicates the presence or absence of x-ray satellites, and whether the stripe period corresponds to the lower (F_α) or higher (F_β) value. The phase without satellites at low T we label F_γ. We believe that F_γ is the monodomain ferroelectric phase, since it has the lattice parameter of the ferroelectric phase but no satellites, indicating that the depolarizing field has been compensated without 180° stripe domain formation. The transition from F_β to F_γ is cooling-rate dependent in thinner films. For example, stripe domains were observed in a 10.5 nm film cooled relatively quickly (5 seconds per degree) to room temperature (similar to Figure 8.5 (c)), but not in the same film if cooled more slowly (160 seconds per degree). The observation of three ferroelectric phases can be qualitatively understood using existing concepts. The F_α phase observed just below T_C has a smaller stripe period and more diffuse domain walls than the F_β phase. It has been proposed that polarization-wave uctuations in the paraelectric phase become unstable at T_C, forming a 180° stripe domain phase with diffuse walls [17, 21]. A polarization gradient energy term is used to represent the domain wall energy. The predicted stripe period for F_α is similar to or smaller than that for

8.3 Observation of nanoscale 180° stripe domains

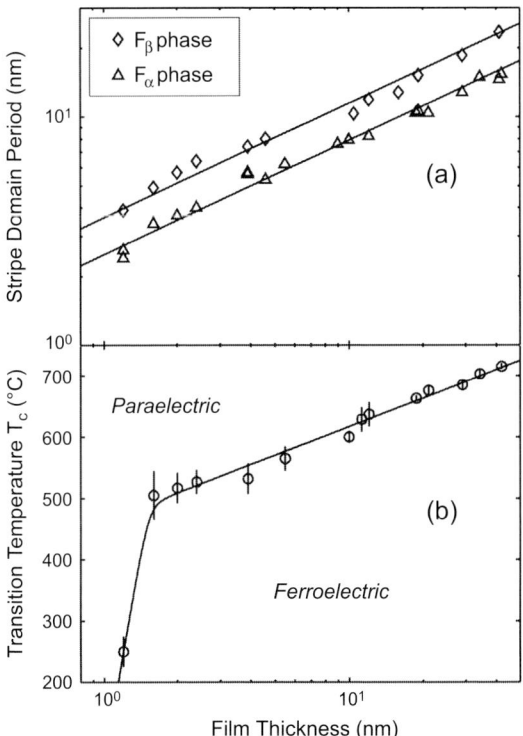

Figure 8.7: Upper plot: Ferroelectric transition temperature versus film thickness. Lower plot: Stripe period versus film thickness. Diamonds: phase F_β at $T = T_c - 250$ K. Triangles: phase F_α at $T = T_c - 50$ K. Lines fits to parabolic dependece.

F_β, as observed. At lower T, non-linearity in the free energy functional is expected to favor sharp domain walls [21]. The transition to the monodomain ferroelectric phase F_γ has been predicted by treating the ferroelectric as a semiconductor, in which carriers can be created by the field effect [16]. This transition can be calculated to occur at $T_C - T = 325\,K$ for $d = 10$ nm [4], in qualitative agreement with our observations. The rich phase diagram we have observed in this simple system makes it an excellent test case for more quantitative development of these concepts.

Acknowledgement

We would like to acknowledge dedicated and capable experimental assistance from L. Thompson, G. R. Bai, M. E. M. Aanerud, and the BESSRC beamline staff. Work supported by the US Department of Energy, BES-DMS under contract W-31-109-ENG-38, and the State of Illinois under HECA.

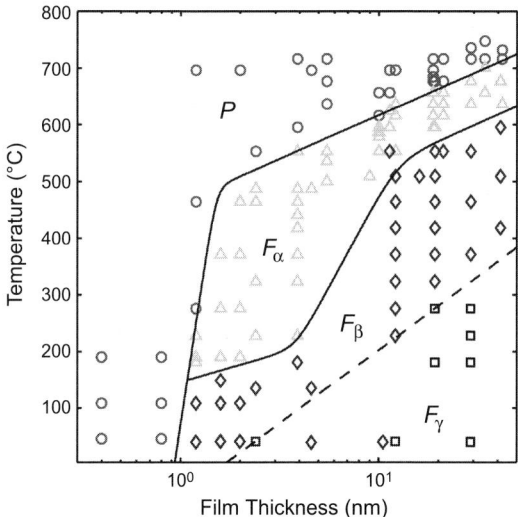

Figure 8.8: Phase diagram for epitaxial PbTiO$_3$ thin films on SrTiO$_3$. Circles: no satellites, paraelectric phase P. Triangles: ferroelectric stripe phase F_α. Diamonds: ferroelectric stripe phase F_β. Squares: ferroelectric monodomain phase F_γ.

Bibliography

[1] G. B. Stephenson, J. A. Eastman, O. Auciello, A. Munkholm, Carol Thompson, P. Fuoss, P. Fini, S. P. Den-Baars, and J. S. Speck, MRS *Bull.* **24**, 21 (1999).

[2] M. V. Ramana Murty, S. K. Streiffer, G. B. Stephenson, J. A. Eastman, G. R. Bai, A. Munkholm, O. Auciello, and Carol Thompson, *Appl. Phys. Lett.* **80**, 1809 (2002).

[3] A. Munkholm, S. K. Streffer, M. V. R. Murty, J. A. Eastman, Carol Thompson, O. Auciello, L. Thompson, J. F. Moore, and G. B. Stephenson, *Phys. Rev. Lett.* **88**, 016101 (2002).

[4] S. K. Streiffer, J. A. Eastman, D. D. Fong, Carol Thompson, A. Munkholm, M. V. Ramana Murty, O. Auciello, G. R. Bai, and G. B. Stephenson, *Phys. Rev. Lett.* **89**, 067601 (2002).

[5] D. D. Fong, G. B. Stephenson, S. K. Streiffer, J. A. Eastman, O. Auciello, P. H. Fuoss, and Carol Thompson, unpublished (2003).

[6] G. J. M. Dormans, P. J. van Veldhoven, and M. de Keijser, *J. Cryst. Growth* **123**, 537 (1992).

[7] I. Barin, *Thermochemical Data of Pure Substances*, 3rd. edn., (VCH, New York, 1995).

[8] G. B. Stephenson, D. D. Fong, M. V. Ramana Murty, S. K. Streiffer, J. A. Eastman, O. Auciello, P. H. Fuoss, A. Munkholm, M. E. M. Aanerud, and Carol Thompson, to appear in *Physica* B (2003).

[9] N. A. Pertsev and V. G. Koukhar, *Phys. Rev. Lett.* **84**, 3722 (2000).

[10] Y. Watanabe, M. Okano, and A. Masuda, *Phys. Rev. Lett.* **86**, 332 (2001).

- [11] Y. G. Wang, W. L. Zhong, and P. L. Zhang, *Phys. Rev.* B **51**, 5311 (1995).
- [12] A. Kopal, T. Bahnik, and J. Fousek, *Ferroelectrics* **202**, 267 (1997).
- [13] A. M. Bratkovsky and A. P. Levanyuk, *Phys. Rev.* B **63**, 132103 (2001); A. Kopal *et al. Ferroelectrics* **223**, 127 (1999).
- [14] A. M. Bratkovsky and A. P. Levanyuk, Phys. Rev. Lett. 84, 3177 (2000); *Phys. Rev. Lett.* **85**, 4614 (2000); *Phys. Rev. Lett.* **87**, 179703 (2001).
- [15] Y. Watanabe and A. Masuda, *Jpn. J. Appl. Phys.* **40**, 5610 (2001); A. M. Bratkovsky and A. P. Levanyuk, *Phys. Rev.* B **61**, 15042 (2000).
- [16] E. V. Chenskii, *Sov. Phys. Sol. State 14*, 1940 (1973).
- [17] E. V. Chenskii and V. V. Tarasenko, *Sov. Phys.* JETP **56**, 618 (1982).
- [18] T. Mitsui and J. Furuichi, *Phys. Rev.* **90**, 193 (1953).
- [19] R. Kretschmer and K. Binder, *Phy. Rev.* B **20**, 1065 (1979).
- [20] A. G. Zembilgotov, N. A. Pertsev, H. Kohlstedt, and R. Waser, *J. Appl. Phys.* **91**, 2247 (2002).
- [21] P. N. Timonin, JETP **83**, 503 (1996).

9 Characterization of Polar Oxides by Photo-Induced Light Scattering

M. Imlau[1], M. Goulkov[2], M. Fally[3], Th. Woike[4]

1. Fachbereich Physik, Universität Osnabrück, Barbarastr. 7, 49069 Osnabrück, Germany
2. Institute of Physics, National Academy of Sciences of Ukraine, Science Ave 46, 03650 Kiev, Ukraine
3. Institut für Experimentalphysik, Universität Wien, Boltzmanngasse 5, 1090 Wien, Austria
4. Institut für Mineralogie und Geochemie, Universität zu Köln, 50674 Köln, Germany

Abstract

The characterization of polar oxides using the photorefractive phenomenon of photo-induced light scattering is demonstrated. Relaxor-ferroelectric properties are studied exemplarily at crystals of strontium-barium-niobate, $Sr_{0.61}Ba_{0.39}Nb_2O_6$, doped with Cerium. It is shown that a deep insight into the polar structure can be gained, especially with respect to the phase transition behavior of ferroelectric domains with different sizes. In principle, the tool of photo-induced light scattering is applicable to all polar media presumed the existence of any kind of photorefractive response.

9.1 Introduction

The electrical characterization of polar media is crucial to investigate their suitability for ferroelectric memories, piezo- or pyroelectric devices and many other ferroelectric applications (see Chapter 3). Optical characterization of polar media is fundamental to investigate their servicability for electro-optic devices or applications in the field of nonlinear optics (see Chapter 4). Additionally there are intentions to characterize polar media with a combination of both, electrical and optical methods, such as to understand ferroelectric phenomena that are influenced by the action of light.

In this review the characterization of polar oxides by photo-induced light scattering is demonstrated. This is shown in particular at the example of the relaxor-ferroelectric strontium-barium-niobate, $Sr_{0.61}Ba_{0.39}Nb_2O_6$ (SBN), exhibiting a spontaneous polarization parallel to the c-axis in the ferroelectric phase. Ferroelectric and optical parameters are revealed by contact-free measurements. Furthermore the method of photo-induced light scattering offers a deep insight into the polar structure and domain dynamics. It is therefore a comprehensive and interesting tool in the field of ferroelectrics. In principle, it is applicable to any polar medium among the 21 crystallographic point groups without center of inversion presumed any kind of photorefractive response.

Polar Oxides: Properties, Characterization, and Imaging
Edited by R. Waser, U. Böttger, and S. Tiedke
Copyright © 2005 WILEY-VCH Verlag GmbH & Co. KGaA, Weinheim
ISBN: 3-527-40532-1

In SBN, the phenomenon of photo-induced light scattering originates from the photorefractive effect, i.e. the change of the refractive index upon light exposure in electro-optic media. The field of photorefraction started in 1966 with the observation of an optically induced inhomogeneity of the refractive index in LiNbO$_3$-crystals during exposure to coherent laser light [1]. The light pattern induces a space-charge field by photo-excited charge carriers, and the electric field is transferred into a change of the refractive index via the linear electro-optic effect. It is remarkable that amplitudes of the refractive-index change of $\Delta n \approx 1 \cdot 10^{-3}$ are obtained even with moderate intensities ($I_p \approx 100\,\text{mW/cm}^2$). The effect offers some unique properties and several advantages, which are useful for applications in the field of technical optics, such as self-pumped phase conjugation [2], volume holographic memories [3] or optical detection of ultrasonic waves [4]. The photorefractive effect allows to record volume phase gratings by spatially induced changes of the refractive index [5]. Recording is stamped by a remarkable feedback, i.e. the amplitudes and/or phases of the recording beams are changed by diffraction at the already recorded grating, which in turn modifies the recording process. This feedback results in an effective light amplification. It is commonly expressed by the holographic gain coefficient Γ with values up to $\approx 100\,\text{cm}^{-1}$. It is sufficient to considerably enhance even the coherent optical noise appearing from a laser beam propagating through the crystal volume. As a result an extremely rich variety of bright scattering patterns of different shapes and fashion appears on a viewing screen behind the crystal. The particular light scattering pattern may consist of wide-angle lobes as well as of sharp rings, lines and spots. It is obvious to designate this effect as *photo-induced light scattering* [6].

There is a strong correlation between photorefractive and ferroelectric properties via the dependence of the linear electro-optic effect on the spontaneous polarization. This allows to control photorefractive phenomena by the polar structure. As a consequence the phenomenon of photo-induced light scattering can be either enhanced or suppressed. Obviously, suppression is of importance for applications of photorefractive crystals in technical optics. However, it should be emphasized that photo-induced light scattering is a clear fingerprint of the processes of photo-excitation and charge transport, so that it can be utilized for contact-less material characterization. In particular photo-induced light scattering is an explicitly promising tool for the study of the domain structure as well as of the phase transition from the ferroelectric to the paraelectric phase of polar media. Passing the phase transition temperature, the polar structure exhibits its strongest changes and thermal decay of ferroelectric domains results in the disappearance of the photorefractive response and thereby of the scattering. Furthermore, scattering centers, which are responsible for the coherent noise seeding the scattering, are incorporated into the polar structure. Thus the properties of scattering centers strongly depend on the state of the polar structure. As a result there is a pronounced sensitivity of photo-induced light scattering on changes in the polar structure.

This article shows up the possibilities to study ferroelectric media by photo-induced light scattering. In particular the optical characterization of SBN by photo-induced light scattering is reviewed including the determination of optical parameters [7], the investigations of the relaxor-kind phase transition [8] and the analysis of the polar structure [9, 10].

9.2 Fundamentals

In this chapter the properties of the relaxor-ferroelectric SBN as well as observation and modelling of photo-induced light scattering in SBN are introduced.

9.2.1 The relaxor-ferroelectric $Sr_xBa_{1-x}Nb_2O_6$

Strontium barium niobate, $Sr_xBa_{1-x}Nb_2O_6$, is a polar oxide and crystallizes in the range of $0.25 \leq x \leq 0.75$ in the tetragonal tungsten-bronze structure [11–13] with a congruently melting point at $x = 0.61$. Characteristic for this structure, schematically shown in a view along the [001]-direction in Figure 9.1, are the edge-coupled niobium-oxygen-octahedra.

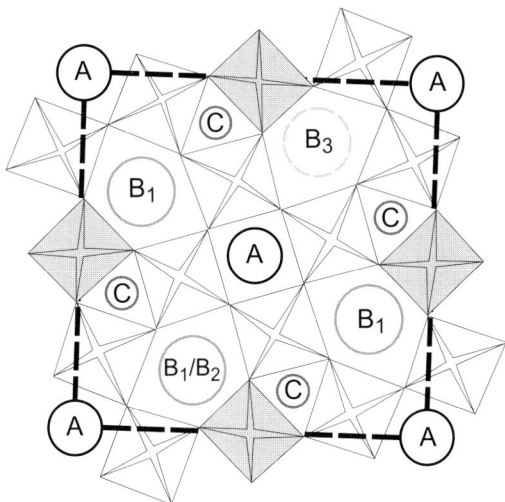

Figure 9.1: Tetragonal tungsten-bronze structure of $Sr_xBa_{1-x}Nb_2O_6$ (Strontium barium niobate, SBN). Characteristic are the corner-linked niobium-oxygen-octahedra and 3 types of vacancies A,$B_{1,2,3}$ and C. Sr^{2+}-ions are embedded in the A- and B_2-positions and Ba^{2+} in the B_1-positions according to their ionic radii. The C-positions and B_3-position remain empty.

Further the structure offers 3 types of vacancies denoted by type A,$B_{1,2,3}$ and C, where the A-positions clearly mark the tetragonal system with its fourfold axis. Sr^{2+}-ions are embedded into the A- and B_2-positions and Ba^{2+} into the B_1-positions according to their ionic radii. The C-position remains empty. That applies to the B_3-position, too, since there are 5 Sr- and Ba-ions to be distributed onto 6 A- and B-vacancies. The structure therefore is designated to as *open tungsten-bronze* [14].

All cations are located within the equatorial plane of the O^{2-}-octahedra in the paraelectric high-temperature phase [15]. In this case the crystal belongs to the centrosymmetric space group P4/mbm. In the low-temperature phase the metal atoms are displaced out of this plane. As a result the crystal structure is transferred into the polar space group P4bm and the polar axis coincides with the crystallographic c-axis. A non-vanishing spontaneous polarization, in particular a ionic polarization (see Chapter 1), and 180°-domains appear. Furthermore this displacement causes a distortion of the covalent Nb-O bonding in the octahedra resulting in the large linear electro-optic coefficients.

In the low-temperature phase strontium barium niobate behaves like a ferroelectric medium (see Chapter 1). It shows a ferroelectric hysteresis, so that the sign of the remanent polarization can be adjusted by an externally applied electric field. In addition there is a phase transition into the paraelectric phase whereby the dielectric susceptibility χ increases with increasing temperature. However, there are three remarkable particularities: the hysteresis shows a so-called slim-loop (see Figure 15.1), the spontaneous polarization decreases smoothly through the phase transition and a sharp peak or a discontinuity of the susceptibility χ is not observed - in contrast χ shows a broad, strongly frequency dependent maximum [16]. Thus and according to Chapter 15 strontium-barium-niobate belongs to the group of *relaxor-ferroelectrics*, in particular to *uniaxial relaxors* (see Chapter 15.6).

There are three relevant models to explain the relaxor behavior. At first Smolenskii proposed the model of a smeared phase transition [17], that explains the relaxor behavior by a superposition of a multitude of critical phase transitions having slightly different phase transition temperatures T_c due to the imperfect atomic positions of the different elements in the crystal. Secondly, the model of a glassy transition was introduced by Burns et al. [18]. Here, the model of dipolar glasses is adapted for low temperatures and the existence of polar clusters is predicted for high temperatures. Consequently, in SBN rather a freezing temperature than a real phase transition temperature exists. In contrast the *spherical random-bond random-field model* (SRBRFM) [19] assumes a real phase transition from the paraelectric to the ferroelectric phase and introduces a long range order between polar clusters. In this case the dynamic transition temperature T_m is equated with the breakdown of this long range order. All models have in common that polar clusters survive at higher temperatures. Thus, the low-temperature behavior of the material is decisive for selecting the appropriate model. SBN fulfills all necessary conditions to be described by the ferroic random field Ising model (RFIM) [20–22], which is a simplification of the SRBRFM (see Chapter 15). In the following we refer to the RFIM, which predicts a formation of ferroelectric domains, an important foundation, for the explanation of the physics being the origin of our observations. For reasons of simplicity we will refer to T_m as phase transition temperature.

Ferroelectric domains have been visualized in the ferroelectric phase in SBN with high resolution piezo-response force microscopy (see Figure 15.8) [23]. The domains are found to be needlelike with lengths in the range of 10 to 500 nm and are oriented along the polar c-axis. The dynamics of the domain walls under externally applied electric fields or heating are expected to influence the polarization especially at low frequencies (see *domain wall polarization*, Chapter 1) [24].

The phase transition temperature of undoped SBN is found at about $T_m = 85\,°C$ [20, 25] and depends strongly on composition, doping element and doping concentration [26–28]. A deviation of the composition from the congruently melting one lowers the transition temperature with decreasing Ba-concentration [26, 29, 30]. The same holds for doping with different elements [31–33]. In the experimental section below SBN crystals doped with 0.66 mol% Cerium are investigated. Here the phase transition temperature is about 52 °C and the Ce^{3+}-ions occupy Sr^{2+} sites in off-center positions [34, 35]. Doping with Cerium enhances the

9.2 Fundamentals

photorefractive response of SBN. Non-uniformly photo-excited electrons are redistributed by diffusion as the dominating charge transport mechanism. In addition SBN exhibits a large linear electro-optic effect in the low-temperature phase [36, 37].

9.2.2 Observation of photo-induced light scattering in SBN

A single crystal of SBN doped with 0.66 mol% of Cerium is exposed to extraordinarily polarized coherent light of a helium-neon laser ($\lambda = 632.8$ nm, $P = 10$ mW) with its wave vector normal to the crystal surface. Figure 9.2 shows the steady-state far-field scattering pattern observed on a screen in transmission. A scattering distribution into a wide solid angle (scattering lobe) appears beside the directly transmitted laser beam. Consequently the intensity of the pump beam strongly decreases. The scattering lobe builds-up into a preferred direction antiparallel to the crystallographic c-axis. Its light polarization is conserved with respect to the polarization of the pump beam. Thus the scattering is designated to as polarization-isotropic photo-induced light scattering.

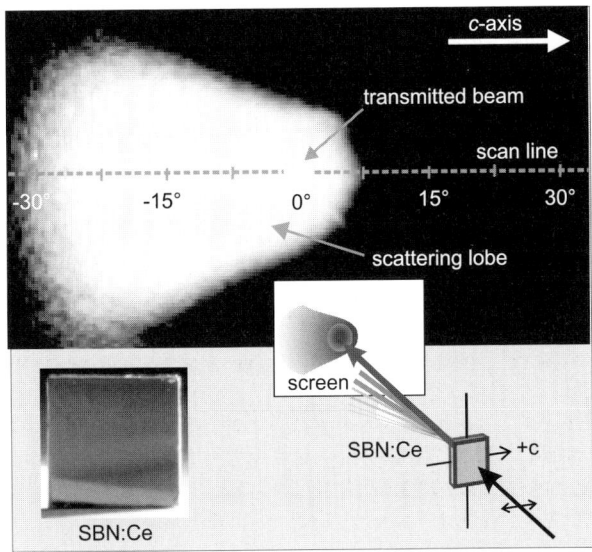

Figure 9.2: Polarization-isotropic photo-induced light scattering in strontium-barium-niobate doped with 0.66 mol% Cerium (SBN61:Ce). Strongly amplified scattered light occurs in a wide solid angle opposite to the polar c-axis. The label denotes the apex angle measured with respect to the directly transmitted laser beam in air. $\lambda_p = 632.8$ nm, $\mathbf{E_p} \parallel c$. The insets show a photograph of the SBN crystal (left) and the schematic setup of the experiment (right), respectively.

A characteristic property of photo-induced light scattering is its dependence on exposure. Figure 9.3 shows the scattering lobe observed at different times of illumination for the same SBN:Ce crystal but with $\lambda_p = 488$ nm. From the beginning of exposure the scattering lobe builds-up in the direction antiparallel to the c-axis accompanied by an increase of the overall scattered intensity. The following chapter explains the appearance of the scattering pattern, its asymmetry and dependence on exposure by employing a simplified photorefractive model.

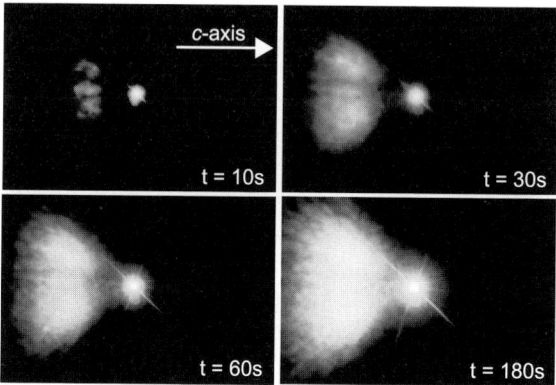

Figure 9.3: Polarization-isotropic photo-induced light scattering in strontium-barium-niobate doped with 0.66 mol% Cerium at different times of illumination. Here, the wavelength $\lambda_p = 488$ nm was chosen and $\mathbf{E_p} \parallel c$.

9.2.3 Description of photo-induced light scattering in SBN

Polarization-isotropic wide-angle photo-induced light scattering of a single pump beam in SBN is usually interpreted as resulting from a multitude of beam-coupling processes between scattered and transmitted parts of the incident beam [6, 38]. In this section we describe the appearance of scattering within a simplified photorefractive model employing the following assumptions reasonable for SBN:

- diffusion of photocarriers is the dominating charge transport mechanism in SBN in the absence of an external electric field, since the photovoltaic effect is extremely small in Cerium-doped SBN [39] at the temperatures used in the experiments,

- electrons give the major contribution to photoinduced currents [40],

- an undepleted pump approximation [5, 41] can be used in the case of the comparatively weak light-induced scattering,

- the absorption coefficient for SBN:Ce ($\alpha = 4 \text{cm}^{-1}$ at $\lambda = 632.8$ nm) is assumed to be zero in order to simplify the model equations and the numerical treatment of the experimental results.

9.2 Fundamentals

The recording of parasitic holograms leading to photo-induced light scattering is schematically illustrated in Figure 9.4 in the reciprocal space and the plane of pump beam and polar c-axis. The beam of a coherent light source with wave vector $\mathbf{k_p}$, commonly called pump beam, propagates through the SBN-crystal and is partially scattered from optical inhomogeneities and imperfections in the crystal (see Figure 9.4 (a)).

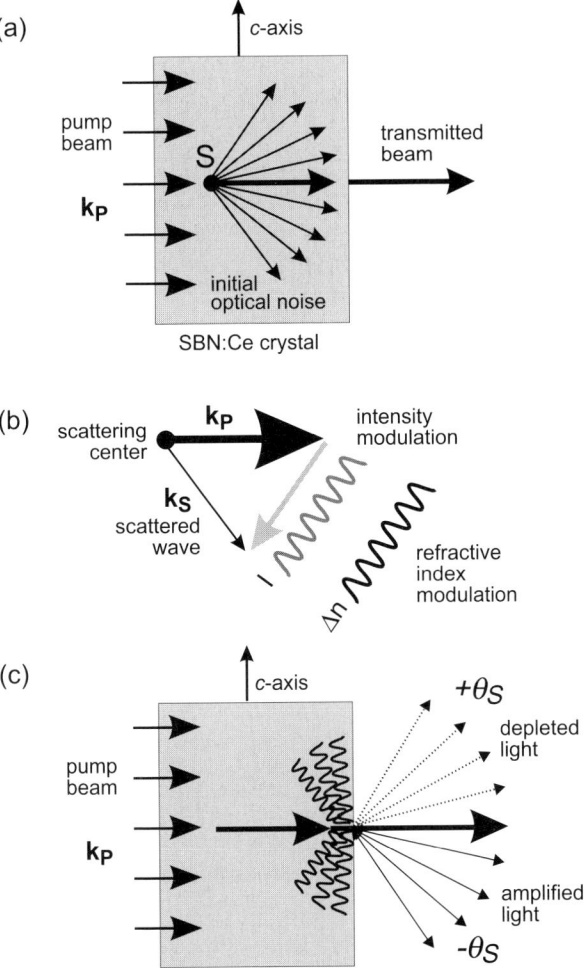

Figure 9.4: (a) The incoming pump beam $\mathbf{k_p}$ is scattered at the scattering center S. (b) The scattered wave $\mathbf{k_s}$ interferes with the propagating pump beam. A sinusoidal light interference pattern $I(\mathbf{r})$ occurs, which is transferred into a refractive-index modulation $\Delta n(\mathbf{r})$ via the photorefractive effect. (c) The pump beam is diffracted at the recorded refractive index modulation. Initially scattered light in direction of the polar axis is depleted and amplified in the opposite direction.

This initial optical noise consists of seed waves propagating at angles $+\theta_s$ and $-\theta_s$ with respect to the direction of the pump beam, i.e. in and opposite to the direction of the polar c-axis. Each seed wave $\mathbf{k_s}$ interferes with the pump wave $\mathbf{k_p}$ (9.4(b)) and forms an elementary light interference pattern

$$I(\mathbf{r}) = I_0[1 + m\cos(\mathbf{K}\cdot\mathbf{r})] \tag{9.1}$$

with a spatial period of $\Lambda = \lambda/2\sin\theta_s = 2\pi/|\mathbf{K}|$, where λ is the wavelength of the incident light, $\mathbf{K} = \mathbf{k_s} - \mathbf{k_p}$ is the grating vector, $m = 2\sqrt{I_s \cdot I_p}/I_0$ is the modulation depth and $I_0 = I_s + I_p$ is the total intensity of the recording beams. Due to processes of thermal diffusion (and drift in an externally applied field), photoexcited electrons migrate from bright to dark regions, where they are trapped. This results in the periodically modulated space-charge field

$$E_{sc}(\mathbf{r}) = mE_{sc}^o\cos(\mathbf{K}\cdot\mathbf{r} + \Phi), \tag{9.2}$$

where E_{sc}^o is the amplitude of the spatially varying field and Φ its phase shift with respect to the incoming light interference pattern [5, 42]. (A detailed study of transport models and photo-ionization processes can be found in [39, 43].)

In the case of SBN the phase shift is always exactly $\Phi = \pi/2$ provided that an external electric field is absent ($E_o = 0$) and there is no frequency shift between the pump and scattered waves. (In the experiments described below, where $E_o \neq 0$, a change of the phase shift Φ of up to 5% results, which can be neglected. The increase of the space charge field at the maximum field strength of $E_o = 4\,\mathrm{kV/cm}$ is only 15% in comparison to zero field.)

The linear electro-optic effect (see Chapter 4) transfers the periodically modulated space-charge field into a refractive index grating:

$$\Delta n(\mathbf{r}) = (\Delta n)_o \sin(\mathbf{K}\cdot\mathbf{r})$$

$$= -\frac{1}{2}r_{\mathrm{eff}} n_{\mathrm{eff}}^3 m E_{sc}^o \sin(\mathbf{K}\cdot\mathbf{r})$$

where n_{eff} is the effective refractive index and r_{eff} is the effective electro-optic coefficient. The spatial $\Lambda/4$-shift between the refractive grating and the light pattern causes an effective stationary energy exchange between pump and seed waves, resulting in an exponential change of the intensity:

$$I_s(\theta_s) = I_{so}(\theta_s)\exp[\Gamma(\theta_s)d], \tag{9.3}$$

where d is the crystal thickness, and the exponential increment

$$\Gamma(\theta_s) = \frac{4\pi(\Delta n)_o}{\lambda m \cos\theta_s} \tag{9.4}$$

is the gain coefficient describing the efficiency of the direct coupling between the waves $\mathbf{k_p}$ and $\mathbf{k_s}$. When using the corresponding expression for the amplitude of the diffusion field [42] and taking into account the particular conditions of our experiment, the coupling coefficient Γ can be written for the polarization-isotropic wide-angle scattering in SBN as:

$$\Gamma(\theta_s) = \frac{q\,r_{33}\,\zeta\,n_e^3\sin(\theta_s/2)\cos(\theta_s/2)}{\dfrac{\lambda^2 q^2}{8\pi^2 k_B T} + \dfrac{2\epsilon_{33}\epsilon_0 \sin^2(\theta_s/2)}{N_{\mathrm{eff}}}}, \tag{9.5}$$

9.2 Fundamentals

where q is the charge of the photocarriers, k_B is Boltzmann's constant, T is the absolute temperature, n_e the refractive index for extraordinarily polarized light, ϵ_{33} is the relative permittivity of SBN and ϵ_0 the permittivity of free space. $0 < \zeta \leq 1$ is introduced to account for electron-hole competition. The effective trap density N_{eff} is measured to 2.2×10^{23} m^{-3} for SBN doped with 0.66 mol% Cerium at room temperature [37]. The sign of r_{33} was measured to be positive [37, 44] and electrons are identified as dominating charge carriers in SBN, i.e., $q = -e$, where e is the elementary charge. Hence, it follows from Equation (9.5) that the sign of Γ is negative for $+\theta_s$ and positive for $-\theta_s$. Employing Equation (9.3) initial optical noise is amplified in the $-c$-direction and is depleted in the $+c$-direction (see Figure 9.4 (c)). A spatially asymmetric intensity distribution of scattered light results as presented in Figure 9.2.

In order to demonstrate both, depletion and amplification of initially scattered waves, Figure 9.5 shows the temporal development of the scattered light I_S at angles in air of $\theta_s = \pm 15°$, i.e., symmetrically with respect to the directly transmitted laser beam and within the plane of the polar c-axis and the incoming laser beam.

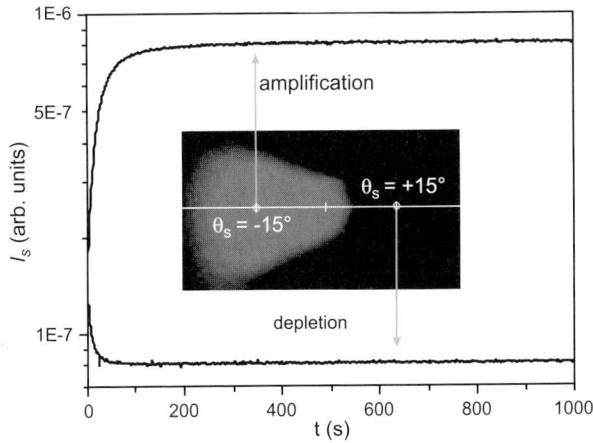

Figure 9.5: Kinetics of the scattered light as a function of exposure. Amplification and depletion of the initially scattered light is observed measured at apex angles of $\theta_s = \pm 15°$, i.e., symmetrically with respect to the directly transmitted laser beam and within the plane of the polar c-axis and the incoming laser beam.

Obviously, the intensity measured in $-c$-direction $I_S(-15°)$ increases whereas the intensity measured along the polar axis $I_S(+15°)$ decreases with time. Note, that comparable intensities are measured at the very beginning of exposure, that probably originate from light initially scattered at defects in the crystal volume or the surface. Thus beside effective *light amplification* the phenomenon of photo-induced light scattering additionally includes processes of *light damping*, i.e., the dark side of Figure 9.2 in direction of the polar c-axis resembles depletion of initially scattered light. (Note that all experiments described below are performed in the steady state.)

From this simplified photorefractive model it is straightforward to use the concept of photo-induced light scattering for the optical characterization of polar oxides: In SBN, the relation between r_{33} and the spontaneous polarization $P_3 = P_s$ can be written as [15]:

$$r_{33} = 2g_{33} P_s \epsilon_{33} \epsilon_0, \tag{9.6}$$

where g_{33} is the quadratic electro-optic coefficient. Thus, there is a relation between the orientation of P_s and the sign of Γ (see Equation (9.5)). High electric fields E_o larger than the coercive field E_c should result in a spatial inversion of the macroscopic polarization P_s of the sample and thereby in a corresponding inversion for the sign of the gain coefficient Γ (see hysteresis behavior below). As a result the direction where the scattering lobe appears will be inverted, which is successfully applied to SBN:Ce in order to investigate its ferroelectric properties. Further, according to Equation (9.5), we expect a strong temperature dependence of the scattering, which in particular is used for investigations of the phase transition.

9.3 Experimental

It follows from the model of photo-induced light scattering that the spatial intensity distribution of the scattered light originates from the angular distribution of the initially scattered light as well as from the dependence of the gain factor $\Gamma(\theta_s)$. Sign and value of Γ especially depend on the direction and value of the spontaneous polarization. This fundamental relation allows to investigate the polar structure of the relaxor-ferroelectric SBN by investigating the intensity distribution of photo-induced light scattering as will be demonstrated in this section.

9.3.1 Experimental setup

The experimental setup used for our investigations is shown schematically in Figure 9.6.

The beam of a low-power He-Ne-laser (5 mW) serving as the pump beam with a wavelength of $\lambda = 632.8$ nm is directed normal to the large a-surface of the sample with $\mathbf{E_p} \parallel c$. The intensity of the pump beam is adjusted to 70 mW/cm^2 using a half-wave retarder plate and a Glan-Thomson prism in order to prevent nonlinear effects other than photo-induced light scattering. The pump beam has a Gaussian intensity distribution with a FWHM of 0.8 mm. A small fraction of the pump beam is directed onto a photodiode PD1 by a beam-splitter BS to monitor the laser intensity. Photodiode PD2 was placed behind the sample at a distance of 5.5 cm and is mounted onto a rotation stage driven by an electronic motion controller. When moving, photodiode PD2 drives an exact half-circle around the sample and determines the intensity distribution in the plane of incidence parallel to the c-axis. The scattered light is measured in the angular range $+90° \leq \theta_s \leq -90°$. At $\theta_s = 0°$ the photodiode crosses the pump beam directly behind the crystal. The aperture of the diaphragm on PD2 limits the apex angle of the measured scattered light to 0.5°. The entire setup is enclosed in an opaque box (represented by the dotted rectangle in the figure) with only a small opening for the pump beam to minimize the noise due to external light sources. To obtain a baseline curve, the intensity distribution of the laser was measured without a sample in the holder.

For our investigations we use a single crystal of SBN doped with 0.66 mol% Cerium. It was grown by the Czochralski technique from the congruently melting composition by

9.3 Experimental

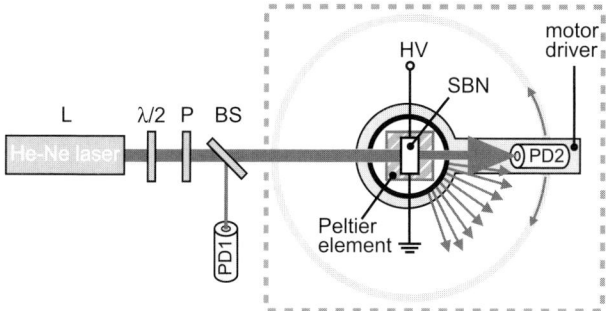

Figure 9.6: Experimental setup for measuring the angular distribution of the scattered light at different temperatures and externally applied electric fields. L is a He-Ne-laser, λ/2 a half-wave retarder plate, P a Glan-Thomson prism, BS a beam splitter, PD1 and PD2 are photodiodes and HV the high voltage amplifier. The SBN sample with 0.66 mol% Cerium is placed on a stack of Peltier-elements to control the temperature.

Pankrath from the crystal growth department of the University of Osnabrück. The boule is cut parallel to the crystallographic axes into a rectangular parallelepiped with dimensions of $0.90 \times 7.15 \times 6.20$ mm^3 along the a-, b- and c-axis, respectively. (*Note, $a = b$ in the tetragonal crystal system of* SBN.) The sample is electrically poled by heating up to 140 °C, applying an external electric field of 350 V/mm along the crystallographic c-axis and then slowly cooling down to room temperature before removing the field. This procedure results in a sample where practically all existing domains are aligned according to the external field [45]. Then the sample is placed onto a holder and fixed on a thermoelectric element. A temperature controller allows to adjust the sample temperature from 10 °C to 150 °C with an absolute accuracy of 0.3 °C. The faces of the sample normal to the crystallographic c-axis are short-circuited in order to prevent any influence of the pyroelectric effect when the sample is heated. A high-voltage amplifier is used for measurements with externally applied electric field. The maximum amplitude is ±3.5 kV, which can be adjusted with a resolution of 7 V.

9.3.2 Spatial distribution of the scattering intensity

Figure 9.7 shows the one-dimensional angular intensity distribution of the steady-state scattering pattern (crystal temperature is T = 28 °C) as a solid curve (a). The sharp intensity peak in the central part of the scan corresponds to the directly transmitted pump beam. Because of the small aperture of the diaphragm on the photodiode PD2, this peak is only a section of the beam cut by the diaphragm. The dashed curve (b) represents the angular profile of the pump beam without the sample. Both intensity profiles are normalized to their maximum intensities at $\theta_s = 0°$. The angular distribution of the light pattern as well as the pattern on the photo in Figure 9.2 are strongly asymmetric: The main part of the scattered light is found at negative scattering angles $-\theta_s$. The scattering in the $-c$-direction exhibits a broad and weakly pronounced maximum at $-\theta_s^{max} = -(30 \pm 5)°$. The scattering measured along the c-axis is

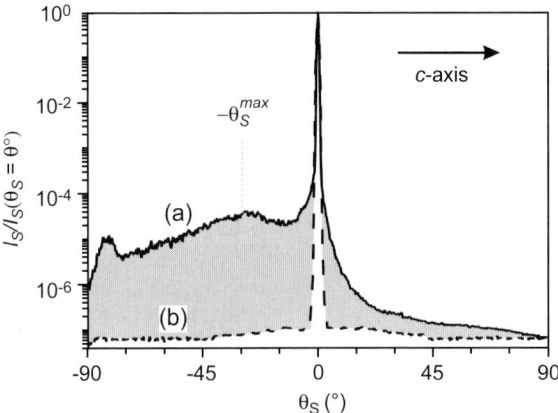

Figure 9.7: (a) Typical angular distribution of the scattered light in SBN at T = 28 °C in the plane spanned by the c−axis and the laser beam. The central peak is the transmitted laser beam. The two grey areas to the left and to the right of the peak are used to derive the integral intensities in the $-c$ and $+c$ directions of the scattering profile, respectively. $-\theta_s^{max}$ defines the maximum of the intensity distribution of the scattering pattern. (b) Beam profile without a sample, representing the optical noise of the system.

rather weak. To get a quantitative measure of this asymmetry of the scattering distribution, we determined the ratio of the light scattered in the $-c$-direction and in the $+c$-direction

$$R = \frac{\int_{-\frac{\pi}{2}}^{-\delta} I_s(\theta_s) d\theta_s}{\int_{+\delta}^{+\frac{\pi}{2}} I_s(\theta_s) d\theta_s}. \tag{9.7}$$

The integrated angular ranges are shown in Figure 9.7 by the two crosshatched areas below the intensity curve. The angular interval $|\delta| = 2°$ is excluded from the integrals in order to take into account the scattered light only. As a result for the asymmetry of the scattering profile shown in Figure 9.7 we get a ratio of $R = 14.3$.

9.3.3 Investigating the relaxor-kind phase transition

In the following we will present photo-induced light scattering as a function of temperature [8]. The spatial intensity distribution of the scattered light was studied at different temperatures in the range from +15 to +148 °C. Figure 9.8 shows three beam-fanning profiles measured with the same SBN sample at different temperatures: $T = 28$ °C (ferroelectric phase), $T = 52$ °C ($T \approx T_m$) and $T = 130$ °C (paraelectric phase). The central intensity peak at $\theta_s = 0°$ is again formed by the transmitted pump beam and the shadowed area marks the angular interval where the pump peak influences the scattering distribution. The basic properties of the temperature dependence of photo-induced light scattering in SBN are:

9.3 Experimental

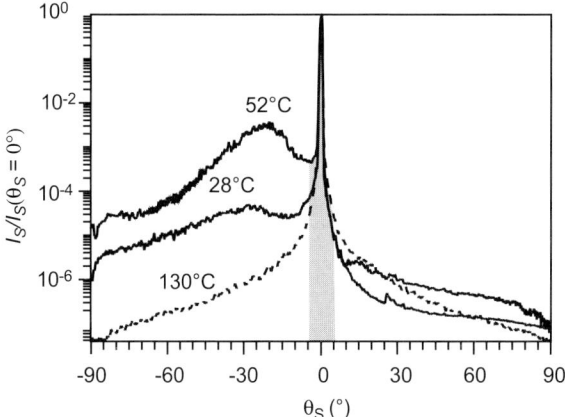

Figure 9.8: Angular distributions of scattered light measured along the c-axis at zero external field and different temperatures: T = 28 °C, T = 52 °C and T = 130 °C. The central peak corresponds to the directly transmitted pump beam. The shadowed area displays the angular interval influenced by the transmitted pump beam.

- In the ferroelectric phase, the scattering pattern of SBN is strongly asymmetric, and most of the scattered light is observed in the direction opposite to the c-axis of the crystal (negative scattering angles θ_s in Figure 9.8).

- The strong increase of ϵ_{33} in the vicinity of the phase transition temperature results in a strong increase of the gain factor (see Equation (9.5)). Thus the total scattering intensity significantly increases when heating up to $T_m = 52\,°C$. At the same time the asymmetry of the angular light distribution becomes more pronounced.

- Because of the relaxor-kind behavior of SBN, quite strong light-induced scattering is observed even at $T > T_m$ due to the presence of polar clusters.

- In the paraelectric phase at $T \gg T_m$, where polar clusters vanish and beam-coupling is not possible because the polar macrostructure is no longer present, only the weak seed scattering is observed. Due to the drastic changes in the domain structure at the phase transition, the seed scattering at 130 °C differs from that at 28 °C both in the total amount and in the angular distribution.

These properties demonstrate that the study of photo-induced light scattering can be used to investigate the relaxor-kind phase transition by an optical method. Hereby, various measures can be used, like the temperature dependence of the intensity of the directly transmitted laser beam, of the integrated scattered light and of the intensity ratio R, respectively, as shown in Figure 9.9.

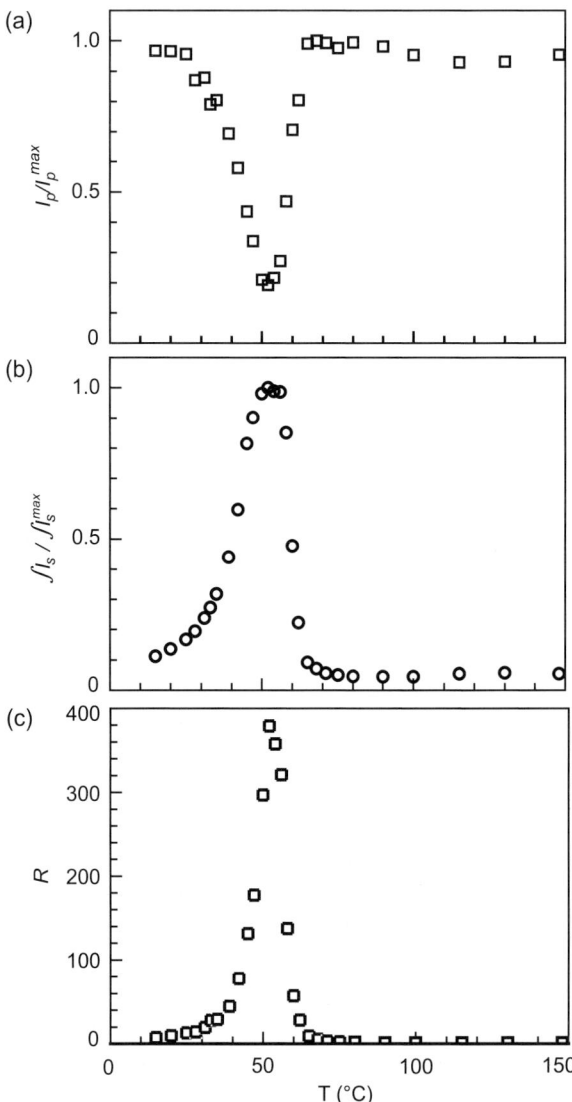

Figure 9.9: (a) Temperature dependence of the transmitted pump intensity. (b) Temperature dependence of the total scattered intensity. Both data sets have been normalized to their maximum values. (c) Dependence of the intensity ratio R.

9.3 Experimental

Figure 9.9 (a) displays the transmitted pump intensity I_p, Figure 9.9 (b) the integrated scattering intensity $\int I_s(\theta_s)d\theta_s$ and Figure 9.9 (c) the intensity ratio R according to Equation (9.7) as a function of the temperature. The integrated scattering intensity is determined from the angular intensity scans by

$$\int I_s(\theta_s)d\theta_s = \int_{-\frac{\pi}{2}}^{-\delta} I_s(\theta_s)d\theta_s + \int_{+\delta}^{+\frac{\pi}{2}} I_s(\theta_s)d\theta_s \qquad (9.8)$$

To focus on the temperature dependence, both intensity curves have been normalized to their maximum values. In the low temperature regime up to about 52 °C, the total scattered intensity increases nonlinearly. Additionally, the transmitted pump beam decreases. The intensity of the scattering grows in the $-c$-direction and drops in the $+c$-direction, leading to a sharp increase of R. At about T $\geq$ 52 °C up to T = 65 °C the total scattering intensity starts to decrease, while the intensity of the transmitted pump beam increases. The decrease of the scattered intensity is especially pronounced for large angles in both directions, while the scattered intensity for small angles still shows a pronounced maximum. In this temperature regime, the fast decay of the asymmetry of the scattering pattern takes place, the intensity of the scattered light decreases in $-c$-direction and increases in $+c$-direction, resulting in a very pronounced maximum of the intensity ratio R. The scattering intensity and R are reduced to the values comparable to those at room temperature. When the temperature is further increased to T = 100 °C the photo-induced scattering vanishes in the background. The intensity ratio R in the paraelectric phase becomes much smaller as compared to the values at room temperature and finally approaches unity for T $\geq$ 100 °C, indicating a perfectly symmetric angular distribution of the scattered light. It should be noted that the scattering exhibits a reversibility with respect to the heating and cooling procedures: When the crystal is cooled back to room temperature, the scattering maximum reappears and the light pattern becomes asymmetric again. The evolution of the scattering in this case passes the above mentioned three regimes in reverse order.

These results clearly demonstrate that the relaxor-kind phase transition determines the appearance of photo-induced light scattering, and therefore its temperature dependence can be used as a method for optical investigation of phase transitions. E.g., the transition temperature can be determined by the asymmetry parameter as a measure for the symmetry of the scattering pattern or by the comparably simple measurement of the directly transmitted laser beam.

9.3.4 Determination of material parameters: gain factor Γ, effective electro-optic coefficient ($\zeta \cdot r_{\text{eff}}$) and effective trap density N_{eff}

The scattering pattern contains additional information about the scattering process itself as well as for the determination of material parameters. We will demonstrate in the following, that both the angular dependence of the holographic gain and the product of the effective linear electro-optic coefficient with the electron-hole competition factor as well as the effective trap density can be determined from the spatial distribution of the scattering pattern as a function of temperature [7].

We assume that the intensity of the initially scattered light from a single scattering center is the same for symmetric scattering angles $\pm\theta_s$. This assumption does not imply that the intensity distribution of the initially scattered light shows a homogeneous angular distribution [9]. The values of Γ for two symmetric angles $+\theta_s$ and $-\theta_s$ differ only in sign. Thus the logarithm of the corresponding ratio of the two scattered intensities $I_s(+\theta_s) = I_{so} \exp(-|\Gamma|d)$ and $I_s(-\theta_s) = I_{so} \exp(+|\Gamma|d)$ will give us twice the absolute value of the two-beam coupling gain:

$$2|\Gamma|d = \ln(I_s(-\theta_s)/I_s(+\theta_s)) \tag{9.9}$$

Figure 9.10 shows the values of $|\Gamma|$ versus the scattering angle θ_s determined from the intensity scans for $T = 20\,^\circ\text{C}$ (circles), $T = 45\,^\circ\text{C}$ (squares) and $T = 65\,^\circ\text{C}$ (crosses).

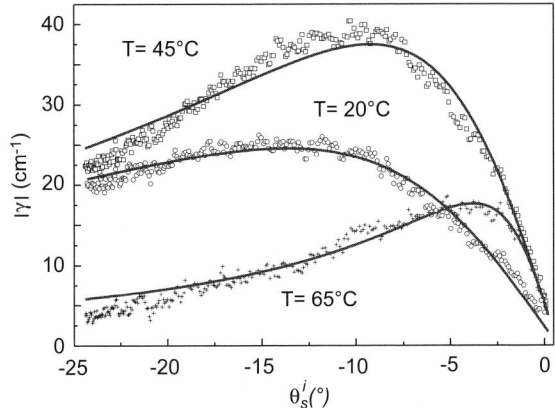

Figure 9.10: Absolute value of the two beam coupling gain Γ versus the scattering angle θ_s for $T = 20\,^\circ\text{C}$ (circles), $T = 45\,^\circ\text{C}$ (squares) and $T = 65\,^\circ\text{C}$ (crosses). The solid lines are fits according to Equation (9.5).

Note that the internal scattering angle θ_s^i displayed in this figure ranges from about $0° \leq \theta_s^i \leq -25°$, corresponding to externally observed angles in the range of $0° \leq \theta_s \leq -90°$, due to the large refractive index of $n_e = 2.281$ (for undoped SBN [46]). Now, Equation (9.5) can be fitted to the experimental data, which allows to determine the effective electro-optic coefficient r_{33} and the effective trap density N_{eff} since the material parameters n_e and ϵ_{33} are known at different temperatures [20, 47, 48]. Fits are represented by the continuous lines in Figure 9.10. A good agreement of the fitting curve with the Γ-dependence is apparent and shows that the simple model considered above describes the angular distribution of the beam fanning in SBN quite successfully. We would like to draw attention to the fact that the accurate fitting of the original I_s-curves is not possible without knowledge of the angular behavior of the seed scattering $I_{so}(\theta_s)$. As mentioned above, the initial scattering exhibits a very strong dependence on the angle θ_s [9], and this can essentially influence the intensity distribution of the scattered light from sample to sample.

The extraction made from Figure 9.10 for $T = 20\,^\circ\text{C}$ gives $\zeta r_{33} = (324 \pm 16)$ pm/V and $N_{\text{eff}} = (1.9 \pm 0.2) \cdot 10^{23}$ m^{-3}. In comparison, the standard interferometric method for the

9.3 Experimental

same SBN sample gives a value of the electro-optic coefficient as $r_{33} = (354 \pm 2)$ pm/V [49]. Thus, one can deduce that the electron-hole competition factor is $\zeta \approx (0.92 \pm 0.04)$ and the product ζr_{33} can be substituted simply by r_{33} in a good approximation. It is obvious, that this analysis can be performed for different temperatures, so that a determination of the mentioned parameters as a function of the temperature is possible with sufficient accuracy.

The advantage of this method is apparent, if compared to commonly used holographic or interferometric setups. At first the complexity of the experimental setup is marginal and the requirements to the mechanical stability are negligible. The second point belongs to the effort which is necessary to determine the desired parameters. In holographic two-beam coupling experiments the recording beams have to be adjusted for each single Bragg angle θ_s, whereas the complete angular range over $180°$ is directly determined by the steady state intensity distribution of the scattering pattern.

9.3.5 Investigating ferroelectric properties

We will demonstrate in the following, that photo-induced light scattering can be used for the optical determination of the ferroelectric hysteresis behavior [9]. In contrast to the electrical detection of the ferroelectric polarization P by measuring surface charges per area, the optical determination by photo-induced light scattering is based on the determination of the asymmetry parameter as a function of the externally applied electric field. Here, it should be stressed that an advantage of the scattering method is the determination of the ferroelectric properties in the crystal bulk and thus the method is not restricted to surface properties.

The starting point of this method is again the close relationship between the sign of the gain factor and the direction of the polar axis, i.e., the inversion of the direction of the spontaneous polarization will result in a change of the sign of the electro-optic coefficient (according to Equation (9.6)) and, thus, of the sign of the gain factor. The consequence is a spatial inversion of the scattering lobe, which is visualized by the asymmetry parameter.

The procedure to determine the scattering hysteresis is performed as follows: An external field E_o is applied to the crystal along the polar axis for a duration of 10 seconds without illumination. Then the field is switched off. After a relaxation time of $1-2$ minutes, which is long enough to reach the steady state of the spontaneous polarization in SBN [45], the crystal is illuminated with the pump beam. The asymmetric scattering pattern builds up and in accordance to the procedure shown above the spatial distribution of the scattered light is investigated in the steady state.

Figures 9.11 (a), (b) show the experimental results of the electric field dependency of the asymmetry coefficient m_s

$$m_s = \frac{R-1}{R+1} = \frac{\int_{-\frac{\pi}{2}}^{-\delta} I_s(\theta_s)d\theta_s - \int_{+\delta}^{+\frac{\pi}{2}} I_s(\theta_s)d\theta_s}{\int I_s(\theta_s)d\theta_s} \qquad (9.10)$$

as well as of the integrated scattering intensity $\int I_s(\theta_s)d\theta_s$ normalized to its value at zero electric field.

The arrows indicate the sequence of changes of E_o during the experiment. It is obvious from Figure 9.11 (a) that the appearance of the scattering lobe can be switched from negative to positive angles θ_s by the application of an external voltage antiparallel to the direction of the polar axis with amplitudes $E_o > -1.53$ kV/cm. A subsequent increase from -4 kV/cm to positive fields does not change the scattering pattern, i.e. the asymmetry coefficient m_s, until a field of $E_o = +1.2$ kV/cm is exceeded. Then m_s reverses its sign and approaches $m_s = 0.96$ measured at $E_o = +4$ kV/cm. It is obvious that the value E_o at which the m_s-curve crosses the abscissa is different for the descending (left) and ascending (right) parts of the scattering hysteresis. This allows us to determine the coercive field E_c to -1.53 kV/cm and 1.22 kV/cm if the external field is antiparallel and parallel to the initial direction of P_s after the poling of the sample at high temperatures, respectively. At the same time the total scattering intensity shown in Figure 9.11 (b) shows a pronounced minimum when applying an external field antiparallel to the polar axis. Subsequent switching of the field direction displays a slight increase of the scattering intensity at negative fields of $E_o \approx -4$ kV/cm, and a saturation near to zero field. The scattering intensity drops into a second minimum at $E_o \approx +1.3$ kV/cm and then increases again at high positive fields. When the externally applied electric field is finally reduced to zero, i.e. the hysteresis loop is closed, the coefficient m_s remains nearly constant and the total scattering intensity saturates with a value by a factor of 2.2 higher than that at the beginning. The hysteresis-like behavior of m_s demonstrates the ability of photo-induced light scattering to reverse its orientation in space perfectly when an external field is applied to the SBN sample. These features indicate that the domain structure in SBN is disordered at $E_o = E_c$ and the long-range order restores itself in the new direction at $E_o > E_c$. Further, the macroscopic polarization of the crystal increases again, and most of the domains are aligned along the external electric field at $E_o \gg E_c$, where P_s reaches a maximum.

The different values of E_c found on the descending (left) and the ascending (right) parts of the m_s-E_o hysteresis can be attributed to the memory effect appearing in the field-cooled samples. During the poling procedure electric fields are induced in the crystal volume at high temperatures by the external field E_o and then frozen during the cooling process. Such random fields cause a predisposition of the macroscopic polarization to adopt its initial orientation if the crystal then is repeatedly repoled by an alternating external field at low temperatures. A similar memory effect has been observed in measurements of the ferroelectric hysteresis [45].

From the general relation Equation (9.6) between the linear electro-optic effect and the spontaneous polarization of a photorefractive crystal, we have shown qualitatively that the light-induced scattering hysteresis is unambiguously defined by the relevant $P_s(E_o)$ hysteresis. Quantitative consideration of this question requires further development of the model of photorefractive scattering, particularly a separate study of the origin of the initial scattering and the subsequent process of nonlinear amplification (gain coefficient Γ) of the scattered light. In the following section we give a first attempt to the origin of the seed scattering.

9.3.6 Investigating the polar structure

In this section we show that photo-induced light scattering is a powerful tool, especially to obtain information about the polar structure in the crystal bulk. The polar structure in SBN:Ce can be considered as the composition of different periodical and/or quasi-periodical assemblies of ferroelectric 180°-domains distributed in the bulk aligned along the c-axis. The existence of

9.3 Experimental

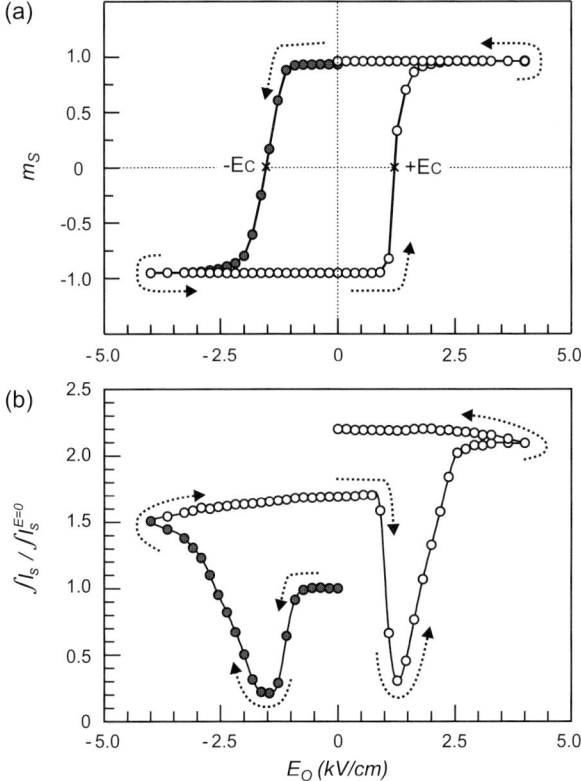

Figure 9.11: Results of the scattering hysteresis experiment with an external electric field E_o: a) asymmetrie coefficient m_s versus E_o; b) total scattering intensity $\int I_s$ normalized on the scattering intensity at zero field versus E_o. The direction of the arrows denote the sequence of the field changes.

such bulk domains has been proven by Fogarty *et al.* [50]. Their size and arrangement depend strongly on the pretreatment of the crystal (such as the poling procedure) and result in an accordant macroscopic polarization [15]. The internal random fields postulated in the RFIM for the explanation of the relaxor-like phase transition in SBN [23] play a decisive role in the formation of the domain structure [51]. Here, it should be noted that a multitude of ferroelectric domains is present even in SBN crystals with maximum macroscopic polarization.

When investigating the polar structure by photo-induced light scattering we assume that the largest contribution to the initial optical noise is due to diffraction of the pump beam on optical inhomogeneities located at boundaries of ferroelectric domains [9]. Figure 9.12 illustrates this concept schematically. Internal electric fields E_i (random fields) yield local perturbations δn of the index of refraction via the linear electro-optic effect $\delta n = -\frac{1}{2}n_e^3 r_{\text{eff}} E_i$.

Such refractive index anomalies are built in along with the domain structure and remain unchanged if the crystal is not influenced externally, e.g., by thermal treatment, optical illumi-

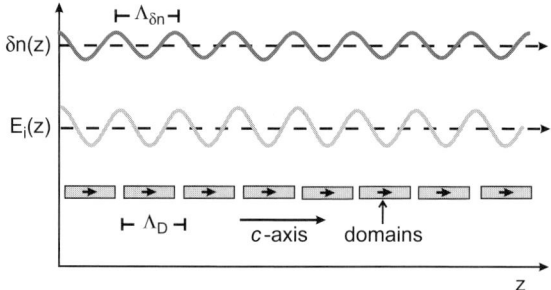

Figure 9.12: Seed scattering at refractive index modulations induced by localized internal random fields via the electro-optic effect. The internal fields are also responsible for the formation of a rich ferroelectric domain structure. Here, a periodic sequence of domains with lengths Λ_D is shown. Note, that the grating period of the refractive index modulation $\Lambda_{\delta n}$ is equal to the lengths of the ferroelectric domains.

nation or application of external fields. The most efficient modulations of the refractive index appear in the $\pm c$-directions, because r_{33} is the largest electro-optic coefficient. Any periodic sequence of domains Λ_D, i.e. any periodic sequence of random fields, will result in a periodic modulation of the refractive index $\Lambda_{\delta n} = \Lambda_D$. The incoming laser beam will be diffracted at this grating, whereby the efficiency depends on the amplitude δn and the angle of diffraction θ_s on the grating period $\Lambda_{\delta n}$:

$$\theta_s = 2\arcsin\left(\frac{\lambda}{2\Lambda_{\delta n}}\right) \tag{9.11}$$

Obviously, there are close connections between such optical noise and the domain structure:

(1) The amplified optical noise in the direction of θ_s corresponds to a refractive index modulation with grating period $\Lambda_{\delta n}$ and, consequently, can be associated with a sequence of ferroelectric domains of length Λ_D. According to the accessible angular range of our setup, structures with spatial dimensions in the range from 0.6 μm to 7.3 μm can be investigated.

(2) The diffraction of the pump beam from the δn-perturbations associated with a large-scale domain structure results in small-angle seed scattering, while δn-perturbations on the assemblies composed from small domains result in seed components propagating at large θ_s (see Equation (9.11)).

(3) Large-scale domains feature a larger spontaneous polarization P_s than small ones. Thus larger amplitudes of δn, and larger intensities of the initial noise result.

(4) The scattering properties of such built-in refractive-index perturbations depend strongly on their spatial regularity. The most efficient scattering occurs if the perturbations are arranged with high regularity. Any deviation or a low spatial regularity will lead to a decrease of the scattered intensity.

9.3 Experimental

Since the symmetry properties of SBN cause the alignment of polar domains either along the $-c$- or the $+c$-direction, we assume that the angular distribution of the seed scattering in these directions is symmetric, as mentioned above. This assumption allows to extract the seed scattering along the c-axis from the corresponding intensity profile shown in Figure 9.7 via the relation:

$$I_{so}(\theta_s) = \sqrt{I_s(+\theta_s)I_s(-\theta_s)}, \tag{9.12}$$

Applying Equation (9.12) to the results of our scattering hysteresis experiment, we receive the seed scattering amplitude I_{so} as functions of the scattering angle θ_s and of the external field (Figure 9.13), respectively.

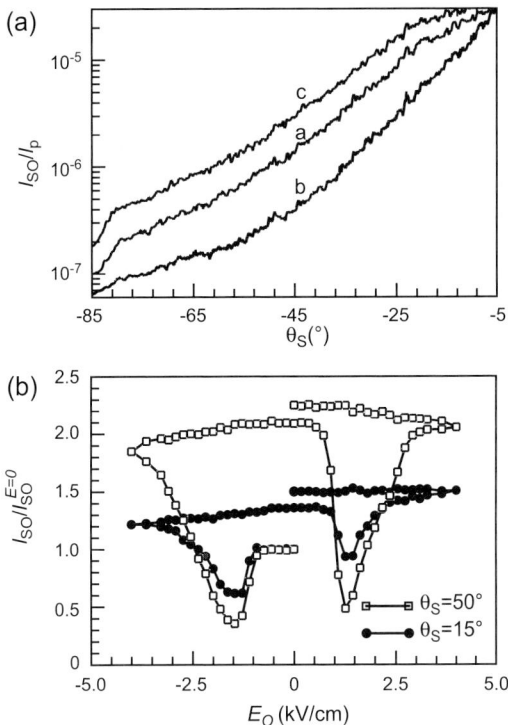

Figure 9.13: (a) Initial scattering amplitude I_{so} as functions of the scattering angle θ_s determined from single spatial intensity distributions (Figure 9.7) after application of electric fields. a: $E_o = 0\,\text{kV/cm}$, b: $-1.5\,\text{kV/cm}$, and c: $-4\,\text{kV/cm}$. The intensity is normalized to the pump beam intensity I_p. (b) Initial scattering amplitude I_{so} as functions of the externally applied electric field E_o for $\theta_s = 50°$ and $15°$, respectively, and normalized to the intensity without external field $I_{so}^{E=0}$.

Figure 9.13 (a) shows the angular distribution of the initial scattering after application of external electric fields with amplitudes $E_o = 0\,\text{kV/cm}$, $-1.5\,\text{kV/cm}$, and $-4\,\text{kV/cm}$. Thus the angular distributions refer to scattering at the beginning, in the vicinity of the coercive field and for saturation fields of the scattering hysteresis (Figure 9.11). The initially scattered light at the beginning of the hysteresis measurement, i.e. without pretreatment by an external electric field, strongly increases with decreasing angle. According to the considerations (1),(2) and (3) of our model small angles are related to large ferroelectric domains featuring a large spontaneous polarization. This in turn leads to a pronounced refractive index amplitude and thereby explains the high intensity of the seed scattering especially at small angles.

After application of an external field of $E_o = -1.5\,\text{kV/cm}$ the initially scattered light decreases over the entire angular range, i.e. the spatial regularity of the built-in refractive-index perturbations is reduced (see consideration (4)). According to our model such reduction occurs, if there is any disorder in the domain structure. Since curves a and b depend nearly similar on the angle over a large range and seem to converge at small angles it is obvious that large domains suffer less disordering than small domains. The seed scattering increases again over the entire angular range if $E_o \gg E_c$ (curve c).

This different contribution of large and small domains to the initial scattering noise especially becomes visible in the dependence of the seed scattering intensity I_{so} on the external field shown in Figure 9.13 (b) for small and large scattering angles, $\theta_s = 15°$ and $50°$, respectively. Two minima in the vicinity of the coercive field are observed, which are caused by the domain disordering at $E_o = E_c$. Incomplete domain reversal at $E_o = E_c$ and redistribution of photoelectrons to new locations during the illumination break the spatial order in the structure of the local fields and consequently in the δn-structures. This distortion of the built-in noisy gratings results in a strong decrease of the seed scattering. When the long-range order reemerges at $E_o > E_c$, it causes a new order for the δn-structures with respect to the modified spatial alignment of domains. According to Figure 9.13 (b), multiple repetition of coherent illumination of the crystal with subsequent application of an external field results in a considerable increase of the seed scattering. This can be explained as follows: An accumulation of photoexcited electrons at positive tips of domains and a respective exhaustion of electrons at negative tips results in an increase of the local charges located on these domains and in an enhancement of the δn-structures. A drain of electric carriers under the external field along domain walls to domain tips yields further changes in local fields. Such redistribution of electric charges reduces the local fields and should assist the further reversal of the 180°-domains. Obviously this will result in a refinement of the polar structure in SBN and therefore to an increase of P_s.

Thus – beside the electrical pretreatment – photo-excited charge carriers play an important role during the determination of the optical hysteresis. Consequently it is a decisive experimental requirement that illumination is performed when a steady state of P_s is reached after the application of an external electric field E_o, and in addition, that the scattered light is detected in the steady state of the light scattering process. Both requirements are met in our experiments.

At this point it is evident that photo-induced light scattering offers the possibility to investigate the behavior of ferroelectric domains of different sizes when passing the phase transition [10]. Figure 9.14 (a) shows the initially scattered light as a function of the temperature for three scattering angles: $\theta_s = 10°$, $30°$ and $60°$. In all cases and comparable to our findings in

9.3 Experimental

Figure 9.9 (b) the seed scattering increases as a function of temperature, reaches a maximum and subsequently decreases.

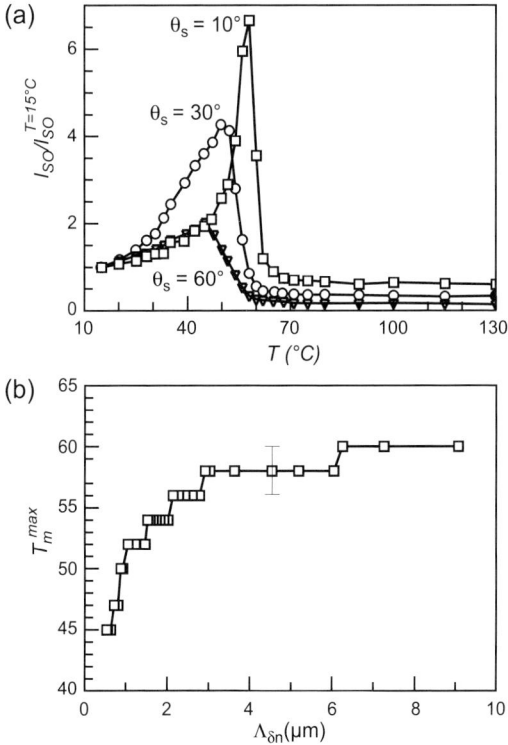

Figure 9.14: (a) Initially scattered light as a function of the temperature measured under three scattering angles $\theta_s = 10°, 30°, 60°$. All spectra are normalized to the initially scattered intensity at $T = 15\,°C$. (b) Temperature T_m^{max} of the maximum of $I_{so}(T)$ versus spatial period $\Lambda_{\delta n}$ calculated from corresponding angles θ_s.

However, the temperature, at which the maximum of the initial scattered light occurs, seems to be related to the scattering angle θ_s and thus to the period $\Lambda_{\delta n}$, respectively. Figure 9.14 (b) shows the correspondence between the temperature T_m of maximum intensity $I_{so}^{\theta_s}$ and the spatial period $\Lambda_{\delta n}$. A spatial disorder of the smallest polar structures occurs at $T_m = 45\,°C$, while the spatial orientation of the largest structures remains stable up to $T_m = 60\,°C$. Such big dispersion of the thermal decay of polar structures over $\Lambda_{\delta n}$ unambiguously illustrates the relaxor behavior of SBN. At the same time it is a key point to understand the bandwidth in the determination of the phase transition temperature T_m in SBN from different methods. For example, in SBN doped with 0.66 mol% Cerium, T_m detected from the maximum of the dielectric permittivity ϵ at 100 Hz (ϵ-method) equals $T_m = 67\,°C$ [20]. Determination of T_m from the inflection point of the spontaneous electric polarization P_3

(P-method) [52] gives $T_m \approx 57\,°C$. The smallest value, $T_m = 52\,°C$, has been received by the photorefractive method (PR-method) [8], where T_m is detected from the maximum of the light-induced scattering distribution. We assume that these differences in T_m result from their sensitivity to different scales of polar clusters when determining the phase transition temperature. According to Figure 9.14 (b), the PR-method responds to a spatial disorder in polar structures with $\Lambda \geq 1\,\mu$m, which can also be considered as the spatial limit for this method. The pyroelectric P-method is sensitive to spatial changes in the range $\Lambda \geq 2.85\,\mu$m. The ϵ-method is the least sensitive method. The largest contribution to the dielectric response always is received from a totally fragmented polar structure with a very high concentration of cluster walls in the crystal volume. Thus it measures noticeable changes only when the polar structure is almost shattered and is not rendered by any data in Figure 9.14.

In conclusion, detailed information on changes in the polar structure of SBN:Ce after application of external electric fields as well as during the ferroelectric-paraelectric phase transition is received from the study of the initially scattered light. Strong hints are revealed that the relaxor-kind phase transition in SBN is due to a dispersion of the thermal decay for different spatial scales in the polar structure.

9.4 Summary

In this work we have summarized the most important properties of photo-induced light scattering in Cerium-doped SBN. We have shown, that this specific kind of scattering phenomenon can be used as comprehensive tool for material investigation. Ferroelectric and optical parameters are revealed. Furthermore the method of photo-induced light scattering offers a deep insight into the polar structure and domain dynamics. It is a contact-free method and therefore an interesting tool in the field of ferroelectrics. All investigations refer to a simplified model based on beam-coupling processes between scattered and transmitted parts of the incident beam. To account for the increasing popularity of methods in materials research these results are presented in the frame of applicability for the determination of material parameters and properties.

Acknowledgement

The authors like to thank Prof. Dr. M. Wöhlecke for fruitful discussion on the relaxor ferroelectric SBN and Dr. R. Pankrath for supplying the samples. We thank K. Bastwöste and S. Möller for experimental support and T. Granzow for discussion on pyroelectric measurements. Financial support by the Deutsche Forschungsgemeinschaft within the projects GRK 695, IM37/2-1, OS55/12-1 and Wo618/3-3, by INTAS within the project 01-0173 and by the Austrian Science fund within the project P-15642 is gratefully acknowledged.

Bibliography

[1] A. Ashkin, G. D. Boyd, J. M. Dziedzic, R. G. Smith, A. A. Ballman, A. A. Levinstein, and K. Nassau, *Appl. Phys. Lett.* **9**, 72 (1966).

[2] J. Feinberg, *Opt. Lett.* **7**, 486 (1982).

[3] H. J. Coufal, D. Psaltis, and G. T. Sincerbox, *Holographic Data Storage*, Springer Series in Optical Sciences, **76**, 2000.

[4] M. P. Petrov, I. A. Sokolov, S. I. Stepanov, and G. S. Trofimov, *J. Appl. Phys.* **68**, 2216 (1990).

[5] N. V. Kukhtarev, V. P. Markov, S. G. Odulov, M. S. Soskin, and V. L. Vinetskiĭ, *Ferroelectrics* **22**, 961 (1979).

[6] V. V. Voronov, I. R. Dorosh, Y. S. Kuz'minov, and N. V. Tkachenko, *Sov. J. Quantum Electron* **10**, 1346 (1980).

[7] M. Goulkov, T. Granzow, U. Dörfler, Th. Woike, M. Imlau, R. Pankrath, and W. Kleemann, *Opt. Commun.* **218**, 173 (2003).

[8] M. Goulkov, M. Imlau, R. Pankrath, T. Granzow, U. Dörfler, and Th. Woike, *J. Opt. Soc. Am. B* **20**, 307 (2003).

[9] M. Y. Goulkov, T. Granzow, U. Dörfler, Th. Woike, M. Imlau, and R. Pankrath, *Appl. Phys. B* **76**, 407 (2003).

[10] M. Goulkov, O. Shinkarenko, T. Granzow, Th. Woike, and M. Imlau, *Eur. Phys. Lett.* **66**, 48 (2004).

[11] A. A. Ballman and H. Brown, *J. Cryst. Growth* **1**, 311 (1967).

[12] P. B. Jamieson, S. C. Abrahams, and J. L. Bernstein, *J. Phys. Chem.* **48**, 5048 (1968).

[13] B. N. Savenko, D. Sangaa, and F. Prokert, *Ferroelectrics* **107**, 207 (1990).

[14] H. Bach and J. Liebertz, *Fortschr. Miner.* **55**, 59 (1977).

[15] A. J. Fox, *J. Appl. Phys.* **44**, 254 (1973).

[16] G. A. Smolenskii and A. I. Agranovskaya, *Fiz. Tverd. Tela* **1**, 1562 (1958).

[17] G. A. Smolenskii, V. A. Isupov, A. I. Agranovskaya, and S. N. Popov, *Sov. Phys. Solid State* **2**, 2584 (1961).

[18] G. Burns and F. H. Dacol, *Phys. Rev. B* **28**, 2527 (1983).

[19] R. Pirc and R. Blinc, *Phys. Rev. B* **60**, 13470 (1999).

[20] J. Dec, W. Kleemann, T. Woike, and R. Pankrath, *Eur. Phys. J. B* **14**, 627 (2000).

[21] P. Lehnen, J. Dec, W. Kleemann, T. Woike, and R. Pankrath, *Ferroelectrics* **240**, 1547 (2000).

[22] W. Kleemann, J. Dec, P. Lehnen, R. Blinc, B. Zalar, and R. Pankrath, *Eur. Phys. Lett.* **57**, 14 (2002).

[23] P. Lehnen, W. Kleemann, Th. Woike, and R. Pankrath, *Phys. Rev. B* **64**, 224109 (2001).

[24] T. Nattermann, V. Pokrovsky, and V. M. Vinokur, *Phys. Rev. Lett.* **87**, 197005 (2001).

[25] S. C. Abrahams, P. B. Jamieson, and J. L. Bernstein, *J. Chem. Phys.* **54**, 2355 (1971).

[26] C. David, T. Granzow, A. Tunyagi, M. Wöhlecke, Th. Woike, K. Betzler, M. Ulex, M. Imlau, and R. Pankrath, *Phys. Stat. Sol.* (a) **201**, R49 (2004).

[27] P. Lehnen, W. Kleemann, Th. Woike, and R. Pankrath, *Eur. Phys. J.* B **14**, 633 (2000).

[28] T. Granzow, Th. Woike, W. Rammensee, M. Wöhlecke, M. Imlau, and R. Pankrath, *Phys. Stat. Sol.* (a) **197**, R2-R4 (2003).

[29] A. M. Prokhorov and Y. S. Kuz'minov, *Ferroelectric crystals for laser radiation control*, Adam Hilger, Bristol, 1990.

[30] T. S. Chang, E. Amzallag, and M. Rokni, *Ferroelectrics* **3**, 57 (1971).

[31] S. Kuroda and K. Kubota, *J. Phys. Chem. Solids* **42**, 573 (1981).

[32] J. Perez, H. Amorin, J. Portelles, F. Guerrero, J. C. M'Peko, and J. M. Siqueiros, *J. Electroceramics* **6**, 153 (2001).

[33] T. Volk, L. Ivleva, P. Lykov, N. Polozkov, V. Salobutin, R. Pankrath, and M. Wöhlecke, *Opt. Mat.* **18**, 179 (2001).

[34] T. Woike, G. Weckwerth, H. Palme, and R. Pankrath, *Sol. State Commun.* **102**, 743 (1997).

[35] J. Wingbermühle, M. Meyer, O. F. Schirmer, R. Pankrath, and R. K. Kremer, *J. Phys.: Condens. Matter* **12**, 4277 (2000).

[36] S. Ducharme, J. Feinberg, and R. R. Neurgaonkar, IEEE *J. Quant. Electron.* **23**, 2116 (1987).

[37] U. B. Dörfler, R. Piechatzek, Th. Woike, M. Imlau, V. Wirth, L. Bohatý, T. Volk, R. Pankrath, and M. Wöhlecke, *Appl. Phys.* B **68**, 843 (1999).

[38] J. Feinberg, *J. Opt. Soc. Am.* **72**, 46 (1982).

[39] K. Buse, U. van Stevendaal, R. Pankrath, and E. Krätzig, *J. Opt. Soc. Am.* B **13**, 1461 (1996).

[40] M. D. Ewbank, R. R. Neurgaonkar, W. K. Cory, and J. Feinberg, *J. Appl. Phys.* **62**, 374 (1987).

[41] R. A. Vazquez, F. R. Vachss, R. R. Neurgaonkar, and M. D. Ewbank, *J. Opt. Soc. Am.* B **8**, 1932 (1991).

[42] P. Yeh, *Introduction to photorefractive nonlinear optics*, John Wiley, New York, 1993.

[43] K. Buse, *Appl. Phys.* B **64**, 273 (1997).

[44] J. R. Oliver, R. R. Neurgaonkar, and L. E. Cross, *J. Appl. Phys.* **64**, 37 (1988).

[45] T. Granzow, U. Dörfler, Th. Woike, M. Wöhlecke, R. Pankrath, M. Imlau, and W. Kleemann, *Phys. Rev.* B **63**, 174101 (2001).

[46] Th. Woike, T. Granzow, U. Dörfler, Ch. Pötsch, M. Wöhlecke, and R. Pankrath, *Phys. Stat. Sol.* (a) **186**, R13 (2001).

[47] L. E. Cross, *Ferroelectrics* **76**, 241 (1987).

[48] A. S. Bhalla, R. Guo, L. E. Cross, G. Burns, F. H. Dacol, and R. R. Neurgaonkar, *Phys. Rev.* B **36**, 2030 (1987).

[49] V. Wirth, *Temperaturabhängige elektrooptische und elektrostriktive Untersuchungen an Kristallen mit ferroischen Phasenumwandlungen*, Dissertation Univ. Köln (1999).

[50] G. Fogarty, B. Steiner, M. Cronin-Golomb, U. Laor, M. H. Garrett, J. Martin, and R. Uhrin, *J. Opt. Soc. Am.* B **13**, 2636 (1996).

[51] A. G. Chynoweth, *Phys. Rev.* **117**, 1235 (1960).

[52] T. Granzow, U. Dörfler, Th. Woike, M. Wöhlecke, R. Pankrath, M. Imlau, and W. Kleemann, *Appl. Phys. Lett.* **80**, 470 (2002).

10 Ferroelectric Domain Breakdown: Application to Nano-domain Technology

G. Rosenman[1], *A. Agronin*[1], *D. Dahan*[1], *M. Shvebelman, E. Weinbrandt*[1], *M. Molotskii*[2], *and Y. Rosenwaks*[1]

1. School of Electrical Engineering-Department of Physical Electronics and
2. The Wolfson Materials Research Center, Tel Aviv University, Ramat-Aviv, 69978 Israel

Abstract

The major trends in ferroelectric photonic and electronic devices are based on development of materials with nanoscale features. Piezoelectric, electrooptic, nonlinear optical properties of FE are largely determined by the arrangement of ferroelectric domains. A promising way is a modification of these basic properties by means of tailoring nanodomain and refractive index superlattices.

The physical size of spontaneously polarized regions and modern technological requirements of lateral domain dimensions scale down to 100 nm in millimeter thick ferroelectric bulk crystals; this implies a development of a new approach both in instrumentation and physics of polarization reversal. We review in this paper a new and novel tool for nanodomain tailoring and technology in ferroelectric bulk crystals - high voltage atomic force microscopy. It is shown that application of the switching voltage in the kV range to a nanometer size switching electrode leads to new physical mechanism of polarization-reversal ferroelectric domain breakdown. In this process string-like domains of nanometer radius penetrate into hundreds of micrometers depth of ferroelectric crystals. Experimental results on nanodomain reversal and theory of breakdown phenomenon are analyzed and discussed in details. Pronounced domain breakdown phenomenon has also been observed in $LiNbO_3$ bulk crystals subjected to high-energy electron beams. C^--polar face of the $LiNbO_3$ crystal coated by thick amorphous dielectric layers protects penetration of the beam electrons into the ferroelectric crystal resulting in formation of immobile electron drops. The electron charge localized in a small volume creates high intensity electric fields and causes domain reversal in a nanometer scale with a very high ratio between the domain radius and its length.

A domain shape invariant is defined; this allows obtaining unambiguous criteria for string-like nanodomain generation resulting in a figure of merit of a ferroelectric crystal applicable for nanodomain technology. It is shown that such a FE should possess high spontaneous polarization, low density of domain wall surface energy and small lateral dielectric permittivity.

We report on the application of atomic force microscopy tip arrays for nanodomain engineering in ferroelectric crystals. Using a ten-tip array, it is shown that nanodomain writing results in a regular domain grating that penetrates throughout the bulk crystal as in the case of single tip writing. The developed nanodomain reversal techniques and understanding of physical mechanisms paves the way to new technologies for fabrication of various nanodomain configurations for advanced optoelectronic and microelectronic devices.

10.1 Introduction

Reversible spontaneous electrical polarization observed in ferroelectric (FE) materials made them attractive for development of new generation of microelectronic and optical devices. The field of microelectronic applications is mainly based on FE thin films where switchable polarization in small homogenously polarized regions (FE domains) is the basis for nonvolatile random-access-memories [1]. Another wide application of FE domains is related to nonlinear optics where devices are fabricated in bulk FE crystals. In this case specifically tailored 1-D domain configurations (periodic, aperiodic, quasiperiodic) allow development of a new generation of lasers in the visible and infrared spectral regions for novel TV and electronic cinema, fine gas spectroscopy and functional devices for telecommunication [2, 3]. Recently, some works have been devoted to nonlinear frequency conversion in 2-D-graded nonlinear photonic FE crystals [4–6], which may be applied for multiple nonlinear optical interactions in multiple directions. All these applications are based on micrometer scale domains and domain configurations. Rapid development of high-density computer memory and next generation of domain-based photonic devices need tailoring ferroelectric domains in a nanometer scale [7].

The key issue of the nanoscale technology is the development of a new technology for direct domain writing with either single or an array of nanometer-size switching electrodes. Small tips of Atomic Force Microscope (AFM), an array of tips or electron drops formed by electron beam may serve as a switching mobile nanoelectrode. Domains and domain matrices in nanoscale were tailored in FE thin films by the use of AFM [8] and electron beam [9–11]. The recently developed high voltage AFM (HVAFM) [12] was applied to FE bulk crystals for nanoscale domain reversal and fabrication of one- and two-dimensional nanodomain gratings [13, 14].

A switching electrode having nanometer dimensions leads to profound difference in poling of FE thin films and FE bulk crystals. In the case of FE thin films with a thickness of about 100 nm, a voltage application between the bottom electrode and an AFM tip of 50–100 nm radius forms a quasi-homogeneous electric field. Such a distribution implies a conventional polarization reversal setup as in the case of uniform switching electrodes applied to polar faces of a FE sample.

The electric field magnitude and distribution strongly changes when point-like switching electrodes of dozens nanometer-size is applied to FE bulk crystals. It was shown [13, 14] that an application of a switching voltage in the kilovolt range to the HVAFM tip generates super high electric fields of around 10^8 V/cm in the vicinity of the tip apex, which is by three to four orders of magnitude higher than the field that may be applied by the use of the conventional polarization switching setup. This field rapidly decreases in the crystal bulk and reaches a negligible value at a depth of dozens of micrometers. Another distinguished feature of a nanometer-size switching electrode application to FE bulk crystals is a strong limitation of the external screening process to the depolarization field. The limited screening of the depolarization field and strong inhomogeneous electric field changes the physical mechanism of polarization reversal at the nanometer scale and leads to a new phenomenon [13, 15] when nucleated domains of nanometer scale radius penetrate in practically zero external electric field through hundreds of micrometers of FE crystal bulk in the form of string-like domains [13–15]. *This phenomenon was named Ferroelectric Domain Breakdown (FDB) because such domains resemble electrical breakdown channels.*

In this paper we present a comprehensive experimental and theoretical description of nanodomain reversal in FE bulk crystals: an experimental method for domain switching using HVAFM, results on nanodomain switching using HVAFM and under indirect electron beam exposure, theory of FE domain breakdown and our last development of the fabrication of nanodomain gratings by the use of multiple tip arrays. We show that FDB is a new physical phenomenon observed in bulk FE crystals, and is the basis for the development of nanodomain technology in bulk FE crystals. This technology is required for future photonic, acoustic and microelectronic devices.

10.2 Nanodomain size limitations

Three physical and technology-related fundamental key issues of FE nanodomain reversal and fabrication of nanodomain structures in FE thin films and FE bulk crystals are under theoretical and experimental consideration: (I) technological requirements and the minimal size of FE domains (II) experimental technique for nanodomain reversal and (III) physics of nanodomain switching in FE films and bulk crystals.

10.2.1 Technological demands for FE domain-based devices

Currently developed 64 Mb FE thin film memories contain elementary cells (FE capacitor and transistor) in the micrometer scale. Prospective ultrahigh density 10 Gbit FE storage devices require lateral cell dimensions below 100 nm. The current domain engineered structures tailored in FE bulk crystals for nonlinear optical devices by electrical poling [3] and electron beam methods with domain sizes in the 2–20 μm range provide nonlinear optical interaction in the visible to infrared ($\lambda \sim 425$–6000 nm) spectral regions. Despite the intensive research and development in this field during the last 10 years, to the best of our knowledge deep domain-patterned structures are limited to 2 μm wide domains [3, 16]. This is caused by several physical and technological limitations like: ferroelectric domain nucleation process [17, 18], tangential domain broadening effect [19], anisotropy of domain wall velocity [20], etc.

Nanometer scale domain configurations in FE bulk crystals pave the way for a new class of photonic devices. As an example, preliminary calculations show that a UV laser ($\lambda = 300$ nm) based on second harmonic generation in LiTaO$_3$ crystal requires a periodic nanodomain superlattice with domain widths of around 700 nm. In addition, the current domain gratings in ferroelectric crystals are suitable only for quasi-phase-matched nonlinear interactions in the forward direction, where the pump and generated beams propagate in the same direction. Sub-micron ferroelectric domain gratings are the basis for a new family of devices based on backward nonlinear quasi-phase-matched optical interactions in which the generated beam travels in a reverse or another non-collinear direction to the incident beam. Non-collinear quasi-phase-matched second harmonic waves with large angles between the interacting waves was generated in an RbTiOPO$_4$ bulk crystal with nanodomain gratings period of 590 nm [21].

New promising technologies for future electron-beam lithography applications based on pyroelectrically induced electron emission from LiNbO$_3$ ferroelectrics [22] were recently proposed [23]. The developed system possessing micrometer scale resolution used 1:1 electron beam projection. The needed electron pattern was obtained by means of deposited micrometer-size Ti-spots on the polar face of LiNbO$_3$. Another solution for the high resolution electron lithography may be found in nanodomain patterning of a ferroelectric template.

Another type of FE nanodomain-based devices are refractive index superstructures. Desired configuration of refractive index superlattices may be achieved by selective dry or wet etching of the preliminary fabricated domain structure. The ability to selectively etch ferroelectric domains with opposite direction of spontaneous polarization is a well-known phenomenon. This method allows converting the domain superlattices to refractive index superlattices and developing different kinds of patterned structures with high refractive index contrast. Optoelectronic applications based on refractive index superlattices (photonic band gap crystals, rib waveguides, tunable wavelength filters) should have feature sizes of around $\lambda/2$. Such devices operating in the spectral range of 200–2000 nm will require domain widths of 100–1000 nm. It should be stressed that to date almost all the available domain engineered superlattices are based on two-dimensional domain structures composed of sheet-like (in analogy to multi-quantum well structures) domains. However, some of the advanced devices, photonic band gap crystals for example, will require one-dimensional (strip-like or quantum wire type) domain configurations. This short review shows that the next generation of FE domain-based optical devices such as UV lasers in LiTaO$_3$ crystals, backward optical radiation FE converters, refractive index engineered devices and FE photonic crystals require a domain fabrication technology in sub-micron region scaling down to $r \sim$ 100–700 nm. Thus the advanced FE domain-divided devices fabricated both in thin films and bulk crystals imply development of a new approach allowing dense patterning of FE templates in the nanometer scale.

10.2.2 Physical limit of domain dimensions in FE

The spontaneously polarized FE state is based on alignment of elementary dipoles along the polar axis within a correlation volume. This electrically ordered phase exists in a FE due to electrostatic dipole-dipole interactions. Such a long-range interaction is highly anisotropic resulting in the anisotropic correlation volume which permits formation of long dipole chains in the polar direction P_s. The force elongating the dipole string along P_s is much stronger than that in the direction normal to P_s. So far the correlation volume defines the ultimate limit of the physical size of thermodynamically stable homogeneously polarized regions with length l along P_s and its minimal radius ρ. It is obvious that aforementioned anisotropic considerations imply that $\rho < l$. The characteristic size l of the polar phase in FE Pb(Zr$_{0.5}$Ti$_{0.5}$)O$_3$ was calculated by the use of the mean-field Landau-Ginzburg-Devonshire theory in a three-dimensional base [24]. The theoretical prediction of the critical thickness l was around $l \sim 200$ Å. Studies undertaken recently showed consistent experimental data. Single FE lead titanate grains were fabricated by chemical solution deposition. The critical size of the FE particles was found to be around $l \sim 4$–14 nm [25]. The FE polarization was measured by the use of the combination of electrostatic force microscopy and piezoelectric microscopy in films down to a thickness of

$l \sim 4$ nm [26]. It was also observed that a FE PbTiO$_3$ film as thin as 1.6 nm thick (two unit cells) still exhibits ferroelectric behavior [27].

Another classical factor affecting the domain lateral dimension ρ is a depolarization field energy. Minimization of this energy occurs by splitting of the FE monodomain sample into domains with opposite direction of P_s having the lateral size ρ. The striped domain structure found in the FE film had domain features with $\rho = 3.7$–24 nm [27]. These data show that the physically limited lateral domain size ρ is significantly smaller than technologically defined lateral elementary FE cell dimensions r, $\rho \ll r$. Thus spatial resolution of tailored nanodomain structures is mainly limited by the applied experimental technique. In the next sections we show that the physical mechanism of polarization reversal at the nanoscale in FE bulk crystals changes dramatically.

However in FE bulk crystals nanodomain reversal is governed by mechanisms of the FDB effect. We show below that string-like domain formation under high voltage applied to the tip-electrode of the HVAFM possess domain shape invariant where the aspect ratio (relation between domain radius and its length r/l) depends on basically the FE material parameters: spontaneous polarization, lateral dielectric permittivity and density of domain wall surface energy. The domain shape invariant allows to obtain unambiguous criteria for nanodomain generation resulting in a figure of merit of a ferroelectric crystal applicable for nanodomain technology. Such a ferroelectric should possess high spontaneous polarization, low density of domain wall surface energy and small lateral dielectric permittivity.

10.3 AFM nanodomain tailoring technology

10.3.1 Nanoscale switching electrode

The classical experimental setup developed for FE polarization reversal implies a single-domain FE sample sandwiched between two electrodes [28]. While conventional domain inversion techniques use equal sized electrodes covering the polar faces of FE templates, nanodomain inversion occurs under totally different conditions when the bottom electrode is a uniform plate and the upper one is a point contact. Two different kinds of the upper switching mobile nanoelectrodes may be considered: AFM tip (and/or array of tips) and electron drop formed using electron beam exposure. When a voltage stress is applied to the nanoelectrode, both the electric field intensity and its spatial distribution strongly differ in FE thin films (thin FE crystals) and bulk FE crystals.

Commonly used AFM devices, having a tip radius of curvature of around $R \sim 50$–100 nm used for nanodomain reversal in FE 100 nm thick films, may be considered as a conventional polarization switching setup [28]. It generates a quasi-homogeneous electric field in the FE film providing polarization reversal in accordance with the classical scenario [28,29]. Switching voltage stress of several volts ($V < 10$ Volts) applied to the AFM tip allows to fabricate nanosize ferroelectric domains in FE thin films [30–32]. This low voltage AFM technique was recently applied to thin monodomain FE crystals such as near stoichiometric 5 μm thick LiNbO$_3$ [33] and congruent 70–150 nm thick LiTaO$_3$ crystals [34, 35]. High density arrays with a domain size $R \sim 10$ nm were tailored in congruent LiTaO$_3$ crystals applying 10–15 V to the AFM tip; this rewritable nanodomain structure was developed towards ultrahigh memory storage system reaching 1 Tbit/in^2 [34, 35].

The switching electrode of the same radius of curvature creates a new physical process in FE bulk crystals. A 0.5 mm thick FE crystal exceeds the tip-electrode radius $R \sim 50$ nm by a factor of 10^4. As a result, the switching electric field generated in the vicinity of the apex of the point-like AFM tip is very high, its value reaches $E \sim 10^5$ V/cm for low voltage $V \sim 10$ V and it may be as high as 10^7–10^8 V/cm when a high voltage stress in kV range is applied [13–15]. The most pronounced difference is a switching field distribution in FE bulk crystals compared to FE thin films and thin FE crystals. The field intensity decreases from the very high value at the polar face of a FE sample to almost zero at the depth of about dozens of micrometers (Figure 10.1) [13–15].

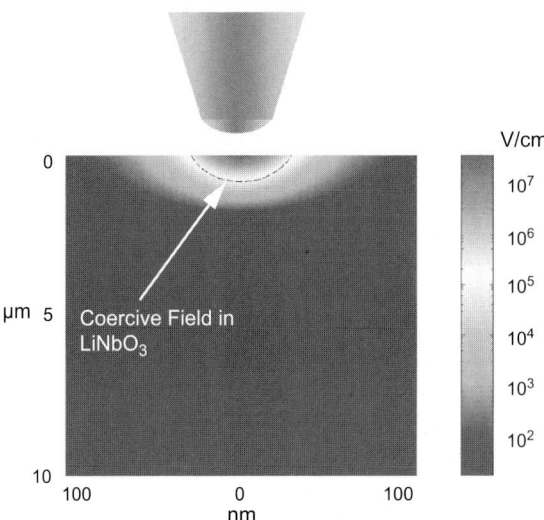

Figure 10.1: Calculated electric field distribution in FE bulk LiNbO$_3$ crystals, tip radius of 50 nm, and applied voltage of 3 kV.

Another factor that influences strongly the evolution of polarization reversal in FE bulk crystals is the limited minimization of the depolarization field by means of screening charges. In the case of the AFM tip of $R \sim 50$ nm the screening charge is limited both by the size of the tip and its non-ohmic contact with the sample surface. Such an effect is especially pronounced in the case of indirect electron beam exposure method described in this paper. The strong limitation of the screening phenomenon provides a necessary condition for the observed effect of FDB [13–15].

10.3.2 Low and high voltage AFM for nanodomain reversal in FE bulk crystals

Low voltage AFM was used for nanodomain reversal in different FE bulk crystals [36–39]. Ferroelectric GASH crystals (50–200 μm thick) were subjected to a pulsed voltage (-5 V to +5 V) applied using a conductive AFM tip [40]. No domain growth throughout the crystal bulk was observed; the reversed domains were circular with cross-sections of 100–150 nm in diameter. Pulses of 10 V applied to 500 μm thick BaTiO$_3$ crystals resulted in spike-like domains of opposite polarization propagating only 5–10 μm deep into the crystal bulk [38]. The domains were very unstable and they disappeared after a short time. Much higher voltage was used for domain reversal of 8 μm thick BaTiO$_3$ crystals [41]; the applied voltage pulses (10 μs in duration and 80 V amplitude) caused a domain reversal to a depth of about 100 nm; the domain lifetime was only two days [41]. So far the commonly used low voltage AFM did not allow to tailor a long-term stable through domains in bulk FE crystals in nanometer scale.

A breakthrough in the field emerged with the development of the high voltage atomic force microscope HVAFM [12]. Diverse stable domain configurations were fabricated in several FE crystals like LiNbO$_3$ (LN), RbTiOAsO$_4$ and RbTiOPO$_4$ (RTP) [13, 15]. The application of the switching voltage in the kV range to a point-like AFM tip generates an extremely high electric field exceeding the conventional switching fields by 2–3 orders of magnitude. These super-high electric fields change drastically the domain nucleation process making it activationless, induce a string-like domain penetration from the ferroelectric surface deep into the crystal bulk, a process which is governed by the FDB effect [13, 15].

The High Voltage AFM was produced by the modification of a commercial AFM (Auto-probe CP, Veeco, Inc.) including a specifically designed stage with sample-holder and high voltage module. The latter allows to apply both dc and pulse bipolar repolarization voltages larger than 10 kV [12]. For HVAFM which is schematically shown in Figure 10.2, simple commutation converts the HVAFM to the commonly used AFM where the reversed domains are imaged by the use of well-known contact and noncontact modes. A heavily boron-doped silicon cantilever with a tip having a radius of curvature not larger than 50 nm was used for domain tailoring. Two different modes of domain tailoring were employed. Local polarization reversal was performed by applying a pulsed high voltage (both single pulses and pulse trains) to a stationary HVAFM tip. A dc voltage was used in the fabrication of the strip-like domain gratings where the domains were written by a scanning AFM tip.

Diverse domain configurations (including one- and two-dimensional structures) were tailored in monodomain LiNbO$_3$ (congruent composition), RbTiOAsO$_4$ and RbTiOPO$_4$ ferroelectric plates 150–250 μm thick. Figure 10.3 shows individual ferroelectric domains tailored in 150 μm thick LiNbO$_3$ crystal by application of single rectangular voltage pulses ($U = 3.2$ kV, $\tau \sim 40$ msec). The domains were imaged using a piezoresponse force microscopy [42, 43] mode where an AC voltage $V = V_{ac} \sin \omega t$ with $V_{ac} = 15$ V, $\omega = 1.5$ kHz, was applied between the HVAFM tip and the bottom electrode. Domains of hexagonal and triangular forms were observed; the triangular domains are strongly oriented towards the Y-axis of the crystal. Chemical etching revealed the tailored domains at the bottom side of the sample indicating a complete penetration of the reversed domains through the sample. This experiment demonstrates a very low domain ratio $r/l = 9.3 \times 10^{-3}$. Decreasing the applied voltage leads to reduction of the domain radius r and its depth of penetration l.

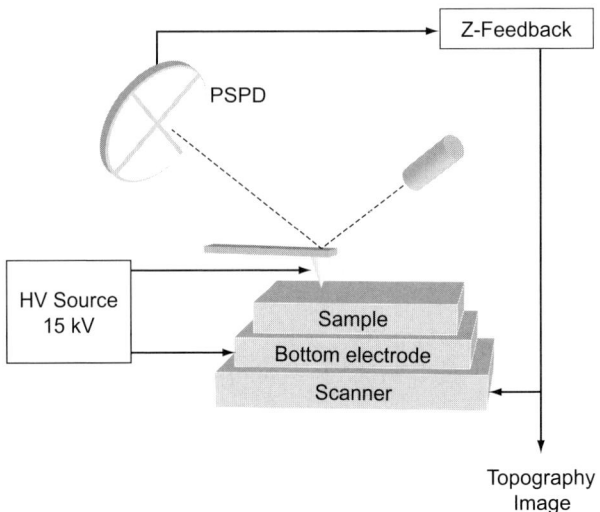

Figure 10.2: Schematic of the high voltage atomic force microscope (HVAFM) setup.

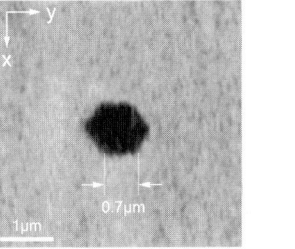 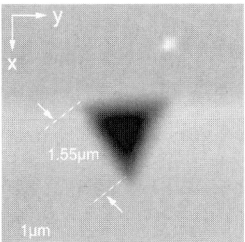

Figure 10.3: Hexagonal and triangular domains tailored in congruent LiNbO$_3$ crystals by HVAFM.

Detailed studies of the domain shapes and their dimensions as a function of the applied high switching voltage showed generation of domains around 200–300 nm [44]. Figure 10.4 demonstrates two sorts of nanodomain gratings. The first grating (domain period is 410 nm) was tailored in a LiNbO$_3$ crystal by application of dc voltage to an HVAFM tip ($U = 2.0$ kV). Applied chemical etching is needed for integrated optical device structures with a periodic refractive index. Another domain grating (domain period is 1180 nm) was fabricated in the RbTiOPO$_4$ crystal for a non-collinear quasi-phase-matched nonlinear optical converter.

These nanodomain superlattices were written by moving the HVAFM tip along the polar Z$^+$-face with a velocity of around 50 μm/sec. The width of the tailored strips was greatly affected by the tip velocity and was measured to be between 50 and 1500 nm; a larger tip velocity resulted in narrower domains.

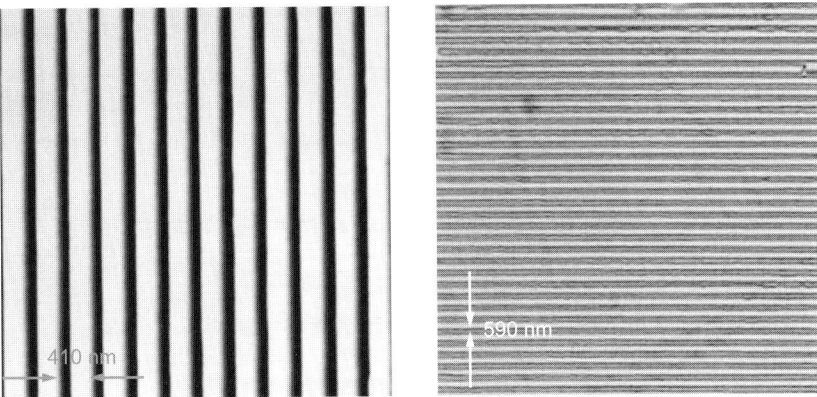

Figure 10.4: Nanodomain grating (domain period is 410 nm) tailored for integrated optical device in LiNbO$_3$ crystal by application of dc voltage ($U = 2.0$ kV). (b) Domain grating (domain period is 1180 nm) fabricated in the RbTiOPO$_4$ crystal for non-collinear quasi-phase-matched nonlinear optical converter.

Figure 10.5 shows a two-dimensional domain structure tailored in a RbTiOPO$_4$ single crystal. A first one-dimensional domain configuration was fabricated in a RbTiOPO$_4$ crystal using a high voltage stress of about 0.5 kV between the tip and the sample by moving the HVAFM tip first along the Y-axis; then a voltage of the opposite polarity was applied to the tip and scanning was performed along the X-axis. The polarization reversal field of the opposite polarity inverted the previously tailored domains only where the tip crossed the Y-oriented strips (Figure 10.5). The resulting structure represents a one-dimensional $1 \times 1\,\mu\mathrm{m}^2$ domain array. Measurements of the fabricated domain configurations one month later resulted in an identical image; this is an indication of a very stable domain structure. Thorough inspection of the tailored individual domain and domain superlattices performed by optical microscopy following chemical etching at the opposite polar face showed complete penetration of the domains throughout the crystal bulk without any significant change in their width.

Figure 10.5: Two-dimensional domain grating tailored in a RbTiOPO$_4$ crystal.

10.3.3 Indirect electron beam induced ferroelectric domain breakdown

In recent years there has been a great interest in electron beam (EB) lithography methods for periodical poling of FE bulk crystals since it is a well established processing tool which provides a high lateral resolution ability controlled by a computer aided design application. It has been reported that EB scanning of the C^--polar face of $LiNbO_3$ induced domain inversion at room temperature [45–47]. EB was also used to invert domain regions with microscale periodicity in other bulk FE crystals like $KTiOPO_4$, $Ti:LiNbO_3$, and $LiTaO_3$ [48–50]. This technique provided stable domain structures with inverted regions penetrating throughout the entire crystal. It should be stressed that despite the long-term studies there is no model describing this effect. We show in this section that the EB-induced domain reversal represents a pronounced FDB effect where FE domains of small lateral size propagate through hundreds of micrometers of FE bulk crystals in almost zero electric field created by electron drop. EB irradiated structures were found to be space limited [51], and with surface cracking at high EB exposure energies or high electron current densities [48]; in addition, domain shapes had poor consistency and reproducibility. It was difficult to reverse domains to be formed in an ideal linear pattern [52] and the domain widths became larger and/or dot-shaped in the crystal depth at the C^+-face compared with the exposed C^--plane [4, 52, 53]. It was also mentioned that etching the crystal before EB exposure might result in much more regular and reproducible domain structures [54]. From technological point of view nanosize domains were not reported before. In Refs. [46, 47] it was found that there is a low limit for domain grating periodicity of 2.5 μm. To the best of our knowledge most of the recent works in the subject still experience the same difficulties as mentioned above [55]. To overcome the hurdles presented by conventional EB poling, it was reasonable to immitate the High Voltage AFM (HVAFM) system, where a mobile active center, the AFM tip (tip array) with an apex of nanometer radius, creates an enormous highly inhomogenous field at the vicinity of the FE crystal surface. This active center induces domain reversal under its location; such domains tend to favor string-like formations in accordance with the FDB phenomenon [13–15]. A simple experimental setup was proposed; a thin dielectric layer with a large deep trap concentration, such as an amorphous photo-resist material [56, 57], was spin-coated on the C^--face of $LiNbO_3$ (Figure 10.6). The thickness of the dielectric layer is carefully selected so that the majority of the incident electrons would stay trapped in the photo-resist layer and create localized electron-drops active centers above the FE crystal, analogously to the AFM tip (tip array).

Calculations of the incident electron penetration depth into the dielectric layer is a well understood phenomenon [58, 59]; in recent years many Monte Carlo simulation tools were developed to study it. In our case it allowed an easy calculation of the photoresist layer thickness for different exposure parameters of the EB. For example if the EB exposure is done with $V_0 = 15$ kV accelerating voltage and the dielectric layer is selected to be polymethylmethacrylate (PMMA), we estimate the penetration depth by Kanaya and Okayama's [58] expression:

$$R(\mu m) = \frac{0.0276 A V_0^{1.66}}{Z^{0.89} \rho} \tag{10.1}$$

Where V_0 is the accelerating voltage in kV, A is the average atomic weight in g/mol, ρ is the density in g/cm^3 and Z is the average atomic number. For PMMA having the parameters

10.3 AFM *nanodomain tailoring technology*

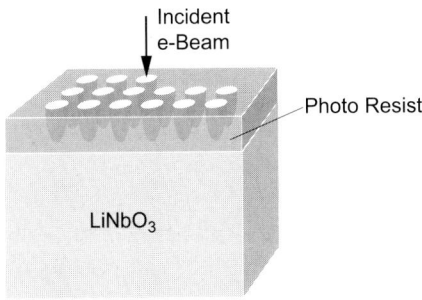

Figure 10.6: An experimental setup for localizing active charge centers (electron drops) in the amorphous dielectric layer to induce FDB effect in the LiNbO$_3$ crystal.

$A = 6.7\,\text{g/mol}$, $Z = 3.6$ and $\rho = 1.2\,\text{g/cm}^3$ [60] the penetration depth is $R \approx 4.4\,\mu\text{m}$ for $V_0 = 15\,\text{kV}$. Alternatively the penetration depth may be found by using Monte Carlo simulations as shown in Figure 10.7.

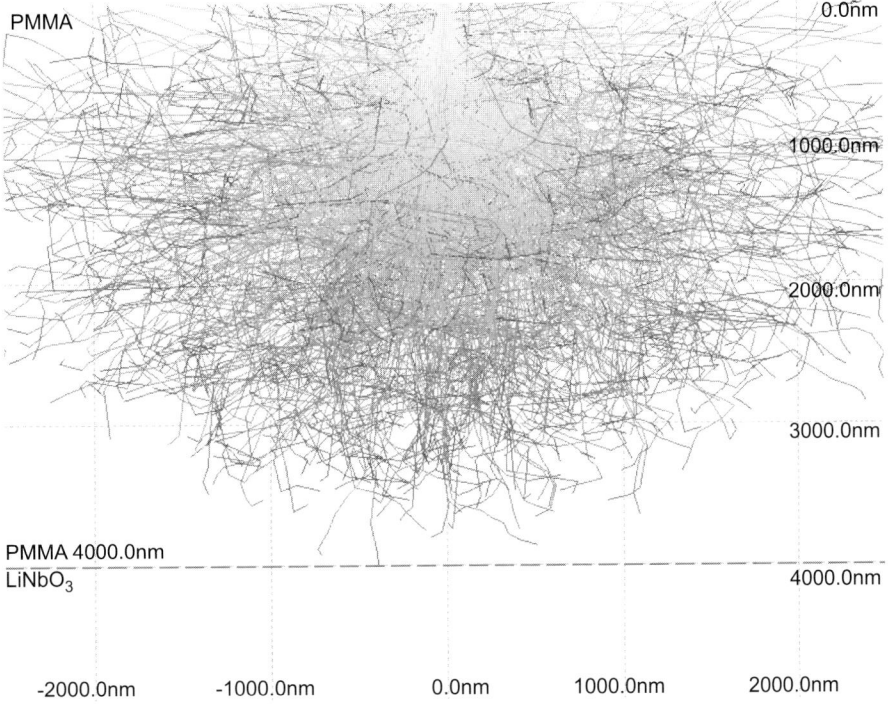

Figure 10.7: Monte Carlo simulation of EB irradiation with an accelerating voltage $V_0 = 15\,\text{kV}$.

It can be seen that most of the electrons are localized in the 3.5 μm thick layer and only a few of them penetrate to a depth of 3.8 μm. In our experiments a 0.5 mm thick LiNbO$_3$ crystal was spin-coated with 3.5 μm thick photoresist layer (Shipley 1818). The prepared sandwiched structure was exposed using a commercial EB lithography system (ELPHY Plus) adapted to a JEOL JSM 6400 scanning electron microscope on the C$^-$-face of the LiNbO$_3$ sample under various accelerating voltages and surface charge densities, as shown in Table 10.1.

Parameter	Value
Accelerating Voltage V_0	$8 \sim 20$ kV
EB Current	0.5 nA
Charge Density η	$100 \sim 400$ μC/cm^2
Scanning Step Size	0.1 μm

Table 10.1: Electron beam exposure parameters

At the next step the photoresist layer was removed and following an HF etching had revealed highly regular clustered 2-D domain structures as seen in Figure 10.8.

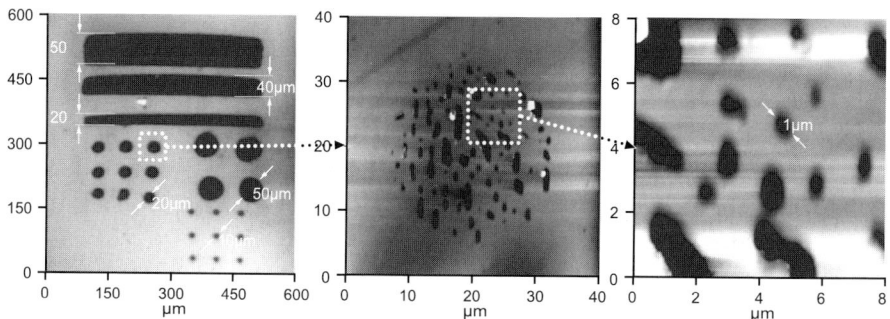

Figure 10.8: FE domains induced by indirect EB writing in a LiNbO$_3$ crystal ($V_0 = 15$ kV, $\eta = 300$ μC/cm^2).

Further in-depth analysis showed that these clusters grow into the LiNbO$_3$ bulk as string-like domain groups rather than bond into a single domain. For $V_0 = 15$ kV the domain clusters penetrated 350 μm deep into the bulk. Even more amazingly, when the beam accelerating voltage was $V_0 = 20$ kV, some of the domains exhibited "cracks" on the surface as seen in Figure 10.9, and bonded into single domains inside the bulk, just like in direct EB exposure when the FE crystal is irradiated directly by the EB. The reason for this is that significant part of the electrons penetrates through the thin photo-resist layer under higher accelerating voltage as if it had not existed.

It was also found that there is a close correlation between the accelerating voltage and surface charge density to the penetration depth of the cluster and its nanodomain density. For

10.3 AFM nanodomain tailoring technology

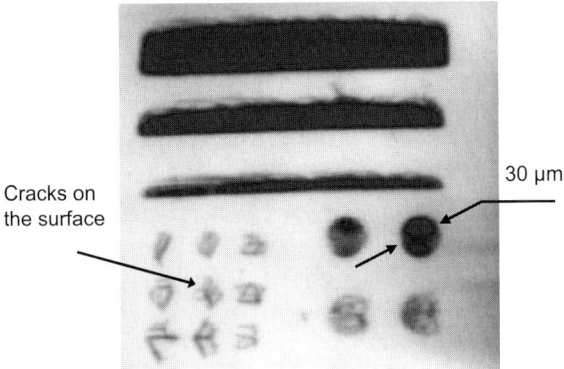

Figure 10.9: Cracks exhibited by indirect EB writing in a LiNbO$_3$ crystal with $V_0 = 20$ kV and $\eta = 200\,\mu$C/cm^2.

example when using smaller accelerating voltages such as $V_0 = 8$ kV the domain clusters were less dense and disappeared in the depth of 15 μm. More complex domain configurations were also produced for optical second harmonic generation measurements on $\lambda = 1542$ nm wavelength, such as "chess board" domains and periodic circles over large areas like $800 \times 800\,\mu^2$m, (Figure 10.10).

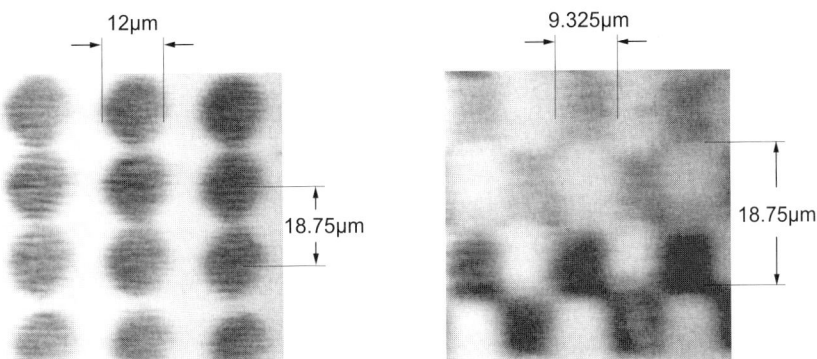

Figure 10.10: Optical micrographs of etched domain-inverted patterns on the C$^-$-face, exposed to electron beam $V_0 = 15$ kV and $\eta = 300\,\mu$C/cm^2. (a) Circles with 12 μm diameter and period of 18.75 μm, (b) chess board pattern with $9.325 \times 9.325\,\mu^2$m squares and period of 18.75 μm.

Thus two different EB methods may be considered. The first method is direct EB domain writing when uncoated FE crystal is directly subjected to the electron beam [45–48]. The previously published data and our results on direct EB domain tailoring showed that inversed domains in LiNbO$_3$ crystals occupy micrometer-size homogeneously polarized areas.

Another method presented in this paper is the indirect EB method when the C^--face of a LiNbO$_3$ ferroelectric is preliminary coated by a highly defective layer of the amorphous photo-resist material PMMA. The thickness of this dielectric layer is large enough to protect the LiNbO$_3$ from penetration of high energy electrons into the bulk. In the presented calculations and simulation a very limited number of electrons penetrated into the LiNbO$_3$ crystal, so most of the injected electron charge remains trapped in the PMMA layer.

The most distinguished feature of this indirect EB domain tailoring method is that induced domain clusters are not homogenously polarized regions as it occurs in the case of direct injection of electrons inside the LiNbO$_3$ crystal. In the experimental data (Figure 10.8) the inverted regions represent domain cluster consisting of individual spatially separated string-like domains of nanometer size radius and hundreds of micrometers length providing high domain aspect ratio $r/l \sim 5 \times 10^{-4}$ which is close to the value found for nanodomains reversed in LiNbO$_3$ by HVAFM.

In the case of the indirect EB method the electron charge is localized in a small region at trap levels of the upper dielectric amorphous layer. Our calculations showed that such localization of electrons led to the formation of a dense electron quasi-drop with a radius about 1 μm. The immobile electron drop generates a high inhomogeneous field; according to Ref. [61] the electric field under such a dielectric coating subjected to the electron irradiation may reach 10^7 V/cm and higher. This field stress is close to the field generated by the HVAFM and may cause domain inversion in bulk FE crystals. The immobile electron-charge quasi-drop formed in the amorphous layer serves as the sharp tip of the HVAFM. It cannot contribute to the screening of the depolarization field, thus it provides the required condition for FDB where the growing domain has a high aspect ratio: nanometer size radius and length of hundreds micrometers.

Thus observation of ferroelectric domain reversal induced by indirect electron beam irradiation in LiNbO$_3$ bulk crystals represents a pronounced FDB effect. The reversed structure represents domain clusters consisting of multiple string-like domains with nanometer radius and 350 μm lengths. Fabricated one- and two-dimensional domain configurations could be used for a broad range of new 2-D as well as potential 3-D nonlinear photonic devices.

10.4 Ferroelectric domain breakdown

10.4.1 Domain shapes under FDB

It is obvious that the HVAFM tip generates a highly inhomogeneous electric field, which decreases very rapidly from 10^7–10^8 V/cm to almost zero at a depth of dozens of micrometers [13–15] in samples of millimeter-size. This implies that the domain forward growth in HVAFM does not follow the electric field (between the tip and the bottom electrode) lines as it occurs in the case of classical polarization reversal field when the domain grows in the homogeneous electric field [28, 29]. Below we show that the super-high and inhomogeneous field generated by the HVAFM and the very limited minimization of the depolarization field by screening charges flowing through the nanometer size tip, lead to generation of string-like domain growth; such domains penetrate a very long distance which maybe as long as 500 μm without any change in their initial lateral dimensions.

10.4 Ferroelectric domain breakdown

Despite pronounced difference in the domain evolution occurring in highly inhomogeneous and uniform electric fields, similar consequent stages of this process take place in both cases. However, the character of these stages in the field induced by the nanometer-size electrode may significantly vary relative to that in homogeneous fields [13–15, 62].

The polarization reversal starts with the nucleation of a new domain near the AFM tip. In contrast to the reversal performed in the homogeneous field the new nuclei formation in the electric field of a AFM tip does not hamper the switching process because of the extremely high field intensity induced by this sharp point-like electrode even under low applied voltages. The voltage, of about several kV in HVAFM, applied to the tip generates an electric field of about 10^7–10^8 V/cm at the ferroelectric polar face. Our calculation [15], performed in the framework of Landauer's model [63], showed that under such fields the nuclei have quasiatomic dimensions, and the nucleation activation energy does not exceed 10^{-3} eV. The nucleation time decreases to extremely short time of about 10^{-13} s, which is on the order of the lattice vibrations period; therefore the nucleation time is negligible small relative to the domain switching time. The observed nanodomain reversal does not require any active centers on the ferroelectric surface; it can start at any point of the crystal surface touched by the tip. The sharp AFM tip or electron drop may be considered as a movable active center for the domain nucleation. The newly formed domain then expands by motion of the domain walls. The domain walls move until the inverted domain reaches an equilibrium state. In the case of a long pulse of the voltage applied to the tip the nucleated domains have enough time to reach the final equilibrium state. In a paper by Molotskii [64] the theory of the equilibrium domain shape has been proposed and applied for explaining the domain size. This theory has been used in [15] to explain the shapes of the domains generated during the FDB process.

A spherical model was used in Ref. [15] in order to obtain the shape of the domains, reversed under the FDB conditions. This model was widely applied for studies of different processes that take place in the field of AFM tip (see Ref. [65]), including ferroelectric polarization reversal [66–69]. In this model the field of the tip apex is supposed to coincide with a field of a metallic sphere, the radius of which is equal to the radius of curvature of the tip apex. Using a simple approximation it may be supposed that the tip charge is concentrated in the center of the sphere [15, 64–69]. We will take into account a more general model and check the accuracy of the simple spherical model application to the ferroelectric domain breakdown condition.

The domain energy W consists of the depolarization field energy W_d, the energy of the domain walls W_s, and the energy, W_t, of the interaction between the electric field of the tip and the spontaneous polarization of the domain. Following Landauer [63] hereafter we assume that the domain has a shape of an elongated half-ellipsoid in the polar direction.

The expression for the domain wall energy and depolarization field energy of the elongated half-ellipsoidal domain has been obtained by Landauer [63]:

$$W_s(r,l) = brl, \quad \text{and} \quad W_d(r,l) = \frac{cr^4}{l} \tag{10.2}$$

The constant

$$b = \frac{\pi^2}{2}\sigma_w \tag{10.3}$$

is proportional to the surface energy density of the domain wall σ_w, and the factor

$$c = \frac{16\pi^2 P_s^2}{3\varepsilon_a} \left[\ln\left(\frac{2l}{r}\sqrt{\frac{\varepsilon_a}{\varepsilon_c}}\right) - 1 \right], \tag{10.4}$$

related to the energy of depolarization field W_d; r and l are the base radius and the length of the domain half-ellipsoid, respectively, ε_c and ε_a are dielectric constants in directions parallel and perpendicular to the polar axis, respectively.

Let us consider the real charge distribution in the vicinity of the tip, in contrast to the simple spherical model, where it was supposed that the overall charge is concentrated in the center of the spherical curvature of the tip apex. The charge $q_0 = RU$ is generated in the center of the spherical curvature of the tip apex under an applied voltage U. Moreover, an infinite series of image charges q_n is located at a distance r_n from the center of the sphere [70, 71]:

$$q_{n+1} = RU \left(\frac{\sqrt{\varepsilon_c \varepsilon_a} - 1}{\sqrt{\varepsilon_c \varepsilon_a} + 1}\right)^n \frac{\sinh \alpha}{\sinh(n+1)\alpha} \tag{10.5}$$

$$r_{n+1} = \frac{R^2}{2(R+\delta) - r_n}, \quad r_0 = 0 \tag{10.6}$$

where R is the radius of curvature of the tip apex, δ is the distance between the tip apex and the sample surface, and α is defined by: $cosh(a) = 1 + \delta/R$.

An expression for the normal component of the electric field produced by a charge q_n concentrated in a tip apex is found by differentiating the expression for a potential which was obtained by Mele [72]:

$$E_{nz}^{apex}(\rho, z) = \frac{2 \cdot q_n}{\gamma \left(\sqrt{\varepsilon_c \varepsilon_a} + 1\right)} \frac{\frac{z}{\gamma} + R + \delta - r_n}{\left(\left(\frac{z}{\gamma} + R + \delta - r_n\right)^2 + \rho^2\right)^{3/2}}, \tag{10.7}$$

where ρ and z are cylindrical coordinates, $\gamma = \sqrt{\varepsilon_c/\varepsilon_a}$. The interaction energy W_{tn} of the field E_n generated by the charge q_n was derived in Ref. [64] taking into account the local change of the spontaneous polarization inside the domain:

10.4 Ferroelectric domain breakdown

$$W_{tn}(r,l) = -\frac{8\pi q_n P_s}{(\sqrt{\varepsilon_c \varepsilon_a}+1)\gamma} \times$$

$$\left[l - \frac{s_n + \frac{l}{\gamma} - \sqrt{s_n^2 + r^2}}{\gamma\left(\frac{1}{\gamma^2} - \frac{r^2}{l^2}\right)} + \frac{\frac{s_n r^2}{l^2}}{\left(\frac{1}{\gamma^2} - \frac{r^2}{l^2}\right)^{3/2}} \times \right.$$

$$\left. \ln \frac{\sqrt{\frac{1}{\gamma^2} - \frac{r^2}{l^2}}\left(s_n + \frac{l}{\gamma}\right) + l\left(\frac{1}{\gamma^2} - \frac{r^2}{l^2}\right) + \frac{s_n}{\gamma}}{\sqrt{\left(\frac{1}{\gamma^2} - \frac{r^2}{l^2}\right)(r^2 + l^2)} + \frac{s_n}{\gamma}} \right] \quad (10.8)$$

where $s_n = R + \delta - r_n$ is the distance between the charge q_n and the sample surface. Ferroelectric breakdown leads to very high aspect ratio domains where the domain radius r is significantly smaller than their length l. In this case it is possible to neglect the high order terms r^2/l^2. Then Equation (10.8) reduces to:

$$W_{tn}(r,l) = -\frac{8\pi q_n P_s}{(\sqrt{\varepsilon_c \varepsilon_a}+1)}(\sqrt{s_n^2 + r^2} - s_n). \quad (10.9)$$

Under the FBD conditions the domain radius r is larger by an order of magnitude than the radius of the tip apex curvature R. For the tip brought in contact with a surface the distance δ equals approximately to the lattice constant (0.5 nm). Therefore $s_n \ll r$ is satisfied and Equation (10.9) becomes:

$$W_{tn}(r,l) = -\frac{8\pi q_n P_s r}{\sqrt{\varepsilon_c \varepsilon_a}+1} \quad (10.10)$$

Summation of the energy contributions (equation (10.10)) over all the charges from Equation (10.5) results in the total interaction energy of the tip apex charge with the domain:

$$W_t(r,l) = -fr \quad (10.11)$$

where

$$f = \frac{8\pi C_{ts} U P_s}{\sqrt{\varepsilon_c \varepsilon_a}+1}, \quad (10.12)$$

is the effective force acting on the domain. This force is produced by the charges located at the tip apex,

$$C_{ts} = R \sinh\alpha \sum_{n=0}^{\infty} \frac{\left(\frac{\sqrt{\varepsilon_c \varepsilon_a}-1}{\sqrt{\varepsilon_c \varepsilon_a}+1}\right)^n}{\sinh(n+1)\alpha} \quad (10.13)$$

is the tip apex capacitance. Equation (10.11) implies that the field of the tip only affects the radial broadening of the long domains, but it cannot directly cause the domain elongation. As will be shown below the domain elongation is an indirect consequence of the field action; the domain elongation is caused by minimization of the total free energy of the system.

When domain dimensions are much larger than the tip radius R, Equations (11), (12) and (13) become identical to the corresponding equations obtained in the framework of the theory [15, 64] developed using the simple spherical model. In this model the energy of the long domain $W(r, l)$, created within the condition of FDB, equals to the summation of the energies from Equation (10.2) and Equation (10.11)

$$W(r, l) = \frac{cr^4}{l} + brl - fr. \tag{10.14}$$

The minimization of W(r,l) enables to determine the equilibrium domain dimensions [64]

$$r_m = \left(\frac{f}{5\sqrt{bc}}\right)^{2/3} \tag{10.15}$$

and

$$l_m = \frac{f}{5b} \tag{10.16}$$

10.4.2 The domain shape invariant

Let us consider Equations (15) and (16); it can be seen that the ratio $r_m^{3/2}/l_m$ depends logarithmically on the applied voltage and the tip-sample system capacitance yielding:

$$\frac{r_m^{3/2}}{l_m} = \sqrt{\frac{b}{c}}. \tag{10.17}$$

This value may be considered as an invariant of the equilibrium domain shape, which practically does not depend on the AFM parameters, and is defined only by the properties of the ferroelectric material. The concept of the domain shape invariant was introduced in Ref. [64]. We obtained Equation (10.17) for a domain with a large radius ($r \gg R$); it was also shown [64] that Equation (10.17) is valid for the domains with small radius ($r \ll R$) as well. In fact, this equation is valid for various experimental conditions and AFM types because it does not depend on the microscope parameters. Our calculations of the shape invariant were performed in the framework of the spherical model for large and small distances between the tip and the sample, and calculations where we took into account the contribution of the field generated by the conical part of the tip support the above statement. The shape invariant concept allows determining the necessary conditions for the observation of the FDB effect occurring under conditions of highly limited screening of the depolarization field and especially pronounced in high inhomogeneous electric fields generated by the AFM tip or electron drop. Substituting Equations (3) and (4) into Equation (10.17) gives the following expression for the domain shape invariant which is defined by basic material properties:

$$\frac{r_m^{3/2}}{l_m} \propto \frac{\sqrt{\sigma_w \varepsilon_a}}{P_s}. \tag{10.18}$$

10.4 Ferroelectric domain breakdown

The radius of string-like nanodomains observed under the FDB conditions is much smaller than its length, $r_m \ll l_m$. Consequently, for the FDB phenomenon the shape invariant value should also be small. Equation (10.18) shows that the shape invariant has the lowest value if a ferroelectric material satisfies the following conditions:

a. It possesses a high value of spontaneous polarization P_s.

b. The dielectric permittivity normal to the polar axis ε_a is small.

c. It possesses a small value of the domain surface energy density σ_w.

In Table 10.2 below we summarize the calculated values of the shape invariant $r_m^{3/2}/l_m$ and aspect ratio r_m/l_m of string-like domains in different FE crystals formed under the FDB conditions. The experimental parameters used for these calculations are: $U = 3\,\text{kV}$ (applied voltage), $R = 50\,\text{nm}$ (typical radius of curvature of the tip apex), $d = 0.5\,\text{nm}$ (distance between the tip apex and the sample surface). The calculated aspect ratios r_m/l_m are quite different for the chosen FE.

	$(\varepsilon_a; \varepsilon_c)$	P_s [μC/cm^2]	σ_w [mJ/m^2]	$r_m^{3/2}/l_m$ [cm$^{1/2}$]	r_m/l_m
BaTiO$_3$	(2000; 120)	26	7	2.1×10^{-4}	23×10^{-3}
GASH	(5; 6)	0.35	0.3	1.7×10^{-4}	9×10^{-3}
TGS	(8.6; 5.7)	2.8	1	4.4×10^{-5}	3×10^{-3}
LiNbO$_3$	(84; 30)	75	4	9.2×10^{-6}	0.8×10^{-3}
LiTaO$_3$	(51.7; 43.5)	50	22	2.9×10^{-5}	3.5×10^{-3}
Lead Titanate	(140; 120)	75	150	1×10^{-4}	18×10^{-3}

Table 10.2: Shape invariant and aspect ratio of domains grown under FDB effect in different FE materials

The smallest ratio is found for LiNbO$_3$ crystals because aforementioned conditions a to c are fully satisfied by its physical properties: high P_s, small ε_a and relatively small σ_w. The values of P_s and ε_a for lithium niobate were taken from the review of Gopalan et al. [73] while σ_w was found from experimental data and the proposed FDB model. The second group of FE's with intermediate ratio r_m/l_m includes TGS and LiTaO$_3$ crystals. These FE's are very different; however, unexpectedly TGS possessing extremely low P_s compared to LiTaO$_3$, has very small ε_a and σ_w and therefore should demonstrate highly pronounced FDB effect and small ratio r_m/l_m. The third group of FE includes BaTiO$_3$ and GASH. The parameter r_m/l_m for BaTiO$_3$ is the largest and exceeds by 30 times the calculated value for LiNbO$_3$ despite that this FE has a relatively large value of P_s and an intermediate magnitude of surface energy density of domain wall energy. The reason for the large r_m/l_m ratio in BaTiO$_3$ is the unusually high ε_a reaching $\varepsilon_a = 2000$. Calculation of the equilibrium domain dimensions in barium titanate reported in Ref. [74] showed that for the conditions which provide FDB in lithium niobate, the domain radii of these two materials are close, but their lengths differ by more than an order of magnitude - the domain generated under FDB in lithium niobate is much longer. Lead titanate crystals, which possess very high values of spontaneous polarization $P_s = 75\,\mu C/cm^2$ are also of interest. The value of the dielectric permittivity is not too large $\varepsilon_a = 130$–150; these data permit the FDB in lead titanate and a low r_m/l_m ratio. However, r_m/l_m increases in this crystal due to the very high value of the surface energy density $\sigma_w = 130$–$170\,mJ/m^2$ [75]. Our calculation shows that for identical experimental conditions the domains in lead titanate will be shorter by 20 times than those in lithium niobate (Table 10.2).

10.4.3 Theory and experimental data of FDB effect

Figure 10.11 shows the theoretical dependence of the domain radius on the applied voltage, which was compared to an experimental data obtained in lithium niobate. This calculation was performed in the range of voltages between 0 and 1.4 kV, using $P_s = 75\,\mu C/cm^2$, $\varepsilon_c = 30$, $\varepsilon_a = 84$. The surface energy density σ_w was obtained by a fitting procedure where it was a free parameter; as expected, the obtained value $\sigma_w = 4\,mJ/m^2$ was relatively small.

Even though we obtained a good agreement between theory and experiment, the consistent theory of the FDB is still in progress. The theory does not take into account several factors such as an electric field produced by the cone of the tip which contributes significantly to the field further away from the tip apex [76], the possibility of plastic deformation and flattening of the tip apex during the experiment, the free charge injection from the tip to the sample [77] and piezoelectric effects caused by indentation [78] etc..

These effects may affect the domain size in different ways. While the field produced by the cone and the tip apex flattening will lead to a larger domain size, the indentation may decrease it. The injection current from the HVAFM tip may either support or suppress the domain growth: on the one hand it may lead to the formation of a space charge region in the vicinity of the tip-sample contact creating an electric field, which will accelerate the domain forward growth; on the other hand the injected charge carriers contribute to the screening of the applied electric field and prohibit the FDB effect. In such a case the dominating factor will determine the contribution of the injected carriers to the domain growth process. The agreement between the theory and experiment shown in Figure 10.11 may be explained by mutual compensation of the discussed effects in the voltage range under 1.4 kV. Above 1.4 kV the aforementioned condition of the mutual compensation may be broken and the theory is not consistent with the experiment any more.

10.4.4 Ferroelectric domain breakdown mechanism

The electric field generated by the charges located on the tip quickly plummets inside the sample away from the tip apex. Figure 10.12 shows the electric field as a function of distance in the vicinity of the tip where the contribution of the conical part of the AFM tip was also taken into account [76]; the electric field was calculated for an applied voltage $U = 3\,kV$. The field near the tip apex is very large; it reaches $\sim 10^8$ V/cm. However at a distance $z \sim 38\,\mu m$ away from the surface the field decreases to 10^3 V/cm. This is by two orders of magnitude smaller than the coercive field $E_c = 210\,kV/cm$ of lithium niobate [73]. The electric field that may lead to domain wall motion in congruent LiNbO$_3$ crystals should be at least 22 kV/cm [79]. At a distance z, larger than $\sim 6.4\,\mu m$, the electric field is below this value, i.e. in 150 μm thick LiNbO$_3$ where through domains were observed [80] the domains moved $\sim 96\,\%$ of their way

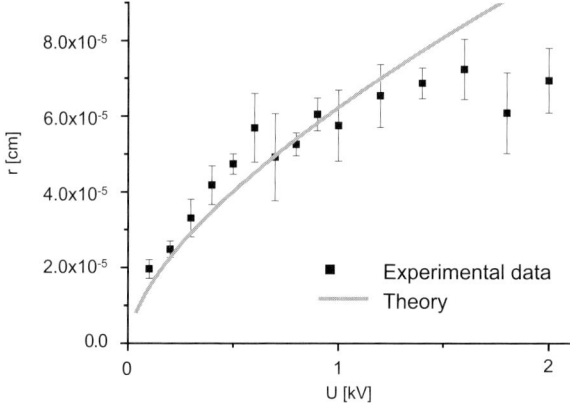

Figure 10.11: Theoretical dependence of the domain radius on the applied voltage obtained by taking of the surface energy density as a free parameter and adjustment to the experimental results.

in a weak field that can not lead to a domain wall motion. Approaching the end of the domain growth throughout the sample, the electric field goes down to ~ 100 V/cm, which is obviously not capable to cause any domain wall motion at the apex of the growing domain.

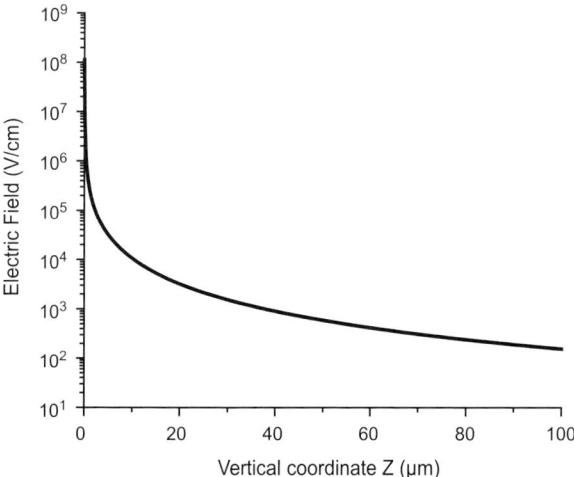

Figure 10.12: Z-component of the inhomogeneous electric field as a function of the distance from the sample.

The question that arises is then: if the electric field in the vicinity of the domain apex does not actually exist, what is the driving force that leads to the domain breakdown? We have shown [15], that the FDB is not related explicitly to the electric field action. This phenomenon may be explained by the action of the internal forces, caused by the depolarization field of the growing domain. The important role of the depolarization field energy in the evolution of ferroelectric domain structures has been pointed out earlier [81–83]. It was shown [84] that the effect of forward motion (high speed of ferroelectric domain wall forward motion in uniform external field in comparison with a sidewise motion of the growing domain) is also dictated by the internal forces action caused by the depolarization field. These forces lead to the accelerated domain elongation and suppress sidewise motion. These forces are also responsible for the ferroelectric domain breakdown. The FDB may be considered as an anomaly of forward motion effect which originates in a strong electric field at the surface when another factor minimizing the depolarization field energy, the screening process, is highly limited due to the nanometer-size switching electrodes.

Let us show that under conditions of FDB the intrinsic depolarization field action may lead to the domain elongation. The driving force for a domain elongation, F_l, may be found from Equation (10.14)

$$F_l = -\frac{\partial W(r,l)}{\partial l} = \frac{cr^4}{l^2} - br.$$

It is seen here that the electric field of the tip does not affect F_l explicitly. The domains are propagating as a result of the depolarization field energy. Evidently, the only reason for the domain elongation is the field of the AFM tip. However, this effect is indirect; it reveals itself through the increase of the domain radius due to the tip field. The field of the tip leads to domain broadening but does not influence its length (see Equation (10.11)). The domain length adjusts itself to the increasing domain radius. It increases until equilibrium size is reached, which is defined by the shape invariant Equation (10.17). As a result of this length adjustment, the domain reaches equilibrium state. In such a state driving forces associated with the increase of the domain surface area br compensate the driving force cr^4/l^2, stipulated by decrease of the depolarization field energy cr^4/l with increase of the domain length.

10.5 Nanodomain superlattices tailored by multiple tip arrays of HVAFM

One- and two-dimensional domain superlattices tailored by HVAFM use a single tip system. A simple estimation shows that the fabrication time of nanodomain gratings occupying the FE template area around $10\,mm^2$ with a tip scanning velocity of around $100\,\mu m$/second is 100 hours. This clearly means that any practical nanodomain writing technology should be based on another solution. Scanning probe microscopy (SPM) technology has been undergoing intensive dynamic growth in the past decade. Presently, most SPM applications involve only a single cantilever and tip. However, many future applications, such as data storage and nanolithography, call for arrayed AFM probes (one- or two-dimensional) to achieve high efficiency and throughput. For example, high-throughput lithography and surface patterning with extremely fine linewidths (e.g. of the order of 10–100 nm) have led to the development of parallel dip-pen nanolithography (DPN) technology, and the demand for miniaturization of chemical sensors to microfabricated cantilever arrays.

In this chapter we report about the use of multiple AFM tip array for parallel domain writing in bulk FE crystals. Fabrication of nanodomain gratings is based on direct writing of nanodomains with small ratio $r/l < 10^{-2} - 10^{-3}$. It is shown that the only mechanism providing such a low ratio is the FDB effect [13–15].

Application of any multiple switching electrode technology leads to strong changes of the applied field due to overlapping of the fields formed by the individual tips. The domain broadening observed under conventional poling using photolithographically patterned strip-like electrodes, limits the domain grating period to 4 μm [3]. This effect of domain propagation into the regions uncoated by the strip electrode was mainly caused by both tangential field and the homogenous switching field in FE crystal bulk [19]. Calculations showed that the homogenous field is formed a crystal depth roughly equal to the distance between the metal strips; that is only several micrometers from the sample face. High voltage application across the tip array-sample system should also change the field distribution in the FE crystal bulk and cause overlapping. However, the generated field results in its inhomogeneous distribution. The limited screening of the depolarization fields by the point-like tip electrodes and inhomogeneous electric field have to lead to FDB effect under each individual tip and generation of string-like domains formation as in polarization reversal using single AFM tips. The results below show a very good agreement with calculated domain dimensions and optimal tailoring conditions using multiple tip arrays are determined. We have used commercially available Si_3N_4 AFM tip arrays (Nanoink Inc.) with a spring constant of ~ 0.2 N/m, tip radius of curvature of about 100 nm similar to the one shown in Figure 10.13.

Figure 10.13: Scanning electron microscope image of the cantilever array, fabricated by Nanoink Inc., used for the domain writing; the insert shows an enlarged view of a single tip at the end of a cantilever.

The cantilevers chip was mounted in a standard cantilever holder and was coated with 27 nm thick Cr to ensure good electrical conductivity. Aligning the probe array parallel to the sample surface was achieved using the internal laser signal feedback control system of the

AFM, and an optical microscope connected to a CCD camera. Figure 10.14 shows a typical domain grating written in 200 μm thick RbTiOPO$_4$ single crystal by applying a DC voltage of 0.7 kV, at a scanning rate of 120 μm/sec; in this case a probe array of 10 cantilevers (30.6 μm apart center to center, 100 μm long, and 25.6 μm wide) was used. Due to the limited scanning range of the AFM scanner, Figure 10.14.a shows a grating image written by 3 out of the ten tips shown schematically above the image. Figure 10.14.b is a zoom on a smaller region which demonstrates that the domains posses perfect geometry (like in HVAFM single tip writing), and do not coalesce. Topography AFM images of the tailored domain configurations following chemical etching showed a complete penetration of the domains throughout the crystal bulk, without any domain broadening exactly like in the case of single tip writing [13, 14].

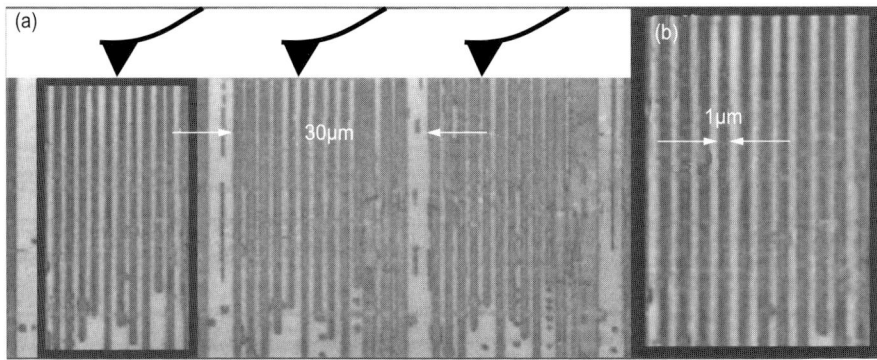

Figure 10.14: (a) Periodic domain structure tailored in a RbTiOAsO$_4$ single crystal by moving the multiple HVAFM tip arrays along the Y-axis direction with a velocity of 120 μm/sec. (b) The insert of a periodic domain structure written by one tip of the arrays.

The calculation of the total free energy in the case of a tip array is explained with the help of Figure 10.15. The figure shows schematics of three AFM tips (with apex radius of curvature R) at a distance ρ_0 apart, located above three inverted domains.

We begin by calculating the contribution of a single tip, located at a distance ρ_0 from a domain located underneath an adjacent tip, to the energy of interaction between the domain and the total external field, $W_{t,neighbor}$. The normal component of the electric field of the tip inside the dielectric is calculated as follows:

$$E_z(\vec{\rho}, \vec{\rho}_0, z) = \frac{2C_t U}{(\sqrt{\varepsilon_c \varepsilon_a} + 1)\gamma} \frac{s + \dfrac{z}{\gamma}}{\left((\vec{\rho}_0 - \vec{\rho})^2 + \left(s + \dfrac{z}{\gamma}\right)^2\right)^{3/2}} \quad (10.19)$$

10.5 Nanodomain superlattices tailored by multiple tip arrays of HVAFM

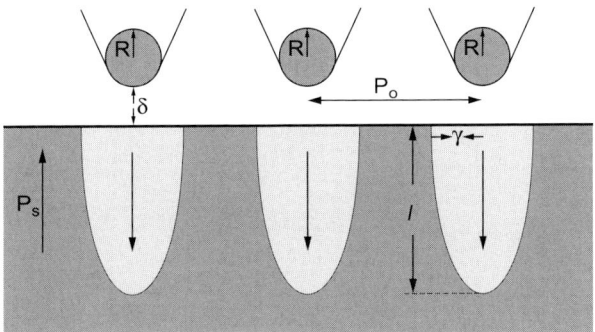

Figure 10.15: Schematic diagram describing the calculation of the electric field generated by the multiple HVAFM tip arrays.

where the vectors $\vec{\rho}$ and $\vec{\rho}_0$ determine the coordinate relative to the domain axis and the domain location relative to the tip correspondingly $\gamma = \sqrt{\varepsilon_c/\varepsilon_a}$, $s = R+\delta$. For a small lateral domain size $r \ll \rho_0$, and for $s \ll \rho_0$ and l, the electric field becomes:

$$E_z(\rho_0, z) = \frac{2 \cdot C_t U \cdot \gamma}{(\sqrt{\varepsilon_c \cdot \varepsilon_a} + 1)} \cdot \frac{z}{[z^2 + (\gamma \cdot \rho_0)^2]^{3/2}} \qquad (10.20)$$

Here, q is the electric charge of the tip $q \cong C_t \cdot U$ and $\gamma = \sqrt{\varepsilon_c/\varepsilon_a}$, where C_t is the capacity related to one tip. The electric field induced by the HVAFM tip array inside RbTiOPO$_4$ single crystal for an applied voltage $U = 0.7$ kV, and tip apex radius of curvature $R = 78$ nm is shown in Figure 10.16. The figure clearly shows that the total external electric field, induced by the whole tip array, decreases to less then 10^3 V/cm at a distance of $\sim 1\,\mu$m from the sample surface, due to the fast decay of the electric field caused by the shape and size of the tip. Thus, although the tip array creates almost a uniform electric field at distances $\rho_0 = 1\,\mu$m from the surface, *this does not change the domain growth relative to the single tip case.*

When the new domain is formed, the spontaneous polarization changes by $2P_s$, and the interaction energy due to the electric field of the neighboring tip becomes:

$$W_{t,neighbor}(\rho_0) = -2P_s \int_V E_z(\vec{r}) dV. \qquad (10.21)$$

The integral is calculated over the volume of the half-ellipsoidal domain, and using Equation (10.20) for the normal component of the field intensity we obtain:

$$W_{t,neighbor}(\rho_0) = -\frac{8\pi \cdot P_s \cdot q \cdot \gamma}{(\sqrt{\varepsilon_c \cdot \varepsilon_a} + 1)} \cdot \int_0^l \frac{z \cdot dz}{[z^2 + (\gamma \cdot \rho_0)^2]^{3/2}} \cdot \int_0^{\rho_m(z)} \rho \cdot d\rho \qquad (10.22)$$

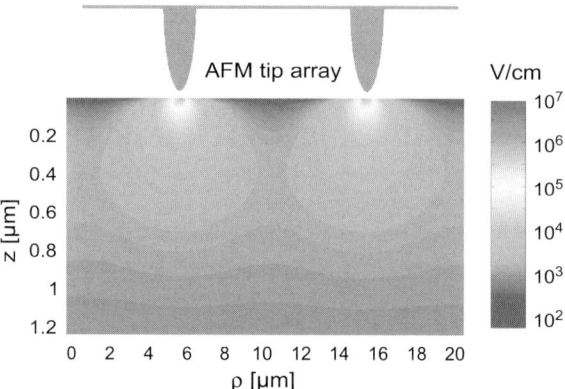

Figure 10.16: The figure clearly shows that the total external electric field, induced by the whole tip array, decreases to less then 103 V/cm at a distance of $\sim 1\,\mu m$ from the sample surface, due to the fast decay of the electric field caused by the shape and size of the tip. Thus, although the tip array creates almost a uniform electric field at distances $\geq 1\,\mu m$ from the surface, *this does not change the domain growth relative to the single tip case.*

where $\rho_m(z)$ is the maximal value of the cylindrical coordinate ρ for a given z coordinate given by $\rho_m(z) = r\sqrt{1 - z^2/l^2}$. Substituting ρ_m in Equation (10.21) we have:

$$W_{t,neighbor}(\rho_0) = -\frac{4\pi \cdot C_t U \cdot P_s \cdot r^2}{(\sqrt{\varepsilon_c \cdot \varepsilon_a} + 1) \cdot \rho_0} \left[1 - \frac{1}{\sqrt{1 + \left(\frac{l}{\gamma \cdot \rho_0}\right)^2}} - \left(\frac{\gamma \cdot \rho_0}{l}\right)^2 \right.$$

$$\left. \times \left(\sqrt{1 + \left(\frac{l}{\gamma \cdot \rho_0}\right)^2} + \frac{1}{\sqrt{1 + \left(\frac{l}{\gamma \cdot \rho_0}\right)^2}} - 2 \right) \right]$$

(10.23)

The total domain energy W_{total} in the case of the tip-array is calculated now as a sum of the total domain energy for a single tip $W_{singletip}$ and the contribution from the nine neighboring tips of a 10-tip array where an i-th tip is located at a distance ρ_{oi} from the domain. It should be mentioned that analytical expressions for the depolarization energy W_d, and the domain surface energy W_s for the case of a tip-array are identical to that of a single tip [15] but the

10.5 Nanodomain superlattices tailored by multiple tip arrays of HVAFM

values of the domain length and radius are different. Thus the total free energy of a single domain inverted by a tip array containing m tips will be:

$$W_{total} = W_{single-tip} + \sum_{i=1}^{m} W_{t,neighbor}(\rho_{oi}) \qquad (10.24)$$

The equilibrium dimensions r_m and l_m of the inverted domain correspond to the minimum of the total free energy of the tip array-sample system, W_{total}. The calculated domain sizes at the energy minimum are shown in Figure 10.17 as a function of the distance ρ_0 between two adjacent tips.

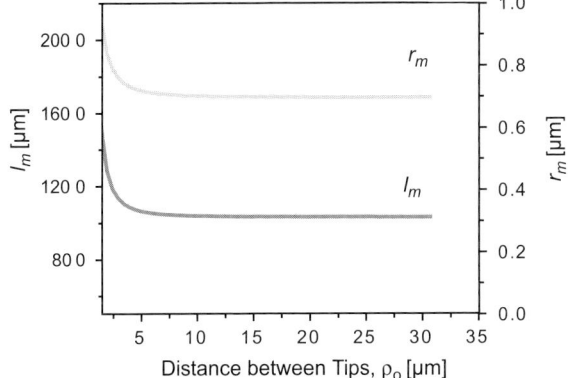

Figure 10.17: The calculation of the equilibrium domain size, r_m (right axis) and l_m, (left axis) as a function of the spacing between tips, for RbTiOAsO$_4$ single crystal ($U = 0.7$ kV, $R = 78$ nm and $\rho_w = 1$ mJ/m^2)

These calculations are based on the following parameters: $\sigma_w = 1$ mJ/m^2, $\varepsilon_c = 25$, $\varepsilon_a = 18$, and $P_s = 30$ mC/cm^2. In this case the domain equilibrium dimension are $r_m = 0.65$–$0.75\,\mu$m and $l_m = 920\,\mu$m. Our measured value of the lateral domain size (1–1.5 μm) is within the calculated range $2r_m = 1.3$–$1.5\,\mu$m.

The most important conclusion from Figure 10.17 is that the equilibrium domain dimensions are almost independent of the distance between adjacent tips down to a distance of about 5 microns; at such a distance our approximation $r \ll \rho_0$ is violated. This conclusion is significant both from scientific and technological points of view. First it shows that the FDB phenomenon is solely due to the strong and inhomogeneous electric field created by the AFM tip and a highly limited effect of the minimization of the depolarization field by external screening charge; under such conditions the domain forward growth velocity is more than two orders of magnitude larger then the lateral one and hence, no domain coalescence takes place. From the technological point of view such a large difference in lateral and forward domain propagation velocity determines the conditions for parallel domain writing using AFM tip arrays without any domain broadening.

Thus in order to obtain sub-micrometer domain gratings in bulk FE crystals it is necessary to use the HVAFM; the tip array method presented here will allow to use the HVAFM in reasonable rates applicable also to real applications.

10.6 Conclusions

1. The physical limits and technological requirements of domain dimensions for a new generation of nanodomain-based devices were considered. It was shown that for both ferroelectric thin films and crystals the achievable domain size is in the range of 100 nm. It is shown that for 100 nm thick ferroelectric films, an application of nanosize electrodes does not make a big difference compared with conventional polarization reversal setups and physical mechanism. However, in the case of bulk ferroelectrics, the use of a switching AFM tip electrode for generation of long domains with a nanometer size radius requires a new approach both for polarization reversal instrumentation and physics of domain inversion.

2. A high voltage atomic force microscope was developed for nanoscale polarization reversal in FE bulk crystals. The switching voltage in the kilovolt range applied to a HVAFM tip generates a superhigh inhomogeneous electric field of about 10^8 V/cm which rapidly decreases in the crystal bulk and reaches a negligible value at the depth of dozens of micrometers. Such a super-high electric field changes the domain nucleation process making it activationless. It induces string-like domains of nanometer radius and hundreds of micrometer length penetrating from the ferroelectric surface into the crystal bulk for a very long distance which may be as long as 500 μm without any change in their initial lateral dimensions, a new phenomenon that was named ferroelectric domain breakdown.

3. A theory for the ferroelectric domain breakdown phenomenon has been developed. In the case of a limited external screening process of the depolarization field the domains propagate as a result of the decrease in the depolarization field energy. The field of the tip influences solely the radial broadening of the long domains. The domain elongation is caused by the total free energy minimization of the system when the domain propagates through the whole crystal thickness without any changes of its radius. Its forward growth continues until forces associated with the increase of the domain surface area compensate the driving force caused by the depolarization field minimization.

4. It was found that string-like domains induced by the ferroelectric domain breakdown effect in HVAFM leads to a domain shape invariant where the aspect ratio (relation between domain radius and its length r/l) of the string-like domains depends only on the basic FE material parameters: spontaneous polarization, lateral dielectric permittivity and density of domain wall surface energy. The domain shape invariant allows to get unambiguous criteria for string-like nanodomain generation resulting in the figure of merit of a ferroelectric crystal applicable for nanodomain technology. Such a ferroelectric should possess a high spontaneous polarization, a low density of domain wall surface energy and a small lateral dielectric permittivity. LiNbO$_3$ has the highest aspect ratio r/l reaching the value of around 10^{-3} which is consistent with the observed experimental data.

5. Multiple AFM tip arrays have been used for parallel domain writing in bulk FE crystals. The fabrication of nanodomain gratings is based on direct writing of nanodomains with small ratio r/l. It is shown that the only mechanism providing such a low ratio is the ferroelectric domain breakdown effect. Following our formalism for the single tip domain reversal the equilibrium domain size i.e. domain length l and base radius r corresponding to the minimum of the total free energy of the tip-array sample system is found resulting in calculation of the minimum distance between adjacent tips in a tip.

6. Pronounce ferroelectric domain breakdown phenomenon has been observed in 500 μm thick $LiNbO_3$ bulk crystals under high energy electron beam exposure. In order to suppress a contribution of the electron screening effect to the minimization of the depolarization field, the C^--polar face of the $LiNbO_3$ crystal is coated by a thick photo-resist layer which avoids a penetration of electrons into the ferroelectric crystal. As a result, the electrons are located at trap levels of high concentration in the amorphous photo-resist layer and form immobile electron drops. This localized electron charge in a small volume in the photoresist layer creates a large electric field and leads to ferroelectric domain reversal with very high domain ratio in agreement with the FDB mechanism.

7. One- and two-dimensional nanodomain configurations have been engineered in $LiNbO_3$, $RbTiOPO_4$ and $RbTiOAsO_4$ bulk FE crystals by the developed HVAFM and indirect electron beam methods for a new generation of photonic and acoustic devices.

Acknowledgement

The work was supported by the Ministry of Science and Technology of Israel, United States-Israel Binational Science Foundation (BSF) and by Tel Aviv University Consortium for Nanoscience & Nanotechnology.

Bibliography

[1] J. F. Scott and C. A. Paz de Araujo, *Science* **246**, 1400 (1989).
[2] L. Myers, R. C. Eckardt, M. M. Fejer, and R. L. Byer, *J. Opt. Soc. Am.* B**12**, 2102 (1995).
[3] G. Rosenman, A. Arie, and A. Skliar, *Ferroelectrics Review* **1**, 4 (1998).
[4] V. Berger, *Phys. Rev. Lett.* **81**, 4136 (1998).
[5] N. G. Broderick R., G. W. Ross, H. L. Offerhaus, D. J. Richardson, and D. C. Hanna, *Phys. Rev. Lett.* **84**, 4345 (2000).
[6] A. Chowdhury, C. Staus, B. F. Boland, T. F. Kuech, and L. McCaughan, *Opt. Lett.* **26**, 1353 (2001).
[7] Y. Rosenwaks, M. Molotski, A. Agronin, P. Urenski, M. Schvebelman, and G. Rosenman, *Nanoscale characterization of ferroelectric materials. Scanning force microscopy approach*, Eds. M. Alexe, A. Guverman, Springer, 2004.
[8] A. Gruverman, J. Hatano, H. Tokumoto, *Jpn. J. Appl. Phys.* **36**, 2207 (1997).
[9] M. Alexe, C. Harnagea, D. Hesse, U. Gosele, *Appl. Phys. Lett.* **75**, 1793 (1999).

[10] S. Kalinin, D. Bonnell, T Alvarez, X. Lei, Z. Hu, J. Ferris, *Nanoletters* **2**, 589 (2002).
[11] J. H. Ferris, D. B. Li, S. V. Kalinin, and D. A. Bonnell, *Appl. Phys. Lett.* **84**, 774 (2004).
[12] G. Rosenman and Y. Rosenwaks, P. Urenski, US Patent (2003).
[13] G. Rosenman, P. Urenski, A. Agronin, Y. Rosenwaks, and M. Molotskii, *Appl. Phys. Lett.* **82**, 103 (2003).
[14] G. Rosenman, P. Urenski, A. Agronin, A. Arie, and Y. Rosenwaks, *Appl. Phys. Lett.* **82**, 3934 (2003).
[15] M. Molotskii, G. Rosenman, P. Urenski, A. Agronin, M. Shvebelman, and Y. Rosenwaks, *Phys. Rev. Lett.* **90**, 107601 (2003).
[16] R. G. Batchko, V. Ya Shur., M. M. Fejer, and R. L. Byer, *Appl. Phys. Lett.* **75**, 1673 (1999).
[17] V. Ya Shur, E. Rumyantsev, E. Nikolaeva, E. Shishkin, D. Fursov, R. Batchko, L. Eyres, M. Fejer, and R. Byer, *Appl. Phys. Lett.* **76**, 143 (2000).
[18] P. Urenski, M. Molotskii, G. Rosenman, *Appl. Phys. Lett.* **79**, 2964 (2001).
[19] G. Rosenman, A. Skliar, D. Eger, M. Oron, M. Katz, *Appl. Phys. Lett.* **73**, 865 (1998).
[20] P. Urenski, M. Lesnykh, Y. Roseaks, G. Rosenman, and M. Molotski, *J. Appl. Phys.* **90**, 1950 (2001).
[21] S. Moscovich, A. Arie, R. Urneski, A. Agronin, G. Rosenman, and Y. Rosenwaks, *Optics Express* (2004).
[22] G. Rosenman, D. Shur, A. Dunaevski, and Y. Krasik, *J. Appl. Phys.* **88**, 6109 (2000).
[23] C. Moon, Kim Dong-Wook, G. Rosenman, T. Ko, and K. Yoo In, *Jpn. J. Appl. Phys.* **42**, 3523 (2003).
[24] S. Li, J. A. Eastman, J. M. Vetrone, C. M. Foster, R. E. Newnham, and L. E. Cross, *Jpn. J. Appl. Phys.* Part 1 **36**, 5169 (1997).
[25] A. Roelofs, T. Schneller, K. Szot, and R. Waser, *Appl. Phys. Lett.* **81**, 5231 (2002).
[26] T. Tybell, C. Ahn, and J. Triscone, *Appl. Phys. Lett.* **75**, 856 (1999).
[27] S. K. Streiffer, J. A. Eastman, D. D. Fong, C. Thompson, A. Munkholm, M. V. Ramana Murty, O. Auciello, G. R. Bai, and G. B. Stephenson, *Phys. Rev. Lett.* **89**, 67601 (2002).
[28] W. Merz, *J. Phys. Rev.* **95**, 690 (1954).
[29] R. J. Landauer, *Appl. Phys.* **28**, 227 (1957).
[30] A. Gruverman, O. Auciello, and H. Tokumoto, *Annu. Rev. Mater. Sci.* **28**, 101 (1998).
[31] T. Hidaka, T. Maruyama, I. Sakai, M. Saitoh, L. A. Wills, R. Hiskes, S. A. Dicarolis, and J. Amano, *Integrated Ferroelectrics* **17**, 319 (1997).
[32] C. Harnagea, A. Pignolet, M. Alexe, K. M. Satyalakshmi, D. Hesse, and U. Goesele, *Jpn. J. Appl. Phys.* Part 2 **38**, L1255 (1999).
[33] K. Terabe, M. Nakamura, S. Takekawa, K. Kitamura, S. Higuchi, Y. Gotoh, and Y. Cho, *Appl. Phys. Lett.* **82**, 433 (2003).
[34] Y. Cho, K. Fujimoto, Y. Hiranaga, Y. Wagatsuma, A. Onoe, K. Terabe, and K. Kitamura, *Appl. Phys. Lett.* **81**, 4401 (2002).
[35] K. Fujimoto and Y. Cho, *Appl. Phys. Lett.* **83**, 5265 (2003).
[36] K. Franke, J. Besold, W. Haessler, and C. Seegebarth, *Surf. Sci. Lett.* **302**, 283 (1994).

[37] A. Gruverman, O. Kolosov, J. Hatano, K. Takahashi, H. Tokumoto, *J. Vac. Sci. Technol.* **B13**, 1095 (1995).
[38] L. Eng, *et al.*, *Ferroelectrics* **222**, 153 (1999).
[39] L. Eng, H. Gumtherodt, G. Rosenman, A. Skliar, M. Oron, M. Katz, and D. Eger, *J. Appl. Phys.* **83**, 5973 (1998).
[40] A. Gruverman, O. Kolosov, J. Hatano, K. Takahashi, and H. Tokumoto, *J. Vac. Sci. Technol.* **B13**, 1095 (1995).
[41] L. M. Eng, M. Bammerlin, C. Loppacher *et al.*, *Ferroelectrics* **222**, 421/163.
[42] K. Franke, J. Besold, W. Haessler, and C. Seegebarth, *Surf. Sci. Lett.* **302**, L283 (1994).
[43] A. Gruverman, O. Auciello, J. Hatano, and H. Tokumoto, *Ferroelectrics* **184**, 11 (1996).
[44] A. Agronin, Y. Rosenwaks, and G. Rosenman, *Appl. Phys. Lett.* submitted (2004).
[45] M. Yamada and K. Kishima, *Elect. Lett.* **27**, 828 (1991).
[46] M. Fujimura, K. Kintaka, T. Suhara, and H. Nishihara, *Elect. Lett.* **28**, 1868 (1992).
[47] H. Ito, C. Takyu, and H. Inaba, *Elect. Lett.* **27**, 1222 (1991).
[48] A. C. G. Nutt, V. Gopalan, and M. C., Gupta *Appl. Phys. Lett.* **60**, 2828 (1992).
[49] C. Restoin, C. Darraud-Taupiac, J. L. Decossas, J. C. Vareille, Hauden J., and A. Martinez, *J. Appl. Phys.* **88**, 6665 (2000).
[50] A. C. G. Nutt, V. Gopalan, M. C. Gupta, US Patent 5,748,361 (1998).
[51] M. Fujimura, T. Suhara, and H. Nishihara, *Elect. Lett.* **28**, 721 (1992).
[52] M. Fujimura, K. Kintaka, T. Suhara, and H. J. Nishihara, *Light. Tech.* **11**, 1360 (1993).
[53] J. He, S. H. Tang, Y. Q. Qin, P. Dong, H. Z. Zhang, C. H. Kang, W. X. Sun, and Z. X. J. Shen, *Appl. Phys.* **93**, 9943 (2003).
[54] Nihei, *et al.*, US Patent 5,668,578 (1997).
[55] C. Restoin, C. Darraud-Taupiac, J. L. Decossas, J. C. Vareille, and J. Hauden, *Mat. Sci. In Semi. Proc.* **3**, 405 (2000).
[56] P. K. Watson, IEEE., *Trans. Dielectr. Electr. Insul.* **2**, 915 (1995).
[57] Z. G. Song, H. Gong, and C. K. Ong, *J. Phys. D: Appl. Phys.* **30**, 1561 (1997).
[58] K. Kanaya and S. Okayama, *J. Phys. D: Appl. Phys.* **44**, 2495 (1972).
[59] J. Mompart, C. Domingo, C. Baixeras, and G. Fernandez, *Nucl. Instrum. Methods* **B107**, 56 (1996).
[60] C. K. Ong, Z. G. Song, and H. Gong, *J. Phys: Condens. Matter* **9**, 9289 (1997).
[61] O. Jbara, S. Fakhfakh, M. Belhaj, J. Cazaux, E. I. Rau, M. Filippov, and M. Andrianov, *Nucl. Instrum. & Methods in Phys. Res.* B **194**, 302 (2002).
[62] M. Molotskii, M. Shvebelman, A. Agronin, G. Rosenman, and Y. Rosenwaks, in *The 10-th European Meeting on Ferroelectricity*, EFM 2003, J. Conf. Absracts **8**, 233 (2003); M. Molotskii, M. Shvebelman, *Ferroelectrics*, to be published(2004).
[63] R. Landauer, *J. Appl. Phys.* **28**, 227 (1957).
[64] M. Molotskii, *J. Appl. Phys.* **93**, 6234 (2003).
[65] R. Erlandsson, G. M. McClelland, C. M. Mate, and S. Chiang, *J. Vac. Sci. Technol.* A**6**, 266 (1988); B. D. Terris, J. E. Stern, D. Rugal, and H. J. Mamin, *Phys. Rev. Lett.* **63**, 2669 (1989); G. Zavala, J. H. Fendler, S. Trolier-McKinstry, *J. Appl. Phys.* **81**, 7480 (1997); K. Goto, K. Hane, *J. Appl. Phys.* **84**, 4043 (1998); F. J. Giessibl, *Rev. Modern Phys.* **75**, 949 (2003).

[66] O. Kolosov, A. Gruverman, J. Hatano, K. Takahashi, and H. Takumoto, *Phys. Rev. Lett.* **74**, 4309 (1995).

[67] A. Gruverman, O. Kolosov, J. Hatano, K. Takahashi, and H. Takumoto, *J. Vac. Sci. Technol.* B**13**, 1095 (1995).

[68] C. Durkan, M. E. Welland, D. P. Chu, and P. Migliorato, *Phys. Rev.* B **60**, 16198 (1999); *Appl. Phys. Lett.* **76**, 366 (2000).

[69] W. Wang and C. H. Woo, *J. Appl. Phys.* **94**, 4053 (2003).

[70] G. Van der Zwan and R. M. Mazo, *J. Chem. Phys.* **82**, 3344 (1985).

[71] S. V. Kalinin and D. A. Bonnell, *Phys. Rev.* B**65**, 125408 (2002).

[72] E. J. Mele, *Am. J. Phys.* **69**, 557 (2001).

[73] V. Gopalan, N. A. Sanford, J. A. Aust, K. Kitamura, and Y. Furukawa, in *Handbook of Advanced Electronic and Photonic Materials and Devices: Ferroelectrics and Dielectrics*, edited by H. S. Nalwa, Academic Press, New York, **4**, 57 (2001).

[74] M. Molotskii and M. Shvebelman (unpublished).

[75] S. Pöykkö and D. J. Chadi, *J. Phys. Chem. Solids.* **61**, 291 (2000); B. Mayer and D. Vanderbilt, *Phys. Rev.* B **65**, 104111 (2002).

[76] H. W. Hao, A. M. Baro, and J. J. Saenz, *J. Vac. Sci. Technol.* B **9**, 1323 (1991); S. Belaidi, P. Girard, and G. Leveque, *J. Appl. Phys.* **81**, 1023 (1997); S. Hudlet, Jean M. Saint, C. Guthmann, J. Berger, *Eur. Phys. J.* B **2**, 5 (1998); *J. Appl. Phys.* **86**, 5245 (1999); S. V. Kalinin, D. A. Bonnell, *Phys. Rev.* B **62**, 10419 (2000).

[77] A. Fein, Y. Zhao, C. A. Peterson, G. E. Jabbour, D. Sarid, *Appl. Phys. Lett.* **79**, 3935 (2001); G. Lubarsky, R. Shikler, N. Ashkenasy, Y. Rosenwaks, *J. Vac. Sci. Technol.* B **20**, 1914 (2002); L. J. Klein, C. C. Williams, *Appl. Phys. Lett.* **81**, 4589 (2002); *J. Appl. Phys.* **95**, 2547 (2004).

[78] M. Abplanalp, J. Fousek, P. Günter, *Phys. Rev. Lett.* **86**, 5799 (2001); S. V. Kalinin, D. A. Bonnell, *Phys. Rev.* B **65**, 125408 (2002).

[79] V. Gopalan, T. E. Mitchell, K. E. Sicakfus, *Solid State Commun* **109**, 111 (1999).

[80] G. Rosenman, P. Urenski, A. Agronin, Y. Rosenwaks, M. Molotskii, *Appl. Phys. Lett.* **82**, 103 (2003).

[81] R. E. Loge and Z. Suo, *Acta mater* **44**, 3429 (1996).

[82] B. Wang and Z. Xiao, *J. Appl. Phys.* **88**, 1464 (2000).

[83] F. Davi and P. M. Mariano, *J. Mech. Phys. Solids* **49**, 1701 (2001)

[84] M. Molotskii, *Philos. Mag. Lett.* **83**, 763 (2003).

[85] B. E. Cooper, S. R. Manalis, H. Fang, H. Dai, K. Matsumoto, S. C. Minne, T. Hunt, and C. F. Quate, *Appl. Phys. Lett.* **75**, 3566 (1999).

[86] M. Zhang, D. Bullen, S.-W. Chung, S. Hong, K. S. Ryu, Z. Fan, C. A. Mirkin, and C. Liu, *J. Nanotechnology* **13**, 212 (2002).

[87] S. Hong and C. A. Mirkin, *Science* **288**, 1808 (2000).

[88] F. M. Battiston, J.-P. Ramseyer, H. P. Lang, M. K. Baller, Ch Gerber., J. K. Gimzewski, E. Meyer, and H.-J. Guntherdot, *Sensors and Actuators* B **77**, 122 (2001).

11 Pyroelectric Ceramics and Thin Films: Characterization, Properties and Selection

Roger W. Whatmore

School of Industrial and Manufacturing Sciences,
Cranfield University, Cranfield, MK43 0AL, UK

Abstract

Pyroelectric devices have become established as one of the most important technologies for the low cost detection of long wavelength infra-red radiation, for a host of applications ranging from intruder sensing, pollution control and people monitoring to uncooled thermal imaging. There are a large number of pyroelectric materials and methods for their characterization and the criteria for their selection in any particular application is an important issue. This paper reviews the physics of pyroelectric detectors from the point-of-view of the selection of the pyroelectric material for a given application and then discusses the methods for characterising those properties. Finally, it discusses and compares the properties of a range of pyroelectric ceramics and thin films.

11.1 Introduction

The use of the pyroelectric effect for the detection of infra-red radiation in the 3–5 μm and 8–12 μm IR bands has been known for many years [1] and has been extensively commercialized in applications such as people sensing, environmental monitoring and gas analysis [2]. There has also been, for about 25 years, a strong interest in the development of pyroelectric devices for thermal imaging [3]. Solid state arrays are of particular interest [4]. These have been made using a thin ferroelectric ceramic wafer which is bonded to a 2D array of amplifiers and multiplexer switches integrated on an application specific silicon integrated circuit (ASIC). Companies such as BAE SYSTEMS Infra-red Ltd. in the UK have developed arrays with 128×256 and 384×288 elements [3], while Raytheon in the USA have developed an array with 320×240 elements [5]. The potential for integrating ferroelectric thin films directly onto silicon substrates has been recogniued as a means for both reducing array fabrication costs and increasing performance through reduced thermal mass and improved thermal isolation. This has encouraged the development of fully integrated 2-D arrays and encouraged the low temperature growth ($< 550°$C) of ferroelectric thin films such as lead zirconate titanate (PZT) directly onto active silicon devices [6]. The purpose of this paper is to review the materials issues involved in pyroelectric infra-red detectors and arrays, with specific reference to the characterization of ferroelectric ceramics and thin films, and to discuss some examples of their fabrication and use.

11.2 The physics of pyroelectric detectors

11.2.1 Pyroelectric response

The pyroelectric coefficient is a vector π, equal to the rate of change of the spontaneous polarization P_S, with temperature T:

$$\Delta P_S = \pi \Delta T \tag{11.1}$$

(Note that most, but not all, detectors are made so that the direction of π is normal to the element electrode plane, i.e. $p = |\pi|$. In the discussion which follows, it will be assumed that this is the case and the term "pyroelectric coefficient" will be applied to p.) The pyroelectric charges can be detected as a current i_p, flowing in an external circuit such that:

$$i_p = ApdT/dt \tag{11.2}$$

dT/dt is the rate of change of the element's temperature with time and thus pyroelectric devices are coupled to the changes of any input energy flux which generates a change in element temperature. This energy input is usually in the form of electromagnetic radiation absorbed in the material (or a coating which is "black" at the wavelength of interest).

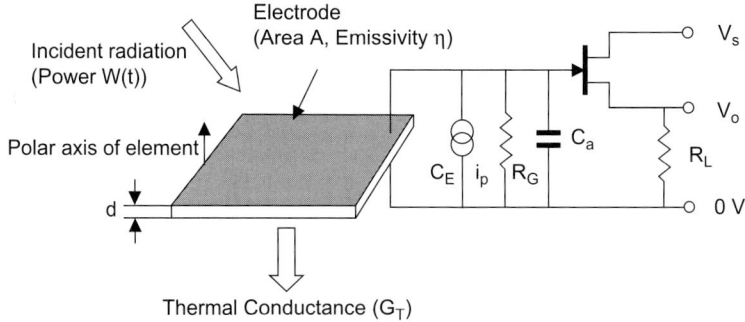

Figure 11.1: Schematic diagram of a pyroelectric detector element

Figure 11.1 shows a schematic diagram of a pyroelectric detector. Its analysis requires consideration of both the thermal and electrical circuits as described by Cooper [1]. Radiation with power $W(t)$ sinusoidally modulated at a frequency ω, so that $W(t) = W_0 e^{i\omega t}$ is incident on the surface of the element (area A and thickness d) which has emissivity η. The element has a thermal capacity H and a thermal conductance to the surroundings G_T giving a thermal time constant $\tau_T = H/G_T$. A high-input-impedance field effect transistor (FET) used as a source follower amplifier amplifies the electrical signal produced. R_G is the gate resistor, giving an

11.2 The physics of pyroelectric detectors

electrical time constant $\tau_E = R_G(C_E + C_A)$ where C_E is the capacitance of the element and C_A the capacitance of the amplifier. The τ_T and τ_E are the fundamental factors which determine the frequency response of the detector. The temperature difference θ between the element and its surroundings is described by the following equation:

$$\eta W(t) = H \frac{d\theta}{dt} + G_T \theta \tag{11.3}$$

which has the solution:

$$\theta = \frac{\eta W_0}{G_T + i\omega H} e^{i\omega t} \tag{11.4}$$

Thus, from Equation (11.2), the pyroelectric current i_p, generated per watt of input power (the current responsivity R_i) is:

$$\frac{i_p}{W_0} = R_i = \frac{\eta p A \omega}{G_T(1 + \omega^2 \tau_T^2)^{1/2}} \tag{11.5}$$

The form of this response is simple. At low frequencies ($\omega \ll 1/\tau_T$) the response is proportional to ω. For frequencies much greater than $1/\tau_T$ the response is constant, being:

$$R_i = \frac{\eta p A}{H} = \frac{\eta p}{c' d} \tag{11.6}$$

where c' is the volume specific heat.

The voltage responsivity of the detector shown in Figure 11.1 is simply derived from the pyroelectric current i_p and the electrical admittance Y presented to it. Ignoring for the moment the AC conductance of the pyroelectric element:

$$Y = \frac{1}{R_G} + i\omega C \tag{11.7}$$

where $C = C_E + C_A$. The pyroelectric voltage V_p generated on the gate of the FET (and therefore the output voltage V_o for a unity amplifier gain) is given by i_p/Y and the voltage responsivity is R_V where:

$$R_V = \frac{i_p}{YW_0} = \frac{R_G \eta p A \omega}{G_T(1 + \omega^2 \tau_T^2)^{1/2}(1 + \omega^2 \tau_E^2)^{1/2}} \tag{11.8}$$

The general form of Equation (11.8) is shown in Figure 11.2 for a 100 μm^2, 25 μm thick ceramic detector element feeding into a typical MOSFET amplifier and with a thermal conductance of 20 mW/K. It is easy to show from this equation that the maximum value of R_V is maximized by minimizing the element's thermal capacity (i.e. making it thin) and by minimizing G_T (i.e. thermally isolating it from its surroundings).

The high-frequency ($\omega \gg 1/\tau_T$ and $\omega \gg 1/\tau_E$) dependence of R_V is given by:

$$R_V = \frac{\eta p}{c' d (C_E + C_A) \omega} \tag{11.9}$$

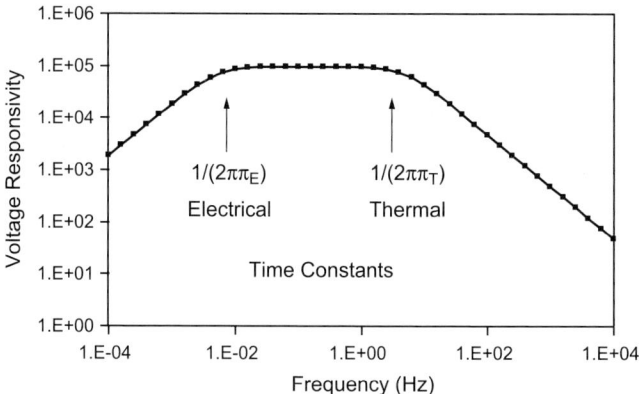

Figure 11.2: Voltage responsivity vs frequency for a typical small ceramic pyroelectric element.

If the element capacitance C_E:

$$C_E = \frac{\varepsilon \varepsilon_o A}{d} \tag{11.10}$$

is large compared with C_A then Equation (11.9) reduces to :

$$R_V = \frac{\eta p}{c' \varepsilon \varepsilon_o A \omega} \tag{11.11}$$

where ε_o is the permittivity of free space and ε is the relative permittivity of the pyroelectric material. For such a detector, the response is proportional to F_V:

$$F_V = \frac{p}{c' \varepsilon \varepsilon_o} \tag{11.12}$$

This only contains parameters describing properties of the pyroelectric material, and is therefore a figure-of-merit, which can be used to compare different materials for their potential voltage responsivities. If $C_A \gg C_E$ the voltage response is proportional to:

$$F_i = \frac{p}{c'} \tag{11.13}$$

which is a different material figure-of-merit. Hence, it can be seen that the choice of the appropriate material will depend both on the amplifier to be used (JFET for example, have higher input capacitances than MOSFET) and on the size of the detector element. Small area detector elements will tend to favour high dielectric constant materials and vice versa, to produce a good match between C_A and C_E. The discussion of figures-of-merit has thus far been confined to single-element detectors. If there are a number of elements or detector points defined on a single plate of material, lateral thermal diffusion in the plate will tend to "smear-out" high spatial frequency information. In this case, it would be appropriate to introduce the thermal diffusivity κ in the denominator of these figures-of-merit, so that lower diffusivity materials will be favoured. However, the introduction of technologies (e.g. "reticulation") to thermally disconnect adjacent elements would eliminate the need for this consideration.

11.2.2 Comparison of noise and signal

It is evidently insufficient to consider only the response of a detector when analysing its usefulness for a particular application. It is generally necessary to analyse both intrinsic and extrinsic noise signals and compare them with the response. The result of this comparison can be expressed in many different ways. One of the most useful is the noise-equivalent power NEP which is the power of an RMS signal input (in watts) required to give a response equal to the total RMS noise voltage ΔV_N. Then:

$$\text{NEP} = \frac{\Delta V_N}{R_V} \quad (11.14)$$

The NEP is usually expressed in units of $\text{WHz}^{-1/2}$ (being specified for a particular frequency and unit bandwidth), or occasionally in W (being the ratio of broadband noise to responsivity at a specified frequency). A figure-of-merit frequently used in the discussion of detector performance is the specific detectivity D^*, which is defined as:

$$D^* = \frac{A^{1/2}}{\text{NEP}} \quad (11.15)$$

normalized to unit bandwidth. (In MKS units D^* is expressed in units of $\text{mHz}^{1/2}\text{W}^{-1}$, but the mixed units $\text{cmHz}^{1/2}\text{W}^{-1}$ are more-commonly used.)

Putley [7] has discussed the primary electronic noise sources for a typical pyroelectric detector circuit. These are: ΔW_T, the statistical fluctuations in the thermal power flow from the detector element to the heat sink; ΔW_J the Johnson noise in the equivalent circuit resistance R_T, Δi_A the amplifier current noise and ΔV_A the amplifier voltage noise. Each of these primary noise sources [8], will have an equivalent voltage generator at the amplifier input (ΔV_T, ΔV_J, ΔV_i and ΔV_A respectively for the above sources). The most important source of electrical noise in the frequency range of interest for most applications, especially thermal imaging, is the Johnson noise, if the amplifier noise sources are minimized. The total electrical conductance g (consisting of the parallel conductances of gate bias resistor R_G and the AC conductance of the detector element g_D) gives rise to the Johnson noise. It can be shown that for frequencies much less than ω_G, where $\omega_G = (R_G C_E \tan \delta)^{-1}$, then ΔV_J is given by:

$$\Delta V_J = (4kTR_G)^{1/2}(1 + \omega^2 \tau_E^2)^{-1/2} \quad (11.16)$$

where k is Boltzmann's constant. This leads to an ω^{-1} dependence at frequencies $\omega \gg \tau_E^{-1}$. For frequencies greater τ_E^{-1} and ω_G the Johnson noise generated by the AC conductance of the detector element frequently dominates in pyroelectric detectors so that:

$$D^* = \frac{R_V A^{1/2}}{\Delta V_J} = \frac{\eta d}{(4kT)^{1/2}} \frac{p}{c'(\varepsilon \varepsilon_o \tan \delta)^{1/2}} \frac{1}{\omega^{1/2}} \quad (11.17)$$

Thus, to maximize D^* in the region where Johnson noise dominates, it is desirable to maximize:

$$F_D = \frac{p}{c'\sqrt{\varepsilon \varepsilon_o \tan \delta}} \quad (11.18)$$

which forms the third figure-of-merit for pyroelectric materials. (This is frequently expressed in units of $Pa^{-1/2}$). The total equivalent input noise ΔV_N is given by taking the RMS summation of the individual components. Porter [9] has considered the relative magnitudes for a detectors with different areas. For a typical detector in which the capacitance of the element is similar to that of the amplifier, the loss-controlled Johnson noise tends to dominate above about 20 Hz, while below this frequency the resistor-controlled Johnson noise and the amplifier current noise contribute almost equally to the total noise.

11.2.3 Other sources of noise

Pyroelectric detectors will give unwanted output voltages due to environmental effects such as rapid changes in ambient temperature or mechanical vibration (which couples via the piezoelectric effect, as discussed below). These effects can be minimized by using a compensated structure in which a reverse-poled element is placed in series with the active detector element, but coated with a reflecting electrode and/or mechanically screened so that it is not subject to the input radiation flux. The compensating element should be placed in a position which makes it thermally and mechanically similar to the detector element. In this case, signals due to temperature changes or mechanical stresses can be cancelled to a large extent. The former is fairly easy to achieve but the latter requires considerable care in mechanical design. Putley [10] has discussed the electrical characteristics of such detectors. If the elements are similar in size so that their capacitances and conductances match, then the compensation will degrade the responsivity and NEP by a factor of between $\sqrt{2}$ and 2, according to the frequency range and the dominant noise mechanism. If, however, the compensation area is arranged to be much greater than the detector area, then the responsivity and NEP will be little affected. Pyroelectrics should also be rigorously screened from sources of electromagnetic interference. This generally entails containing them in an earthed metallic (or metallized) package with, if possible, a conductive, earthed window. Germanium is particularly suitable for infrared detectors as it both transmits in the wavebands of interest and is conductive. It can also be readily coated with multilayer spectral filters.

11.2.4 The piezoelectric effect in pyroelectric detectors

All pyroelectric materials are also piezoelectric (i.e. they generate charge in response to mechanical strain). This has a number of important consequences:

a. The piezoelectric coefficients couple to thermally induced strains so that the pyroelectric coefficients measured under constant stress and constant strain are different. The pyroelectric coefficient at constant strain (when the material is mechanically "clamped") is usually called the "primary" pyroelectric coefficient. This is related to the "unclamped" value measured at constant stress by the addition of a summed product of the piezoelectric, elastic stiffness and thermal expansion tensor coefficients, usually referred to as the "secondary" pyroelectric coefficient [11]. For ferroelectrics, the secondary is generally much smaller in magnitude than the primary effect, although not always of the same sign. For low-frequency detectors in which the elements are free to expand the appropriate coefficient is the sum of the primary and secondary (i.e. the mechanically unclamped) coefficients. However, for detector elements which are totally or partially clamped, it is necessary to take account of the elimination of all or part of the secondary effect. This can occur: (i) for fast detectors where the radiation pulse frequency is higher than one or more of the mechanical resonant frequencies of the detector element; (ii) for thin film detectors [12] where there are appreciable substrate clamping effects; and (iii) for elements in which only a small portion is heated so that the active area is clamped by the rest of the element [13]. The secondary effect can also be important if a detector is subjected to an AC radiation flux whose frequency matches one of the element's mechanical resonant frequencies, or if the pyroelectric material is clamped to a supporting substrate with an appreciably different thermal expansion coefficient. The lateral stress in the pyroelectric caused by bending can generate large piezoelectric contributions to the apparent pyroelectric effect. It is believed that this may be the cause of some discrepancies in pyroelectric coefficients measured on thin film materials mounted in different ways in different laboratories.

b. Pyroelectric detectors used in a high vibration or acoustically noisy environment will produce a microphonic noise signal which can be very significant if steps are not taken to suppress it. Shorrocks *et al.* [14] have discussed methods by which the microphony of pyroelectric arrays may be reduced to a very low level. Minimizing microphony can be achieved by combining various techniques: (i) placing the detector in a very rigid package, designed to minimize flexure; (ii) de-coupling the detector package from deformations in the structure to which it is mounted; (iii) using a compensated detector structure; (iv) mechanically isolating active and compensation areas by reticulation (this reduces lateral stresses due to package flexure and improves the strain match between active and compensation elements). One technique for reduction of microphony involves freely suspending the active elements on a thin polymer film [15], which effectively does (ii) above.

11.3 Measurement of physical parameters

The section above considered the physical properties that are important in determining the performance of a pyroelectric detector, measurement methods used to obtain the physical parameters which determine pyroelectric performance are worthy of critical discussion, as many techniques are reported in the literature.

11.3.1 Dielectric properties

There are a wide variety of measurement systems available that will determine capacitance and loss. These include capacitance bridges (manual and auto-balance), impedance analysers, network analysers etc. It is important for the user to consider the frequency range over which the pyroelectric devices are to be used, as this will largely determine the selection of the instrument to use. As most pyroelectric detectors are used in the range 0.1 to 100 Hz, the instrument should ideally permit measurement over this frequency range. There are very few commercial capacitance and impedance analysers that will work below 20 Hz, and many low cost units will not work below 1 kHz. A further important factor in instrument selection is its ability to measure low losses at low frequencies. As most pyroelectrics are selected to possess low loss, at low frequency this leads to low electrical conductance, which can be outside the specified range for the instrument. It is particularly important to avoid quoting pyroelectric figures-of-merit in which dielectric properties are used that are not measured in the relevant frequency range. Unfortunately, this is all too common in the literature. Ferroelectrics frequently have strongly frequency-dependent dielectric properties. This can be due dielectric relaxation phenomena. Alternatively, at the very low frequencies of interest to pyroelectric devices, one can observe a rise in loss due to free carrier effects. In many cases the dielectric loss is significantly higher at 30 Hz than at 1 kHz, so that dielectric properties measured in the higher frequency ranges tend to give an over-optimistic view of the pyroelectric figures-of-merit, especially F_D. One needs to be conscious of edge effects upon dielectric measurements. These are less important for high permittivity materials if the electrodes come right to the edge of the sample. The use of guard-ring techniques can also help ameliorate such effects.

11.3.2 Pyroelectric properties

The measurement of pyroelectric coefficients is the area which is most likely to lead to problems, which can be inherent to the method used, or due to confusing factors such as the release of thermally stimulated currents (TSC) as the temperature of the sample is changed, or occasionally a permanent loss of ferroelectric polarization due to depoling. The latter is frequently a problem with polycrystalline ferroelectrics as T_C is approached, although sometimes this effect can be troublesome at over 100°C below T_C. In the case of measurements on ferroelectrics which must be electrically poled before measurement, it is a good idea to anneal the specimen for an hour or two after poling at a temperature somewhat above the maximum temperature to be used (although obviously below T_C) with the electrodes short-circuited, prior to making any measurements. An essential aspect of the pyroelectric effect is that it should be reversible, and any method used should take this into account. TSC and depoling currents are unidirectional and frequently decay with time if the sample is held at constant temperature. The most commonly-used methods are those described by Chynoweth [16], Lang and Steckle [17], Glass [18] and Byer and Roundy [19]. The first of these (after Chynoweth) involves illuminating the specimen with a chopped-source of radiation from a hot filament source and measuring the resulting pyroelectric current or charge as a consequence of the temperature changes in the material. This is illustrated schematically in Figure 11.3 (a) for the measurement of pyroelectric current. The method has the great advantage that the temperature of the specimen is continuously cycled by a small amount about an average temperature and the average current produced is reversible. The disadvantage of the method is that the amount of radiation absorbed is dependent upon the emissivity of the electrode, so that it can be difficult to be certain of the actual temperature change in the specimen. Some authors have used chopped visible or near-IR radiation (e.g from a gas or solid state laser) for the same purpose. This has the same inherent problem, but in some cases the results can also be complicated by the release of currents due to photoelectric effects. It should be emphasized that the pyroelectric currents released, especially from small area specimens, can be extremely small (in picoamps) and a good, low noise current amplifier is essential. Care is also required to ensure that the time-waveform of the illuminating radiation is truly sinusoidal, or the effects of the higher frequency components of the illuminating waveform must be taken into consideration. In the case of square-wave illumination, the intensity of the higher harmonics can be considerable, considerably distorting the frequency response curve from the ideal [20].

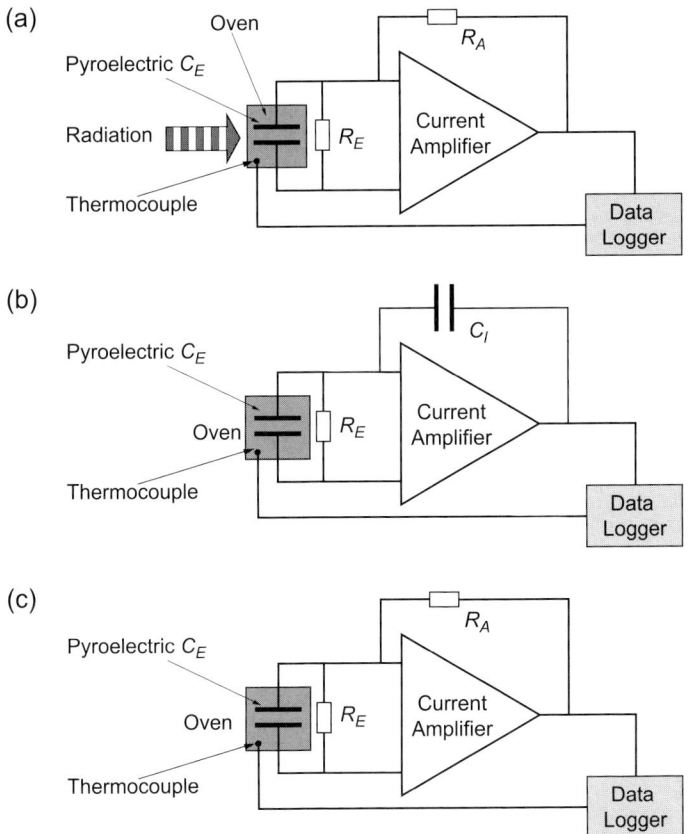

Figure 11.3: Showing schematically three different techniques that can be used to measure the pyroelectric coefficient. (a) After Chynoweth, where the pyroelectric material is illuminated by a chopped source of radiation and the pyroelectric current measured, (b) the charge integration method described by Lang, Steckl and Glass where the pyroelectric charge is integrated using a charge amplifier and, (c) the method described by Byer and Roundy where the sample temperature is changed with time and the poyroelectric current is measured.

In the methods described by Lang and Steckle [17] and Glass [18], the pyroelectric charges released as the temperature of the specimen is changed are integrated either on the capacitance of the specimen or in a charge amplifier (as shown schematically in Figure 11.3(b)). The pyroelectric coefficient can be calculated from the gradient of the charge/temperature curve. This method is easily confused by TSC or depoling currents, and it is essential to measure the charge curves on both heating and cooling. If there is a difference between them, then account must be taken of the disturbing effects. The method described by Byer and Roundy [19] is similar, but relies on heating and cooling the specimen at known rates and measurement of the resulting pyroelectric currents as shown in Figure 11.3(c). Again, it is essential to check reversibility. Sharp and Garn [21, 22] have described a very useful variation on this method which uses a thermoelectric heater/cooler element to cycle a specimen sinusoidally about a mean temperature. Molter et al. [39] have used this method and shown that when the temperature of the pyroelectric sample was oscillated sinusoidally by $\pm 2°C$ about a mean temperature, the pyroelectric coefficient can be computed from the measured the pyroelectric current as follows:

$$p(T) = \frac{i_{0m}}{\Delta T A \omega} \cdot \sqrt{\left(\frac{R_e}{R_S} + 1\right)^2 + \omega^2 R_e^2 C^2} \qquad (11.19)$$

where i_{0m} is the average of the maximum and minimum currents, ω is the angular frequency of temperature oscillation of amplitude ΔT, A is the electrode area, R_e is the resistance of the electrometer (this depends on the instrument range and can be taken from the electrometer specification), and R_S is the resistance of the sample, measured as described below. A typical plot of temperature variation and pyroelectric current generated as functions of time are shown in Figure 11.4.

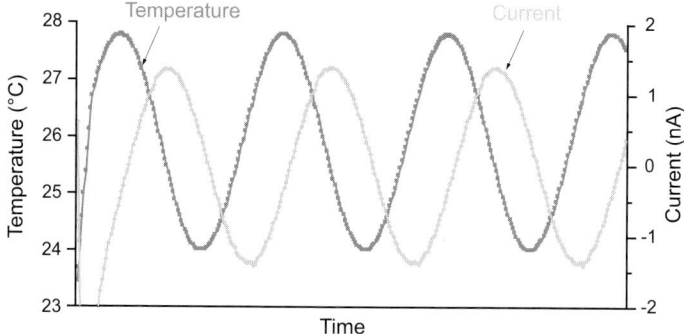

Figure 11.4: Showing the sinusoidal variation of temperature with time in a typical measurement as described in [20]. Note that the temperature and current plots are 90° out of phase.

The advantages of this method in comparison with the Byer-Roundy technique are that it is less susceptible to effects due the thermally stimulated release of trapped charge. Furthermore, it permits consideration of the leakage of charge through the resistance represented by the sample, which can be significant at the high levels of doping.

As a final note on this section, whichever method is used, it is also necessary to consider how the pyroelectric is mechanically mounted in the system. Any mechanical clamping (such as bonding the material to a substrate of different thermal expansion coefficient) can lead to spurious currents due to changes in the stress on the sample, coupling via one or more of the piezoelectric coefficients.

11.3.3 Electrical resistivity

The bias resistor R_G in Figure 11.1 performs an important funtion in the circuit, in that it both sets the electrical time constant and also biases the gate of the FET. For single element detectors, it is possible to include this as an explicit circuit element, in which case a very high value component is needed (ca $10^{12}\,\Omega$ for a 10 s time constant with $C = 10\,\mathrm{pF}$). However, such components are expensive. For large arrays, it is impossible to include a separate resistor for each element. In this case, it is desirable to be able to use the intrinsic DC resistivity ρ of the element to provide the required resistance. In this case, we can set the electrical time constant from $\tau_e = \rho\varepsilon\varepsilon_o$, so that for a ceramic with $\varepsilon = 300$, we need $\rho = 4 \times 10^9\,\Omega\mathrm{m}$. There are thus two issues to address. The first is the best method for measuring the DC resistivity in the range 10^8 to $10^{12}\,\Omega\mathrm{m}$. The second is the method for controlling it. While many excellent commercial electrometers are available that will permit the accurate measurement of high resistances, the accurate measurement of very high DC resistivities is not trivial. Surface moisture is a major problem and the measurements should ideally be carried out under vacuum, after heating the specimen to expel any adsorbed moisture from the surface. For the most-accurate measurements, guard-ring methods should be used. A further consideration in the measurement of resistivity is the relaxation time associated with the polarization of immobile charges in the material. This is particularly important for materials that have only moderately-high resistivities. It leads to a peak in current immediately after applying the DC test voltage, followed by an exponential decay in the current to a stable value. The time constant of this decay can be many minutes long and the best way to get the "DC" value of the resistivity is to measure the current as a function of time and fit it to an equation of the form $i(t) = i_0 e^{1/t}$ where $i(t)$ is the current at time t and i_0 is the current at infinite time. The values of i_0 can be obtained by extrapolation of $\ln(i(t))$ vs $1/t$ and used to compute the DC resistivities at "infinite" time.

11.3.4 Thermal properties

There are many excellent differential scanning calorimeter systems available which can be used to measure the specific heat which, when combined with the sample density can be used to give c'. The thermal diffusivity (which can be important for thermal imaging systems if the target is not reticulated) can be measured directly on a pyroelectric substrate using the laser intensity modulation method described by Lang [23].

11.3.5 Piezoelectric property determination

The most important piezoelectric property from the point-of-view of generating microphonic noise is the d_{31} coefficient. This is because noise is generated due to the flexing of the pyroelectric material/substrate combination, which stresses the pyroelectric in the plane perpendicular to the polar axis. For bulk ceramic materials, d_{31} can be determined by measuring the radial mode coupling factor for a thin flat disk. For thin and thick film materials, a variety of techniques exist that are based upon the flexure of the plate carrying the film. These techniques include those based on the flexure of a whole plate [24] or modified Berlincourt technique based on the flexure of a small section of wafer placed on a ring support [25].

11.4 Pyroelectric materials and their selection

It was noted in the introduction that all polar materials will exhibit the pyroelectric effect, which means that there is potentially a huge range of materials to chose from for use in devices. The material figures-of-merit F_V, F_i and F_D derived the previous section and given in Equations (11.12), (11.13) and (11.18) respectively form a useful guide to the selection of a particular material for a given application. Normally, for a detector element and amplifier with well-matched capacitances, and in which amplifier noise has been minimized so that the device noise is dominated by the dielectric-loss generated Johnson noise in the element, then maximising F_D will maximize the detectivity. However, for frequencies below ω_G, or if one of the other noise sources is dominant and the capacitance of the element dominates the input capacitance of the amplifier, then maximising F_V will maximize the detectivity. On the other hand, if the capacitance of the element is small in comparison with that of the input capacitance of the amplifier, and the noise sources other than that due to the dielectric loss of the element are dominant, then maximising F_i will maximize the detectivity. Hence, it is not sufficient to say that the best strategy is always to choose a material with the highest F_D.

Furthermore, in certain types of device, the physical design of the detector element is constrained by the application and/or fabrication technologies and this can have an effect of the selection of materials. For example, in solid state thermal imaging arrays it is necessary to put a certain number of detector elements into an area which is usually restricted by the size of the electronic chip used to read out the pyroelectric signals, and the size of the optics. This in turn limits the area of the detector elements. (A typical example would be an array of 80 μm square elements on a 100 micron pitch in a 100 × 100 imaging array.) In a well-matched detector we would like the element capacitance to be of the same order as the amplifier input capacitance, which will typically be about 1 pF. It is thus important, if performance is to be optimized, to select a material which possesses a dielectric constant which will give a good capacitance match for a given element thickness d. d is determined by two considerations, one theoretical and one practical. Intuitively, it might be thought best to minimize the element thickness (and therefore its thermal capacity H), to maximize the temperature change for a given energy input. However, it is necessary to consider the balance between H and the thermal conductance G_T at the required operating frequency of the device. For practical reasons, it may not be possible to reduce G_T below a certain value. A practical value for a particular design of array might be perhaps 10 to 20 μWK^{-1}. If the device is to operate at approximately 50 Hz, then a thermal time constant of 20 ms for the device would be appropriate. It is then simple to show that for our example of an 80 μm square element with a volume specific heat of 2.5×10^6 Jm^{-3}K^{-1}, the element thickness must be around 25 μm. Hence, for a C_E of 1 pF, the permittivity of the material should be around 440. A smaller element (say 40 × 40μm) will require a material with an even higher permittivity (approximately 1000), unless the thermal conductance can be lowered by an improvement in the thermal isolation technology. Thus, choosing a material with a higher F_D may not necessarily give a better performance than another with a lower F_D if it has a permittivity much lower than this value. The second consideration determining element thickness is technological. Many of the best pyroelectric materials are brittle oxides, which will have a tendency to break during handling when in very thin sections. The attainable thickness may push the selection of a material towards one with a higher dielectric constant. This simple example serves to show that, while the figures-of-merit listed above can be useful in guiding one in the right general direction for selecting a material, it is essential to consider the full details of any device's operation and manufacturing technology in choosing the most appropriate material to use. These issues are discussed further below. A further consideration in the selection of materials is the frequency dependence of the dielectric properties. The vast majority of pyroelectric devices operate at very low frequencies (usually < 100 Hz, frequently < 10Hz). Most of the so-called "low-frequency" dielectric properties reported in the literature have been measured at 1 kHz or above. For many materials, the loss tangent rises rapidly as the frequency decreases in the range 100 to 0.1 Hz and pyroelectric figures-of-merit based on dielectric data measured in the range of 1 kHz are of questionable utility.

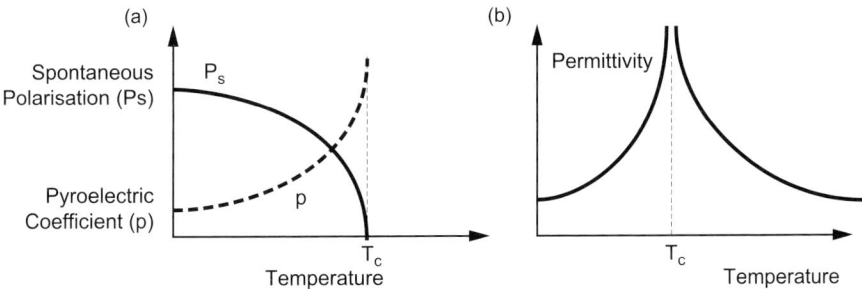

Figure 11.5: (a) Spontaneous polarization P_S and pyroelectric coefficient p and (b) permittivity as functions of temperature in a proper ferroelectric.

It was noted in the introduction that ferroelectrics show the largest pyroelectric coefficients by virtue of the changes in their spontaneous polarizations with temperature as the temperature increases towards T_C. This leads to a maximum in the pyroelectric coefficient at T_C as shown schematically in Figure 11.5 (a). The dielectric constant in a proper ferroelectric also increases and peaks at the Curie Temperature, as shown in Figure 11.5 (b). It is possible to show from Devonshire theory that in a proper ferroelectric in the absence of an electric bias field the ratio $p/\sqrt{\varepsilon}$, and hence the detectivity at constant $\tan\delta$, are constant below T_C, meaning that the devices require no thermal stabilization to maintain a stable performance. The major compromize as the temperature approaches T_C is an increase in dielectric loss due to an increased facility of domain wall movement, which tends to reduce F_D, and a tendency to de-pole the ferroelectric, giving a general loss in activity (especially for ceramics). This means that most ferroelectrics are operated well below their T_C. However, if an electric bias field is applied to the ferroelectric, the polarization and the pyroelectric effect will be stabilized in the region of and above T_C. This effect was first demonstrated by Chynoweth [26] for $BaTiO_3$ above T_C. It is also observed that the dielectric constant tends to reduce, as does the dielectric loss as the domain walls are effectively "swept-out" of the material by the field. The consequence of this is that very large pyroelectric coefficients in conjunction with large dielectric constants and low dielectric losses can be measured in the region of T_C. This is the principle of the "dielectric bolometer" [27] by which the pyroelectric coefficient is given by:

$$p = \left.\frac{dD}{dT}\right|_E \quad (11.20)$$

The following discussion separates pyroelectric materials into 3 groups: "intrinsic" pyroelectrics which are operated well below T_C, dielectric bolometer materials which are operated close to T_C, but with an electrical bias applied and ferroelectric thin films.

11.5 Pyroelectric ceramics and thin films

There are many different types of pyroelectric, including single crystals, polymers, ceramics and thin films and several reviews [2, 3, 28, 29] have considered the properties of many pyroelectric materials in detail, so the discussion here will be confined to a brief review of pyroelectric ceramics and thin films.

Material	T	p	Dielectric Properties		Freq.	c'	T_C	F_V	F_D	Ref.
	°C	$\frac{10^{-4}C}{m^2 K}$	ε	$\tan \delta$	Hz	$\frac{10^6 J}{m^3 K}$	°C	$\frac{m^2}{C}$	$\frac{10^{-5}}{Pa^{1/2}}$	
Mod. PZ (Ceramic)	25	4.0	290	0.003	1000	2.5	230	0.06	5.8	[30]
			300	0.014	33			0.06	2.6	
Mod. PT (Ceramic)	25	3.5	220	0.01	1000	2.5	> 250	0.07	3.2	[2]
			220	0.03	33			0.07	1.8	

Table 11.1: The properties of two pyroelectric ceramic materials

Many of the ceramic ferroelectrics that have been explored for their possible applications to pyroelectric devices are based upon the lead zirconate – lead titanate (PZT) perovskite solid solution series $\{1 - x\}PbZrO_3 - x\ PbTiO_3$. The compositions of greatest interest are located either close to the $PbZrO_3$ end of the phase diagram [30, 31] or close to lead titanate [32]. The dielectric constants and losses of these compositions are lower than the corresponding values close to those at the morphotropic phase boundary at $x = 0.47$, which are commonly used for piezoelectric device applications. Pyroelectric properties typical of these types of materials (referred to as Mod. PZ and Mod. PT respectively) are listed in Table 11.1. The materials are characterized by high dielectric constants (typically 200 to 300), and low, but strongly frequency dependent, dielectric losses. The figures-of-merit (especially F_D) for these materials look especially good if the 1 kHz dielectric properties are taken and the high dielectric constants of these materials make them very suitable for use in arrays of small detectors.

For these materials, it has been possible to quote the dielectric properties at 1 kHz and 33 Hz, to illustrate the differences in figures-of-merit (especially F_D) which are obtained by going to the frequency range of interest for device applications. The F_D figures are reduced by a factor of two. However, low frequency dielectric loss data for the other materials is hard to obtain from the literature, and therefore a direct comparison of the materials at the most relevant frequency range is not possible. This illustrates the difficulty of making valid materials comparisons simply on the basis of available data. Many studies of pyroelectric materials either fail to report the frequency at which measurements have been carried out, or specify a frequency which is well outside the range of interest for most devices.

There have been some attempts [33, 34] to exploit the $F_{R(LT)}$ to $F_{R(HT)}$ phase transition in the rhombohedral PZT ceramics for thermal detector applications. Recent work [35] on MnO_2 doped compositions in the $Pb(Mg_{1/3}Nb_{2/3})O_3$–$PbTiO_3$–$PbZrO_3$ system has shown that it is possible to realize pyroelectric coefficients at this phase transition that are a factor of 10 greater than those measured at room temperature, albeit over a relatively narrow temperature range and with some hysteresis in the transition, meaning that the exploitability of this effect is rather questionable. This system also gave excellent pyrorlectric coefficients and figures-of-merit when measured away from the phase transition.

The control of electrical resistivity in these ceramics is, as noted above, very desirable. There has been considerable success in the use of uranium as a dopant for this function in a variety of compositions [36–38]. Recently, chromium has also been shown to exhibit some promise for this [39].

It was noted above that, by operating a proper ferroelectric close to T_C with an electric bias field applied, very high effective pyroelectric coefficients can be obtained. Several materials have been examined in this mode, including $KTa_xNb_{1-x}O_3$ (KTN) [40], $Pb(Zn_{1/3}Nb_{2/3})O_3$ (PZN) [41], $Ba_xSr_{1-x}TiO_3$ (BST) [28, 42], $(Pb_{0.99}La_{0.01})(Mg_{1/3}Nb_{2/3})O_3$ (La01-PMN) [42] and $PbSc_{1/2}Ta_{1/2}O_3$ (PST) [43]. In characterizing these materials, it is necessary to measure their properties over a wide range of fields and temperatures. A typical set of properties [28] in a BST 67/33 ceramic at 0.6 V/μm bias field is $p = 70 \times 10^{-4} Cm^{-2}K^{-1}$, $\varepsilon = 8800$ and $\tan\delta = 0.004$, giving $F_D = 12.4 \times 10^{-5} Pa^{-1/2}$. The field and temperature dependence of the relevant properties of BST 65/35, La01-PMN [42] and PST [43] ceramics have all been published. In PST the p and ε are strongly peaked with temperature at low bias fields, but the peaks become lower and broader and shifted to higher temperatures at high bias fields. The effect of field on dielectric loss is to dramatically reduce it immediately below T_C and to increase it slightly above.

At low bias fields, the temperature dependence of the response and signal-to noise figures-of-merit of dielectric bolometer materials is strongly temperature dependent in the region of T_C, as would be expected. BST 67/33 [28] is a second-order ferroelectric and the F_D peaks at $14 \times 10^{-5} Pa^{-1/2}$ at 22°C and 0.4 V/μm bias field. Within ± 2°C from this temperature, the value falls to about one half of this figure. F_V varies similarly. This seems to make a tight control of device temperature essential. However, application of higher bias fields allows this requirement to be relaxed. In PST ceramics at 5 V/μm bias there is a much slower temperature variation, with a peak at 40°C which is about $15 \times 10^{-5} Pa^{-1/2}$. Some temperature stabilization would still be required to optimize the performance of the device, but this is not tight and could, for many applications, be achieved by heating alone, eliminating the need for a thermoelectric cooler.

The use of bulk ferroelectrics in pyroelectric devices inevitably leads to a situation where the material must be cut, lapped and polished to make a thin, thermally-sensitive layer. If an array of detectors is required for thermal imaging, this must be metallized on both faces, processed photolithographically and bonded to a silicon read-out circuit to yield a complete hybrid array. Clearly, it would be desirable if the ferroelectric material could be deposited as a thin film, which would remove the requirement for lapping and polishing. If possible the thin film would be directly onto a complete wafer of silicon chips, where it could be processed to yield an array of thin, thermally isolated structures. This requirement places a very important constraint on the temperature at which the ferroelectric is grown. The silicon circuits required for thermal imaging are complex VLSI chips which are manufactured in silicon foundries in wafers of 6" diameter or more. There is no way a general foundry would accept the deposition of the sorts of exotic materials which have been discussed in the previous sections and the ferroelectric layer must be deposited on fully-processed and tested wafers. Experimentally, it has been shown that the interconnect metallization on the chips should not be taken above about 560°C for any length of time, otherwise the yield of working chips is compromized [44]. This places an upper limit on the ferroelectric layer process temperature. The ceramic oxide materials which were discussed above possess good pyroelectric properties (high dielectric constants and high values of F_D) but they are normally manufactured as ceramics with sintering temperatures around 1200°C. Fortunately, many techniques have been researched for ferroelectric thin film deposition, mainly for applications to non-volatile memories. These include chemical solution deposition (CSD) - particularly sol-gel or metal-organic deposition (MOD), RF magnetron sputtering, pulsed laser ablation deposition (PLD) and metal-organic chemical vapour deposition (MOCVD). These methods have all been well reviewed [45] by Paz de Araujo *et al.* and each has its own advantages and disadvantages. Sol-gel type processes tend to be relatively low cost to set-up and simple to apply, with the capability for easy compositional modification, but being wet chemical processes are perhaps less acceptable in a semiconductor device processing context; sputtering and MOCVD are capital intensive processes and difficult to modify for complex film compositions, but are standard in the semiconductor industry and capable of coating onto large-area substrates, PLD is really only suitable for small area substrates but has the advantage of being able to deposit films with complex compositions relatively quickly, as there is good correspondence between the target and the film compositions.

Table 11.2 summarizes the properties as reported recently in the literature of modified PZT thin films deposited onto Pt/Ti on Si substrates where deposition has been at temperatures below 700°C. The best F_D values (in the range 1 to $2 \times 10^{-5} \text{Pa}^{-1/2}$) and lowest deposition temperatures reported to date have been with compositions in the PZT 30/70 region. It has been shown that sol-gel processing can give excellent properties for films grown at 510°C and recent work [46] has shown that the perovskite phase can be crystallized at temperatures as low as 400°C. It can be seen from Table 11.2 that RF magnetron sputtering has been used successfully to produce good quality PZT 25/75 films at 450°C. MOCVD has not yet been used to produce such good films at temperatures below 535°C, mainly because of the problem of finding metal-organic precursors for all the constituent atoms which will decompose at an appropriate rate at such low temperatures. The best films produced to date are those doped with Mn [54], for which the properties are comparable with bulk ceramic materials.

The best films on Pt/Ti/Si substrates are all polycrystalline, but tend to be strongly (111) orientated. At first sight, this may seem to be surprising, as the compositions are all tetragonal, with the polar axis along (001). However, it is found that if films are deposited with the cubic axes of the crystallites normal to the substrate plane, the tensile stresses in the film tend to

11.5 Pyroelectric ceramics and thin films

Material Composition	Dep'n Method	Dep'n Temp. °C	p $\frac{10^{-4}C}{m^2 K}$	Dielectric Properties ε	$\tan \delta$	Freq. Hz	F_V^* $\frac{10^{-2}m^2}{C}$	F_D^* $\frac{10^{-5}}{Pa^{1/2}}$	Ref.
PZT30/70	Sol gel	510	3	380	0.008	33	3.2	2.1	[47]
PZT25/75	Sol gel	510	2.2	350	0.008	33	2.5	1.7	[48]
PZT20/80	Sol gel	510	1.8	260	0.011	33	2.8	1.3	[48]
PL08T	Sol gel	650	1.4	730	0.06	1000	0.8	0.25	[49]
PZT15/85	Sol gel	650	2.2	210	0.015	1000	4.2	1.5	[50]
PZT25/75	Sputtered	450	2	300	0.01	1000	2.6	1.4	[51]
PL07T	MOCVD	535	1.3	414	–	1000	1.3		[53]
PMZ30/70	Sol gel	530	3.0	260	0.006	100	4.8	3.0	[54]

Table 11.2: Pyroelectric properties of several thin film ferroelectrics, $*c' = 2.7 \times 10^6 \, \text{Jm}^{-3}\text{K}^{-1}$

force the polar axis to be in the substrate plane, reducing the net polarization and pyroelectric coefficient.

As has been clearly demonstrated above, there are great benefits in pyroelectric properties, especially for thermal imaging devices, to be gained by using a ferroelectric close to T_C under an applied bias field. This has led to a drive to grow thin films of such materials. Lead scandium tantalate (PST) has been a prime candidate for such work. Complex perovskites such as this tend to be harder to crystallize into the perovskite phase than members of the PZT system close to lead titanate. There have been some extremely promising properties produced from films crystallized at relatively high temperatures on a variety of substrates. Shorrocks et al. [55] have shown that 4 µm thick films of PST can be grown onto sapphire substrates with a 1 µm MgO coating and crystallized at 900°C. These were subsequently released by dissolution of the MgO layer in phosphoric acid [56] to give films with $p > 25 \times 10^{-4} \text{Cm}^{-2}\text{K}^{-1}$, $\varepsilon = 2500$ and F_D up to $11 \times 10^{-5} \text{Pa}^{-1/2}$ at 4.2 V/µm bias field. Watton and Todd [57] have also demonstrated PST films on sapphire substrates annealed at 900°C to yield $p \cong 50 \times 10^{-4} \text{Cm}^{-2}\text{K}^{-1}$, $\varepsilon = 5800$ and $F_D \cong 11 \times 10^{-5} \text{Pa}^{-1/2}$. Cho et al. [58] have demonstrated 0.85 Pb(Mg$_{1/3}$Nb$_{2/3}$)O$_3$ - 0.15 Pb(Zn$_{1/3}$Nb$_{2/3}$)O$_3$ films on (111)Pt-on-(100)MgO substrates. These exhibited $p \cong 15 \times 10^{-4}\text{Cm}^{-2}\text{K}^{-1}$, $\varepsilon = 1150$ and $F_D \cong 3.1 \times 10 - 5\text{Pa}^{-1/2}$ at 3 V/µm bias after annealing at 900°C. Mantese et al. [59] have produced films of KTa$_{1-x}$Nb$_x$O$_3$ on yttria substrates using a MOD method, and after annealing at 1070°C, the films exhibited $p \cong 1.5 \times 10^{-4}\text{Cm}^{-2}\text{K}^{-1}$ at 3 V/µm bias. 10 µm thick films were separated from the substrates by etching and fabricated into working arrays. However, excellent though these pyroelectric properties are (and all at around 15 to 25°C), the films would all require a technique by which the bolometer film can be made separately from the active silicon substrate and subseqyently hybridized with a readout chip to make a working device. Donohue et al. [44] have devised a scheme by which this can be achieved using a silicon interconnect wafer. However, there is still a strong drive to reduce the firing temperature to approximately 500°C so that a process fully-integrated with silicon can be devised. Watton [60] has reported that if PST is sputtered with a large Pb excess (14 to 37%), the perovskite phase can be induced to grow on Pt/Ti/Si substrates at 525 to 575°C, and the films possessed $p \cong 4.4 \times 10^{-4} \text{Cm}^{-2}\text{K}^{-1}$, $\varepsilon = 650$ to 850 and F_D up to $3 \times 10^{-5} \text{Pa}^{-1/2}$ at 10 to 15 V/μm bias, well below the high temperature deposited values, but still useful. Whatmore et al. [61] (1997) have presented evidence to show that the Pb excess in these films resides within the crystallites as Pb^{4+} on the perovskite B-site. Takeishi and Whatmore [62] have also shown that perovskite PST films can be grown using sol-gel processing at temperatures as low as 520°C if a sufficient lead oxide excess is included in the films and hot-plate firing is used.

11.6 Conclusions

In conclusion, it has been demonstrated that the pyroelectric properties of polar materials can be compared relatively simply through the measurement of a few key physical parameters (pyroelectric, dielectric and thermal coefficients) and the judicious use of appropriate figures-of-merit. It is essential that the dielectric properties are measured in the frequency range appropriate for device use, and this is typically in the range of a few to 100 Hz. The properties of many pyroelectric ceramics and thin films have been compared and it has been shown that good pyroelectric properties can be obtained from this films manufactured at relatively low temperatures, a fact that bodes well for their future applications in fully-integrated arrays.

Bibliography

[1] J. Cooper, *Rev. Sci. Instrum.* **33**, 925 (1962).
[2] R. W. Whatmore, *Reports on Progress in Physics* **49**, 1335 (1986).
[3] R. W. Whatmore and R. Watton, *Pyroelectric Materials and Devices*, Published in Infrared Detectors and Emitters: Materials and Devices, P. Capper and C. T. Elliott ed., Chapman and Hall, London, 99, 2000.
[4] M. V. Bennett and I. Matthews, *Proc.* SPIE. **2744**, 549 (1996).
[5] C. M. Hanson, H. R. Beretan, J. F. Belcher, K. R. Udayakumar, and K. L. Soch, *Proc.* SPIE. **60**, 3379 (1998).
[6] R. Watton, P. A. Manning, M. C. J. Perkins, J. P. Gillham, and M. A. Todd, *Proc.* SPIE. **2744**, 486 (1996).
[7] E. H. Putley, *The Pyroelectric Detector in Semiconductors and Semimetals*, R. K. Willardson and A. C. Beer ed., New York: Academic, **5** 259, 1970.
[8] E. H. Putley, *Infra-red Phys.* **21**, 173 (1981).
[9] S. G. Porter, *Ferroelectrics* **33**, 193 (1981).
[10] E. H. Putley, *Infra-red Phys.* **20**, 149 (1980).
[11] J. F. Nye, *Physical properties of crystals*, Oxford: Oxford University Press, 1957.
[12] J. D. Zook and S. T. Liu, *J. Appl. Phys.* **49**, 4604 (1978).
[13] L. B. Schein, P. J. Cressman, and L. E. Cross, *Ferroelectrics* **22**, 937 (1979a).
[14] N. M. Shorrocks, R. W. Whatmore, M. K. Robinson, and S. G. Porter, *Proc.* SPIE. **588**, 44 (1985).
[15] A. A. Turnbull, *Infra-red Phys.* **22**, 299-306 (1982).
[16] A. G. Chynoweth, *J. Appl. Phys.* **27**, 78 (1956).

[17] S. B. Lang and F. Steckle, *Rev. Sci. Instrum.* **36**, 929 (1965).
[18] A. M. Glass, *J. Appl. Phys.* **40**, 4699 (1969).
[19] R. L. Byer and C. B. Roundy, *Ferroelectrics* **3**, and IEEE *Trans. Sonics Ultrason.* SU-**19**, 333 (1972).
[20] K. D. Benjamin, A. F. Armitage, and R. W. Whatmore, Private communications, Napier University, Edinburgh, UK, 1999.
[21] E. J. Sharp and L. E. Garn, *J. Appl. Phys.* **53**, 8980 (1982).
[22] L. E. Garn and E. J. Sharp, *J. Appl. Phys.* **53**, 8974 (1982).
[23] S. B. Lang, *Ferroelectrics* **93**, 87 (1988).
[24] Jr J. F. Shepard, P. J. Moses, and S. Trolier-McKinstry, *Sensors and Actuators A* **71**, 133 (1998).
[25] J. E. A. Southin, S. A. Wilson, S. Schmitt, and R. W. Whatmore, *J. Phys. D:Appl. Phys.* **34**, 1446 (2001).
[26] A. G. Chynoweth, *J. Appl. Phys.* **27**, 78 (1956).
[27] R. A. Hanel, *J. Opt. Soc. Am.* **51**, 220 (1961).
[28] B. M. Kulwicki, A. Amin, H. R. Beratan, and C. M. Hanson, Pyroelectric Imaging, *Proc. 8th International Symposium on Applications of Ferroelectrics*, Greenville, SC, USA Aug 30 to Sept. 2 1992 IEEE Cat. No.90CH3080-9, 1 (1992).
[29] P. Muralt, *Rep. Prog. Phys.* **64**, 1339 (2001).
[30] R. W. Whatmore and F. W. Ainger, *Proc.* SPIE **395**, 261 (1983).
[31] R. W. Whatmore and A. J. Bell, *Ferroelectrics* **35**, 155 (1981).
[32] N. Ichinose, *Am. Ceram. Soc. Bull.* **64**, 1581 (1985).
[33] R. Clarke, A. M. Glazer, F. W. Ainger, D. Appleby, N. J. Poole, and S. G. Porter, *Ferroelectrics* **11**, 359 (1976).
[34] M. Adachi, A. Hachisuka, N. Okumura, T. Shiosaki, and A. Kawabata, *Jpn. J. Appl. Phys. Supplement* **26-2**, 68 (1986).
[35] C. P. Shaw, S. Gupta, S. B. Stringfellow, A. Navarro, J. R. Alcock, and R. W. Whatmore, *J. Europ. Ceram. Soc.* **22**, 2123 (2002).
[36] A. J. Bell and R. W. Whatmore, *Ferroelectrics* **37**, 543 (1981).
[37] R. W. Whatmore, *Ferroelectrics* **49**, 201 (1983).
[38] S. B. Stringfellow, S. Gupta, C. Shaw, J. R. Alcock, and R. W. Whatmore, *J. European Ceram. Soc.* **22**, 573 (2002).
[39] R. W. Whatmore, O. Molter, and C. P. Shaw, *J. Europ. Ceram. Soc.* **5**, 721 (2003).
[40] O. M. Stafsud and M. Y. Pines, *J. Opt. Soc. Am.* **6**, 1153 (1971).
[41] Y. Yokomizo, T. Takahashi, and S. Nomura, *J. Phys. Soc. Japan.* **28**, 1278 (1970).
[42] R. W. Whatmore, P. C. Osbond, and N. M. Shorrocks, *Ferroelectrics* **76**, 351 (1987).
[43] N. M. Shorrocks, R. W. Whatmore, and P. C. Osbond, *Ferroelectrics* **106**, 387 (1990).
[44] P. P. Donohue, M. A. Todd, C. J. Anthony, A. G. Brown, M. A. C. Harper, and R. Watton, *Integrated Ferroelectrics* **41**, 25 (2001).
[45] C. Paz de Araujo, J. F. Scott, G. W. Taylor eds., *Ferroelectric thin films: synthesis and basic properties*, Gordon and Breach, 1996.
[46] Z. Huang, Q. Zhang, and R. W. Whatmore, *J. Appl. Phys.* **86** (3), 1662 (1999).

[47] R. W. Whatmore, P. Kirby, A. Patel, N. M. Shorrocks, T. Bland, M. Walker, *Proc.* NATO *Advanced Research Workshop on Science and Technology of Electroceramic Thin Films*, O. Auciello and R. Waser eds., Kluwer Academic Publishers, Dordrecht, The Netherlands, 383, 1994.

[48] N. M. Shorrocks, A. Patel, M. J. Walker, and A. D. Parsons, *Microelectronic Engineering* **29**, 59 (1995).

[49] M. Alguero, M. L. Calzada, and L. Pardo, *J. Phys France* **8**, Pr9-155 (1998).

[50] M. Kohli, A. Seifert, and P. Muralt, *Integrated Ferroelectrics* **22**, 453 (1998).

[51] P. Padmini, R. Kohler, G. Gerlach, R. Bruchhaus, and G. Hofmann, *J. Phys France* **8**, Pr9-151 (1998).

[52] R. Kohler, P. Padmini, G. Gerlach, G. Hofmann, and R. Bruchhaus, *Integrated Ferroelectrics* **22**, 383 (1998).

[53] J. F. Roeder, I-S. Chen, P. C. Van Buskirk, and H. R. Beretan, *Proc. 10th* IEEE ISAF (IEEE Cat. No. 96CH35948), 227 (1996).

[54] Q. Zhang and R. W. Whatmore, *J. Phys. D: Appl. Phys* **34**, 2296 (2001).

[55] N. M. Shorrocks, A. Patel, and R. W. Whatmore, *Ferroelectrics* **134**, 35 (1992).

[56] A. Patel, N. M. Shorrocks, and R. W. Whatmore, IEEE *Trans on Ultrasonics, Ferroelectrics and Frequency Control.* **38**, 672 (1991),

[57] R. Watton and M. A. Todd, *Ferroelectrics* **118**, 279 (1991) .

[58] M. K. Cho, K. S. Kim, and H. M. Chang, *J. Mater. Res.* **10**, 2631 (1995).

[59] J. V. Mantese, A. L. Micheli, N. W. Schubring, A. B. Catalan, K. L. Soch, K. Ng, S. H. Klapper, R. J. Lopez, and G. Lung, IEEE *Trans. on Electronic Devices*, ED-40, 320 (1993).

[60] R. Watton, *Ferroelectrics* **184**, 141 (1995).

[61] R. W. Whatmore, Z. Huang, and M. A. Todd, *J. Appl. Phys.* **82**, 5686 (1997).

[62] T. Takeishi and R.W. Whatmore, *J. Phys.* IV *France* **8**, Pr9 57 (1998).

12 Nano-inspection of Dielectric and Polarization Properties at Inner and Outer Interfaces in PZT Thin Films

L. M. Eng

Institute of Applied Photophysics, University of Technology Dresden 01062 Dresden, Germany

Abstract

We report on novel approaches using scanning force methods, i. e. piezoresponse force microscopy (PFM), Kelvin probe force microscopy (KPFM) and pull-off force spectroscopy (PFS), in order to deduce the local dielectric and polarization properties of PZT thin films both at outer and inner interfaces with < 50 nm lateral resolution. We show that the polarization profile into the depth of the PZT sample varies dramatically being built up at the bottom Pt electrode over a transition layer of more than 200 nm in thickness. Also this interfacial area shows a different relaxation behavior upon switching. The results are explained both in the view of negatively charged defects pinned at the PZT/Pt interface as well as the possible variation in the local dielectric properties across the film thickness. Investigating the latter made the quantitative deduction of values such as the effective dielectric polarization P_z, the deposited charge density σ, and the surface dielectric constant $\epsilon_{\text{surface}}$ in thin ferroelectric PZT films necessary. We illustrate that such measurements in fact are possible on the nanometer scale revealing quantitative data when combining PFM and PFS.

12.1 Introduction

Ferroelectric thin films considerably gain in interest within the last couple of years due to their potential application in nonvolatile random-accessmemory devices (FeRAM). Among potential candidates, PbZr$_x$Ti$_{1-x}$O$_3$ (PZT) is one of the most promising materials because of its large remanent polarization and low coercive field. However, PZT is also well known for its poor fatigue behavior on metal electrodes [1, 2] and occurrence of size effects [3–5] which are well due to the ferroelectric/electrode interface properties [1–5].

Addressing the inspection to internal interfaces, however, as manifests the ferroelectric film/metal electrode interface, has so far mostly been restricted to investigations using transmission electron microscopy (TEM) [6, 7]. Still the influence of preparation conditions on the nanoscopic properties of as-prepared cross sections for TEM may be debated, in conjunction of how the electron beam does alter the local electronic and physical constitutions.

Therefore we propose here an alternative route to inspect the local dielectric and polarization properties using non-destructive and non-invasive methods based on scanning force microscopy (SFM). Simultaneously, these techniques offer a high resolution in real space being extended down to the atomic scale when inspecting ferroelectric systems under ultra-high vacuum (UHV) conditions [8,9]. Under ambient conditions in air, though, the resolution is limited due to the relatively large capillary forces resulting in weared tips after a few scan lines. Nevertheless, when accurately controlling the tip-sample interaction force (zero force point) the resolution even on ferroelectrics approaches 1 nm in topography and ~ 5 nm for subsurface information [10] deduced for instance with piezoresponse force microscopy (PFM).

12.2 Methods

12.2.1 Piezoresponse force microscopy (PFM)

Figure 12.1 presents such a high resolution imaging of an individual PZT grain. The method applied is PFM the details of which are described elsewhere [8,11]. In principle an AC signal is applied to the tip which consequently, when contacting the sample, activates 3-dimensional local sample vibrations due to the inverse piezoelectric effect. We developed sensitive measurement set-ups in order to deduce the tensorial properties of the piezoelectric constant and hence the three dimensional polarization $\vec{P} = (P_x, P_y, P_z)$, for instance for single crystals [12–16], ceramics [17–19], thin films [10,20–25], etc. down to a 5 nm resolution. Using a highly doped Si cantilever with 70 kHz resonance frequency, it is possible to apply an AC voltage of 2 V_{p-p} in the range between 10–25 kHz (12 kHz in the actual experiment) while still being far from mechanical resonances. Basically, PFM measurements result in deducing the AC contribution S_{AC} of both the polarization and mobile charge which possibly might be accumulated at the sample surface (compensation charge for instance resulting from switching experiments). Differentiation between the two contributions on the nanometer scale, though, so far seemed to be impossible. In fact PFM is sensitive to a signal given by [24]

$$S_{AC} = P_z - \gamma \cdot \sigma \quad (12.1)$$

with P_z showing the effective contribution of dielectric polarization along the z-direction, and σ being the surface charge density due to charge deposition from the tip. As seen both contributions P_z and σ influence the measurement at the same time. The proportionality factor γ so far is unknown and has to be deduced theoretically or during the experiment. Note the subtractive behavior of the two contributions since electrostatic interaction of the tip with the deposited charge is always repulsive while the inverse piezoelectric effect in PFM acts contrary.

As previously shown [26,27] γ covers the regime between $0 < \gamma < 1$. Thus the inverse piezoelectric effect always dominates the overall interaction in PFM. Therefore, PFM effectively reveals the local polarization distribution close to the sample surface [11,21,22,28,29].

12.2.2 Kelvin Probe Force Microscopy (KPFM)

For KPFM the same AC voltage is applied to the cantilever which, however, is now strictly kept away from the sample surface in the true noncontact mode [30,31]. The oscillation amplitude is kept to below 10 nm_{p-p} in order to improve the force resolution and avoiding elastic tip sample interaction due to tapping. The additional AC field exerted by the tip modulates the

12.2 Methods

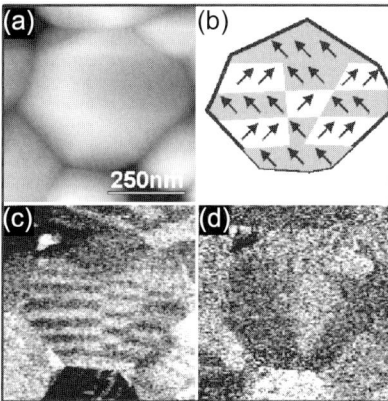

Figure 12.1: Domain pattern in an individual PZT grain: (a) sample topography (total z-scale 10 nm, (b) reconstruction of domain pattern, (c) out-of-plane polarization (OPP) components, and (d) in-plane polarization (IPP) components. Lamellar 90° domain pattern are clearly visible in (c) also revealing domain inversion well inside the grain.

force and allows to minimize all electrostatic interactions by sensing the force minimum [32] due to

$$F = \frac{1}{2}\frac{\partial C}{\partial z}(\Delta\Phi)^2 \qquad (12.2)$$

with C the tip-sample capacitance and z the mean tip sample distance. As indicated in Equation (12.2) both a negative and positive potential difference $\Delta\Phi$ between tip and sample surface affect our measurements similarly (always resulting in an attractive force) showing the quadratic behavior of the Coulomb force term in Equation (12.2). In ordinary noncontact SFM though, both the tip and sample surface potential are electrically not controlled. Moreover when investigating ferroelectrics having a bound surface charge density, any arbitrary tip potential being different from the local sample surface potential directly leads to an additional force term in our non-contact interaction (equ. 12.2). The interpretation of data recorded by EFM or noncontact SFM therefore always suffer from this point, specifically when Coulomb forces dominate the local force contribution. KPFM though enables the most accurate surface potential measurements since all extra forces arising via Equation (12.2) are directly balanced at the tip.

12.2.3 Pull-off force spectroscopy (PFS)

In contrast to PFM, PFS measurements do not allow the two contributions P_z and σ to be separated from their sign, which therefore results in summing up over the two signals rather than building the difference as shown in Equation (12.1) [24]. Since pull-off force experiments are performed with a DC voltage applied to the conducting tip while retracting the tip from the sample surface, we write correspondingly:

$$S_{DC} = P_z + \delta \cdot \sigma \tag{12.3}$$

The positive sign directly results from electrostatic forces always being attractive between tip and (any) sample independent of whether electrons or holes are the majority carriers. The proportionality factor δ is larger than one since the surface charge density σ clearly dominates the polarization induced force on the tip.

In order to separate P_z and σ we therefore have to determine γ and δ either theoretically or in our experiment.

12.3 Materials

We used two different type of PZT samples. For studying the internal interface properties, an RF sputtered 450 nm thick (111)-oriented Pb(Zr$_{0.25}$Ti$_{0.75}$)O$_3$ film deposited onto a Pt(111)/SiO$_2$/Si substrate was used. Such samples are ideal for applications in pyroelectric sensors [33] showing self-polarization with the normal component pointing from the bottom Pt electrode to the top surface [34]. In fact ellipsometry measurements indicate the self-polarization to gradually increase over the first 300 nm [33]. Therefore, such ferroelectric samples were polished under different small angles ranging between 1-6°. This greatly enlarges the cross section of the film making both the surface and inner interface accessible to SFM. For example the 450 nm thick film is enlarged to $\sim 5\,\mu$m as shown in Figure 12.2. The PZT/Pt interface was scanned with the cantilever aligned parallel to the layers. The scan range was large enough ($> 8\,\mu$m) in order to include both the intact PZT film of 450 nm nominal thickness as well as the SiO$_2$ section thus providing good reference marks both for potential and topography measurements. By scanning the tip from left to right over the wedge, the information from different positions on the PZT film corresponding to different thicknesses was directly deduced. The second sample was selected with a Pb(Zr$_{0.53}$Ti$_{0.47}$)O$_3$ concentration having a thickness of 600 nm. Details on sample preparation are given in [35]. The transition layer for the second type of samples is much smaller allowing the preparation of thinner films (down to 20 nm thickness only) and their integration in high-density FeRAM applications.

12.4 Results

12.4.1 Polarization profile across the PZT film

Piezoresponse force microscopy (PFM) [11] and Kelvin probe force microscopy (KPFM) [9] were applied to deduce the polarization and local electric potential distribution over the whole cross section of the PZT sample (see Figure 12.3 and Figure 12.4) under static conditions as well as after switching. The details of our setup are described elsewhere [9, 11].

Figure 12.3 depicts the piezoresponse signal (red curve) and the sample topography (black curve) taken across the wedged sample. Note the different sample regions deduced from the topography profile. From left to right, they are the PZT native surface, the polished PZT

12.4 Results

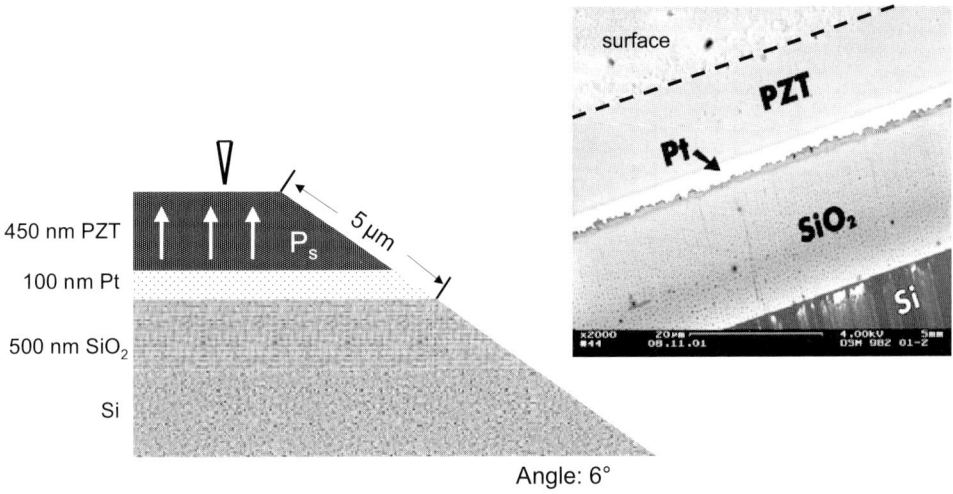

Figure 12.2: Schematic view of the sample system (left), and scanning electron micrograph of the as-polished PZT structure, giving access to all interfaces (right)

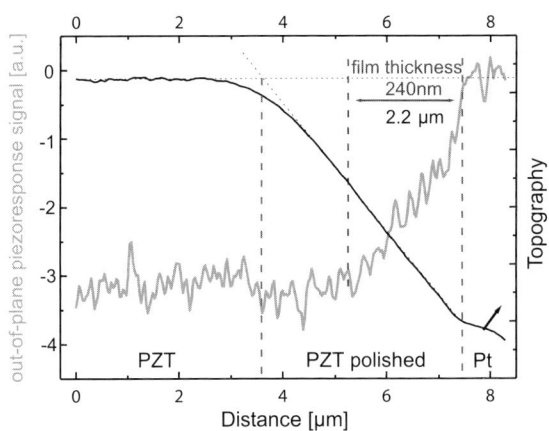

Figure 12.3: PFM and topographic cross-section of the PZT wedge sample. Note that the PFM signal saturates when the sample exceeds a thickness of $\sim$240 nm although the sample still increases in thickness.

wedge, the Pt bottom electrode, and the SiO_2 (not shown). In our setup, the direction of selfpolarization results in a negative response for the PFM signal. We clearly see that on the PZT wedge away from the Pt electrode, the absolute value of the piezoresponse signal first increases gradually with increasing film thickness before then saturating at a distance of 2.2 μm from the Pt, which corresponds to a critical film thickness $t_c \approx 240$ nm. The existence of such a thick transition layer clearly resembles the ellipsometry measurements reported in literature [33]. The overall sign of the polarization is still negative, as expected. Similar experiments were performed with the tip now scanning in non-contact mode. Such KPFM measurements are sensitive to the electrical potential variations induced by charges located both at the surface and within the interior of the film. Figure 12.4 shows a full image scan over the wedged sample area. Due to the built-in polarization of the PZT film the top surface is positively charged while being negatively compensated thus forming a Helmholtz double layer [36]. This is reflected in Figure 12.4 by the negative absolute potential value. Taking into account that long-distance electrostatic forces tend to smear out the experimental potential distribution [37], the observed profile is in good agreement with the transition layer thickness of 240 nm found by PFM and reported in Figure 12.3.

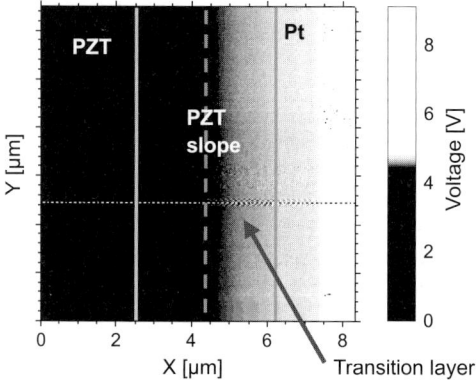

Figure 12.4: KPFM image showing the build-up of the surface potential for the film thickness and increasing from right to left. Note the relative negative value of the surface potential when reaching the nominal sample thickness of 450 nm (@ $x = \sim 4\,\mu$m).

To date many models have been presented in order to explain the effects reported above at the PZT/Pt interface [4, 38–41]. Most of them are related to the charge carrier concentration in conjunction with the Schottky barrier built up at the ferroelectric/metal interface, among which the electron injection scenario [40, 41] seems to be the most appropriate in our case. For ferroelectric films an interface layer usually forms close to the substrate. The band bending in the PZT at the interface leads to a significant lowering of the Schottky barrier [38] between PZT and Pt so that electrons may inject from the Pt electrode into the semiconductor then becoming trapped in the PZT film to form negatively charged defects, for instance Ti^{3+}, or lead vacancies (V_{Pb}^{2-}) [42]. Thus an internal electric field is built up resulting in a permanent poling of the film. This represents one of the most important origins of self-polarization. In order to clarify the existence of such a transition layer, it is necessary also to investigate both the dynamic behavior of the interfacial layer as well as the dielectric properties of the sample, specifically at interfaces.

12.4 Results

12.4.2 Relaxation dynamics within the PZT film

To clarify the switching properties we apply a DC voltage of +20 V to the tip in contact mode while scanning the tip over the wedge at a constant speed of 2 μm/s. The scanning range was chosen to cover the whole PZT slope leaving only a small gap to the Pt electrode to avoid any short circuit (see Figure 12.5). Mapping both the PFM (see Figure 12.5) and KPFM signals directly after switching shows the two signals to decay with a time constant τ being on the order of several hours [23]. The decay is slowest at the center of the switched region (center of the PZT slope) and becomes faster near its borders. This behavior is reasonable since the driving force for back switching is expected to be larger at the edge than for the inner part. Further details are discussed in [23].

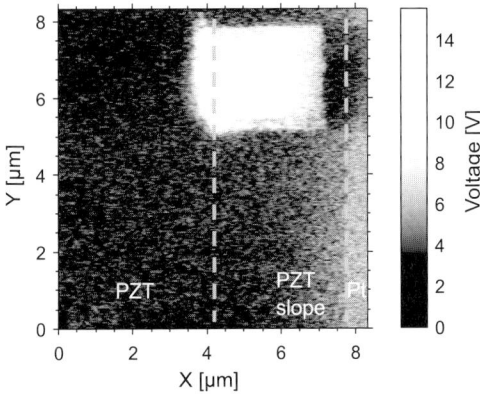

Figure 12.5: Switching dynamics probed across the wedge sample with PFM. Note that the tip did not touch the Pt electrode for switching in order to avoid tip-sample short circuit.

12.4.3 Local dielectric constant at the PZT surface

From Equation (12.1) and (12.3) it follows that both P_z and σ depend on S_{AC} and S_{DC} in the following way:

$$P_z = \frac{S_{AC} + \frac{\gamma}{\sigma} \cdot S_{DC}}{\frac{\gamma + \delta}{\delta}} \tag{12.4}$$

and

$$\sigma = \frac{S_{DC} - S_{AC}}{\gamma + \delta} \tag{12.5}$$

Both S_{AC} and S_{DC} may be deduced from force measurements as denoted by Equation (12.1) and (12.3). In PFS we measure the overall force acting on the tip when pulling the tip back from the surface. Thus such a DC force may be written as:

$$F_{DC} = A_{DC}(\epsilon, U) \cdot U^2 + S_{DC}(P_z, \sigma) \cdot U + C \tag{12.6}$$

where A_{DC} specifies the induced dielectric force component (proportional to U^2), and S_{DC} the force contributions due to polarization and/or mobile charges depending linearly in U. C represents any additional DC force contribution as stems for instance from the capillary force affecting the tip-sample interaction. The coefficient A_{DC} in Equation (12.6) denotes the dielectric properties thus containing information on the tip-sample capacitance. For the geometry chosen in our experiment, the top electrode is our conductive tip while the bottom electrode is laterally extended. In fact one therefore could calculate the field distribution between tip and sample to be that of a point charge in front of an extended dielectric half space having a counter electrode that merges to infinity. For the sake of simplicity, however, and because the PZT film thickness used in this experiment was rather low (~ 100 nm), we still model the tip-sample system in our experiment as a plate capacitor resulting in:

$$A_{DC} = -\frac{1}{2}\epsilon_0 \epsilon(U) \cdot a \tag{12.7}$$

In Equation (12.7), a specifies the geometry while $\epsilon(U)$ now denotes the local dielectric constant probed with the tip.

Experimentally, both the PFM and PFS measurements are performed at one and the same surface spot, and an absolute matching of the two curves is intended using a polynomial fitting. Such an approach is reasonable since it is both P_z and σ which contribute to the AC and DC force terms via Equation (12.4) and (12.5). Our experiment [24] then allows the following results to be deduced:

1.) PFM probes the piezoelectric properties even for tip-sample voltages up to ± 10 V.
2.) Only for larger fields exceeding 10 V in reversed polarization direction will mobile surface charges contribute to the overall signal S_{AC} in PFM (deposited upon switching). In contact experiments, however, such mobile charges are directly eliminated via the conductive tip.
3.) From an absolute matching of PFM and PFS we find $\epsilon_{\text{surface}}$ to measure ~ 140, much smaller than the bulk ϵ value determined from dielectric spectroscopy ($\epsilon_{\text{Bulk}} \approx 500$).

These results suggest, that also the top PZT surface has different properties compared to the bulk values. In fact, various theoretical models already suggested the existence of a pure dielectric surface layer [38] to be present on PZT thin films. Here, for the first time, we have given experimental evidence that this is true for both the inner and outer interfaces in PZT on the nanometer scale

12.5 Conlusion

In conclusion, we reported the investigation of inner and outer interfaces in PZT in order to quantify both the amount of effective ferroelectric polarization and change in dielectric properties. With PFM and KPFM we find a transition layer occurring at the Pt/PZT interface within which the polarization builds up reaching its saturation value for film thicknesses exceeding 240 nm. Its presence was also tested under dynamic switching conditions suggesting that the observed temporal behavior may tentatively be attributed to the influence of negatively charged defects accumulated at that inner surface.

Furthermore, for the voltage regimes used here, no evidence was found that PFM measurements should lack from mobile charge deposition upon switching. In contrast, we prove that the PFM signal purely reflects the measured piezoelectric displacement which hence may be compared to the local polarization distribution.

In addition, the dielectric constant at the PZT top surface was found to be dramatically reduced compared to the bulk value. Our measurements therefore suggest a dead layer to be present. Similar experiments of deducing the local dielectric constant are now necessary for the inner interface.

Bibliography

[1] E. L. Colla, D. V. Taylor, A. K. Tagantsev, and N. Setter, *Appl. Phys. Lett.* **72**, 2478 (1998).

[2] J. J. Lee, C. L. Thio, and S. B. Desu, *J. Appl. Phys.* **78**, 5073 (1995).

[3] M. Alexe, C. Harnagea, D. Hesse and U. Gösele, *Appl. Phys. Lett.* **79**, 242 (2001).

[4] C. H. Lin, P. A. Friddle, C. H. Ma, A. Daga, and H. Chen, *J. Appl. Phys.* **90**, 1509 (2001).

[5] J. F. M. Cillessen, M. W. J. Prins, and R. M. Wolf, *J. Appl. Phys.* **81**, 2777 (1997).

[6] J. C. Jiang, W. Tian, X. Q. Pan, Q. Gan, and C. B. Eom, *Appl. Phys. Lett.* **72**, 2963 (1998).

[7] H. N. Lee, S. Senz, N. D. Zakharov, C. Harnagea, A. Pignolet, D. Hesse, and U. Gösele, *Appl. Phys. Lett.* **77**, 3260 (2000).

[8] L. M. Eng, M. Bammerlin, Ch. Loppacher, M. X Guggisberg, R. Bennewitz, E. Meyer, and H.-J. Güntherodt, *Surf. Interface Analysis* **27**, 422 (1999).

[9] L. M. Eng, M. Bammerlin, Ch. Loppacher, M. Guggisberg, R. Bennewitz, R. Lüthi, E. Meyer, and H.-J. Güntherodt, *Appl. Surf. Sci.* **140**, 253 (1999).

[10] Ch. Loppacher, F. Schlaphof, S. Schneider, U. Zerweck, S. Grafström, L. M. Eng, A. Roelofs, and R. Waser, *Surface Science*, (2003) in press.

[11] L. M. Eng,*Nanotechnology* **10**, 405(1999).

[12] L. M. Eng, G. Rosenman, A. Skliar, M. Oron, M. Katz, and D. Eger, *J. Appl. Phys.* **83**, 5973 (1998).

[13] L. M. Eng, M. Abplanalp, P. Günter, and H.-J. Güntherodt,*J. de Physique* IV **8**, Pr9-201 (1998)

[14] L. M. Eng, M. Abplanalp, and P. Günter, *Appl. Phys.* **A66**, S679 (1998).

[15] M. Abplanalp, L. M. Eng, and P. Günter, *Appl. Phys.* **A66**, S231 (1998).

[16] G. Tarrach, P. Lagos L., R. Hermans Z., Ch. Loppacher, F. Schlaphof, and L. M. Eng, *Appl. Phys. Lett.* **79**, 3152 (2001).

[17] L. M. Eng, H.-J. Güntherodt, G. A. Schneider, U. Köpke, and J. Muñoz Saldaña, *Appl. Phys. Lett.* **74**, 233 (1999).

[18] J. Muñoz-Saldaña, G. A. Schneider, and L. M. Eng, *Surf. Sci.* **480**, L402 (2001).

[19] J. Munoz Saldana, L. M. Eng, and G. A. Schneider, *Sciencia* UANL **3**, 389 (2000).

[20] A. Roelofs, N. A. Pertsev, R. Waser, F. Schlaphof, L. M. Eng, C. Ganpule, and R. Ramesh, *Appl. Phys. Lett.* **80**, 1424 (2002).

[21] A. Roelofs, F. Schlaphof, S. Trogisch, U. Böttger, R. Waser, and L. M. Eng,*Appl. Phys. Lett.* **77**, 3444 (2000).

[22] C. S. Ganpule, V. Nagarajan, B. K. Hill, A. L. Roytburd, E. D. Williams, R. Ramesh, S. P. Alpay, A. Roelofs, R. Waser, and L. M. Eng, *J. Appl. Phys.* **91**, 1477 (2002).

[23] X. M. Lu, F. Schlaphof, C. Loppacher, S. Grafström, L. M. Eng , G. Suchaneck, and G. Gerlach, *Appl. Phys. Lett.* **81**, 3215 (2002).

[24] K. Franke and L. M. Eng,*J. Appl. Phys.* (2003) submitted.

[25] L. M. Eng, F. Schlaphof, S. Trogisch, A. Roelofs, and R. Waser, *Ferroelectrics* **251**, 11 (2001).

[26] K. Franke, *Ferroelec. Lett.* **19**, 35 (1995).

[27] W. Cao and C. Randall, *Solid State Comm.* **86**, 435 (1993).

[28] K. Franke, H. Hülz, M. Weihnacht, W. Hässler, and J. Besold, *Ferroelectrics* **172**, 397 (1995).

[29] K. Franke, H. Huelz, and S. Seifert, *Ferroelec. Lett.* **23**, 1 (1997).

[30] M. Guggisberg, M. Bammerlin, Ch. Loppacher, O. Pfeiffer, A. Abdurixit, V. Barvich, R. Bennewitz, A. Baratoff, E. Meyer, and H.-J. Güntherodt, *Phys. Rev. B* **61**, 11151 (2000).

[31] M. Guggisberg, O. Pfeiffer, S. Schär, V. Barvich, M. Bammerlin, Ch. Loppacher, R. Bennewitz, A. Baratoff, and E. Meyer,*Appl. Phys. A* **72**, S19 (2000).

[32] L. M. Eng, S. Grafström, Ch. Loppacher, F. Schlaphof, S. Trogisch, A. Roelofs, and R. Waser, in:*"Advances in Solid State Physics"*, ed. B. Kramer, Springer, Berlin,**41** (2001), p 287-298.

[33] G. Suchaneck, W.-M. Lin, R. Koehler, T. Sandner, G. Gerlach, R. Krawietz, W. Pompe, A. Deineka, and L. Jastrabik,*Proc. 6th Int. Symp. Sputtering & Plasma Processing* 341 (2001).

[34] R. Köhler, G. Suchaneck, P. Padmini, P. Padmini, T. Sandner, G. Gerlach, and G. Hoffmann, *Ferroelectrics* **225**, 57 (1999).

[35] R. Bruchhaus, H. Huber, D. Pitzer, and W. Wersing, *Ferroelectrics* **127**, 137 (1992).

[36] S. V. Kalinin and D. A. Bonnell, *Phys. Rev. B* **63**, 125411 (2001).

[37] H. O. Jacobs, P. Leuchtmann, O. J. Homan, and A. Stemmer,*J. Appl. Phys.* **84**, 1168 (1998).

[38] A. K. Tagantsev and I. A. Stolichnov, *Appl. Phys. Lett.* **74**, 1326 (1999).

[39] K. Iijima, N. Nagano, T. Takeuchi, I. Ueda, Y. Tomita, R. Takayama,*Mater. Res. Symp. Proc.* **310**, 455 (1993).

[40] G. Suchaneck, G. Gerlach, Yu. Poplavko, A. I. Kosarev, and A. N. Andronov, *Mat. Res. Soc. Symp. Proc.* **655**, CC7.7.1 (2001).

[41] J. F. Scott,*Jpn. J. Appl. Phys., Part 1* **38**, 2272 (1999).

[42] ref from Sandner Thesis.

13 Piezoelectric Relaxation and Nonlinearity investigated by Optical Interferometry and Dynamic Press Technique

D. Damjanovic

Ceramics Laboratory, Swiss Federal Institute of Technology - EPFL, 1015 Lausanne, Switzerland

Abstract

This contribution briefly discusses recent results on piezoelectric relaxation and nonlinearity in ferroelectric materials. The piezoelectric properties are investigated using an optical interferometer and a dynamic press. In the case of the direct effect the hysteresis of the total response is investigated as a function of the driving pressure, temperature and frequency. For the converse effect, the strain-electric field hysteresis and the phase angle and amplitude of the fundamental and the first two harmonics are investigated as a function of the electric field amplitude at room temperature. We demonstrate how these techniques can be used to investigate fine details of the piezoelectric response and thus reveal mechanisms governing the observed behavior. In particular we show validity of Kramers-Kronig relations for the piezoelectric effect in modified lead titanate ceramics, unusual clockwise hysteresis in piezoelectric heterostructures, and discuss frequency dependence of the nonlinear piezoelectric response in disordered systems. The most pertinent details of each technique are discussed.

13.1 Introduction

The term "piezoelectric nonlinearity" is used here to describe relationship between mechanical and electrical fields (charge density D vs. stress σ, strain x vs. electric field E) in which the proportionality constant d, is dependent on the driving field, Figure 13.1. Thus, for the direct piezoelectric effect one may write $D = d(\sigma)\sigma$ and for the converse effect $x = d(E)E$. Similar relationships may be defined for other piezoelectric coefficients (g, h, and e) and combination of electro-mechanical variables. The piezoelectric nonlinearity is usually accompanied by the electro-mechanical (D vs. σ or x vs. E) hysteresis, as shown in Figure 13.2. By hysteresis we shall simply mean, in the first approximation, that there is a phase lag between the driving field and the response. This phase lag may be accompanied by complex nonlinear processes leading to a more general definition of the hysteresis [2].

The nonlinearity and hysteresis have a profound influence on application of piezoelectric sensors and actuators, particularly in high precision devices. Details and additional references can be found in [4].

Polar Oxides: Properties, Characterization, and Imaging
Edited by R. Waser, U. Böttger, and S. Tiedke
Copyright © 2005 WILEY-VCH Verlag GmbH & Co. KGaA, Weinheim
ISBN: 3-527-40532-1

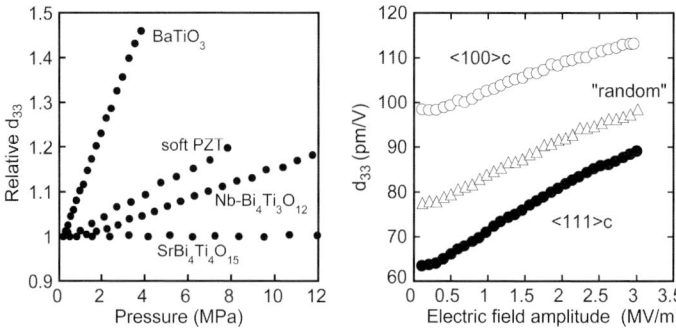

Figure 13.1: Examples of the field dependence of piezoelectric coefficients (a) direct effect in ferroelectric ceramics, measured with a dynamic press (b) converse effect in rhombohedral 60/40 PZT thin films with different orientations, measured with an optical interferometer [1]. $<hkl>$ correspond to pseudocubic axes.

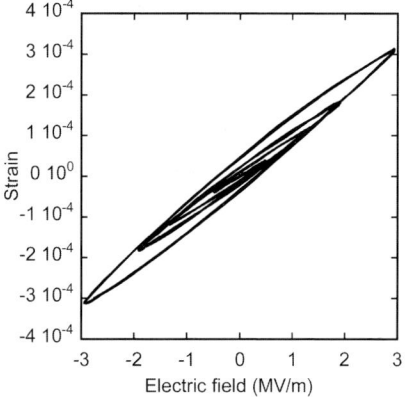

Figure 13.2: Piezoelectric hysteresis for converse piezoelectric effect in a rhombohedral PZT thin film [3].

13.2 Measurement techniques

13.2.1 Optical techniques for measurements of the converse effect

In the converse piezoelectric effect one usually applies voltage V or electric field E on the sample and measures displacement Δl or strain $\Delta l/l$. From relation $\Delta l = d_{33}V$ for the longitudinal effect, we see that even for materials with exceptionally high piezoelectric coefficient ($d_{33} = 2000$ pm/V in PZN-PT) the displacement Δl is only around 2 nm if 1 V is applied on the sample. For the same voltage the displacement is reduced to ≈ 0.2 nm in a typical PZT composition and to only ≈ 2 pm in quartz. The displacement can be increased by application

13.2 Measurement techniques

of high voltages, but this has disadvantages that high voltage signals are often nonlinear in itself (due to nonlinearity of the high voltage generators), can lead to breakdown of samples or lead to polarization switching. Optical interferometric techniques are well suited for measurements of such ultra-small displacements. Theoretically, displacements as small as 10^{-5} nm can be measured [5] while the practical limit is about 10^{-3} nm. However, the final accuracy is mostly limited by the difficulties in optical alignment. The mostly widely used optical dilatometer in the piezoelectric community are Mach-Zehnder [6], Michelson [7], and Fabry-Pérot [8, 9] interferometers. Presently the most popular type, particularly for investigation of thin films [10] is Mach-Zehnder interferometer. It is based on the interference of a laser beam reflected from the piezoelectric sample surface and a reference beam. The interference pattern is modified by an electric field applied on the piezoelectric sample. For measurements of ultrasmall displacements (displacement much smaller than the laser light wavelength) one measures slight variation in the intensity of the central interference spot using a photodiode detector [10]. For high displacement (larger than several wavelengths of the laser light) one needs to measure number of interference fringes which pass the light detector as the sample changes its dimension [7]. In the rest of the text we shall refer only to measurements made using Mach-Zehnder interferometer.

Signal-to-noise ratio in interferometric measurements is rather small [5] so that lock-in techniques need to be used. In practice, routine measurements below about 10 Hz are difficult to make and special feed-back techniques need to be used to remove oscillations of the system due to thermal variations. At high frequencies, measurements are limited by mechanical resonance of the many components in the interferometer set-up. However, the bandwidth of these resonance is relatively small and measurements may be made up to high kilohertz range, and even through the piezoelectric resonance of the sample [11]. For strain vs field hysteresis measurements, where lock-in technique cannot be used, it is necessary to average signal many times (e.g. >200 times in addition to using high resolution features of an oscilloscope) to remove noise [10]. Thus, only the equilibrium hysteresis after many field cycles can be measured and not its evolution on a short time scale. A double beam interferometer is used for measurements of thin films, where bending of the substrate can contribute to the signal displacement [10]. An important recent advancement was use of scanning technique, in which the laser beam is scanned over the whole surface of the top electrode of a thin film sample. Thus variation of the piezoelectric response across the sample and spurious contribution of the substrate can be obtained [12].

For large displacements (above about 10 nm) other optical techniques such as so-called photonic sensor are more convenient to use than an optical interferometer. In one of these techniques two optical fibers are placed next to each other in the proximity of the sample's surface. One fiber emits light and the other receives the light reflected from the sample's surface. The area defined by the overlap of the illuminating field of the emitting fiber and the field of view of the receiving fiber will be a function of the distance between the sample and the fibers. If an ac field is applied on a piezoelectric sample, its surface will move and the change in the overlap area will be detected by a photonic sensor. The advantage of photonic sensor dilatometer is that it is very simple to use for measurements of displacements ranging from 10 nm to 1 mm.

Temperature dependent piezoelectric measurements are difficult to carry out using interferometric optical techniques, although recently some progress has been made in this direction [13].

13.2.2 Dynamic press for the measurements of direct effect

The dynamic press used for all measurements described in this report is based on Berlincourt d_{33} meter. The press exhibits important improvements with respect to a classical d_{33} meter in terms of the frequency, pressure and temperature range, accuracy, and resolution. The schematic is shown in Figure 13.3 The system is pre-stressed by a stepper motor which allows a very precise control of the static compressive prestress. The dynamic force is applied by a PZT actuator driven by a high voltage generator. Due to the static prestress, the equal force is applied on the quartz reference sensor and the sample. The piezoelectric charge signals from the reference and sample are amplified by two identical high-quality charge amplifiers and displayed on an oscilloscope. Since the force generator is a strongly nonlinear PZT actuator, its nonlinearity and accompanying hysteresis are seen by both the sample and the quartz reference sensor. In the first approximation, the charge signals from the sample and reference sensor contain the same information on the phase angle and nonlinearity introduced by the nonlinear force generator. In the plot sample charge vs force ($\propto$ reference sample charge) the identical information will cancel and only the true difference between the sample and reference will be displayed. This can be easily verified by measuring response of a quartz sample, which gives perfectly linear and nonhysteretic relation between the charge and force. When a nonlinear sample is investigated, its nonlinear response will contain nonlinear information that is not included in the linear response of the quartz reference; however, the nonlinear component of the sample and the actuator are at least two to three orders of magnitute lower than their linear components. Therefore, the nonlinear contribution of the actuator to the nonlinearity of the sample can be neglected in most cases.

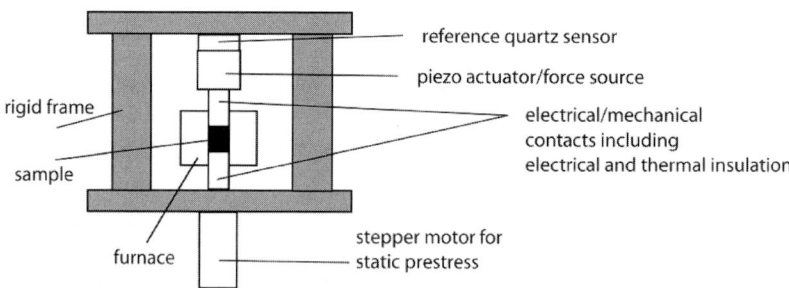

Figure 13.3: Dynamic press for measurements of the direct piezoelectric effect.

The dynamic press allows measurements of the longitudinal, transverse and shear piezoelectric coefficients in the frequency range from 0.01 Hz to about 100 Hz. The lower limit is determined by the insulation resistance of the sample and cables, and charge drifts associated with pyroelectric effect and slight temperature changes. The upper frequency limit is determined by the resonance frequency of the press. The temperature can be varied from ambient to about 500°C. Because the sample is clamped, measurements are strongly dependent on sample aspect ratio that can affect both the magnitude and the apparent field dependence of the piezoelectric coefficients.

In this paper we shall try to demonstrate how these two techniques can be used not only for routine piezoelectric measurements, but to give a deeper insight into different processes operating in ferroelectric materials.

13.3 Investigation of the piezoelectric nonlinearity in PZT thin films using optical interferometry

PZT materials, both ceramics and thin films, exhibit nonlinear and hysteretic piezoelectric response even at fields much lower than the macroscopic switching or coercive field. An example is shown in Figure 13.4.

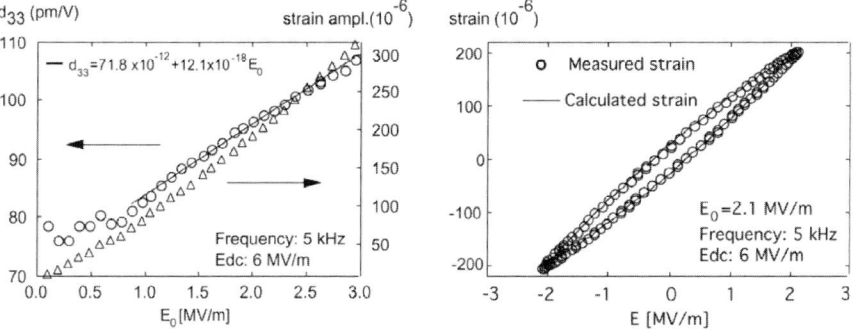

Figure 13.4: Illustration of piezoelectric nonlinearity and hysteresis in a PZT thin film [3].

The origin of the nonlinearity and hysteresis in the films is most likely due to displacement of domain walls [4]. If domain walls move in a medium with a random distribution of pinning center, the response of the material can be described, in the first approximation by Rayleigh relations. We next demonstrate how optical interferometry can be sued to verify whether this particular model applies to the investigated PZT thin film. In the case of the converse piezoelectric effect, when the driving field E is varied between $-E_0$ and E_0, the piezoelectric strain x is hysteretic and can be expressed by the following Rayleigh relations:

$$x(E) = (d_{init} + \alpha_m E_0)E \pm \alpha_m \left(E_0^2 - E^2\right)/2 \qquad (13.1)$$

$$d(E_0) = d_{init} + \alpha_m E_0 \qquad (13.2)$$

In equations above, E is the ac electric field, x the piezoelectric strain, d_{init} designates the value of d at zero field. α_m is called the nonlinear Rayleigh coefficient. The sign $\pm$ corresponds to the upper and lower branches of the hysteresis loop. The practical value of the Rayleigh relations is that they give a simple, unified description of the hysteresis and nonlinearity. Knowledge of either Equation (13.1) or (13.2) allows calculation of the other. Note that both equations must be fulfilled in order to have the Rayleigh case. This is illustrated in Figure 13.4, with data obtained using a double beam optical interferometer.

Interesting properties of the Rayleigh relationships can be obtained from the expansion of Equation (13.1) in Fourier series. For $E = E_0 \sin(\omega t)$, one obtains:

$$x(E) = (d_{init} + \alpha_m E_0)\sigma_0 \sin(\omega t) + \frac{4\alpha_m E_0^2}{3\pi} \cos(\omega t)$$

$$+ \frac{4\alpha E_0^2}{15\pi} \cos(3\omega t) - \frac{4\alpha E_0^2}{105\pi} \cos(5\omega t) + ...$$

(13.3)

The most important features of Equation (13.3) are the following. (i) Only odd-number harmonics have nonzero amplitude. It can be shown that this implies a perfectly symmetrical material response, (so called half-wave symmetry, $x(t + T/2) = -x(t)$, where t is time and T period of the driving field); (ii) all higher harmonics are quadratic functions of the field amplitude, and (iii) the phase angle of all harmonics except the fundamental is $\pm 90°$. These characteristics are essential features of the Rayleigh relations and may be used to discriminate between different nonlinear/hysteretic phenomena as shown by Taylor and Damjanovic [14] for the dielectric response and by Taylor [3] for the piezoelectric response of PZT thin films. An illustration is given in Figure 13.5. The lock-in technique used for the displacement measurements by the interferometer directly gives the needed information.

Discussion on reasons for the deviation of the phase angle from 90° and of the exponent of the third harmonic amplitude from Figure 13.5 can be found in [4].

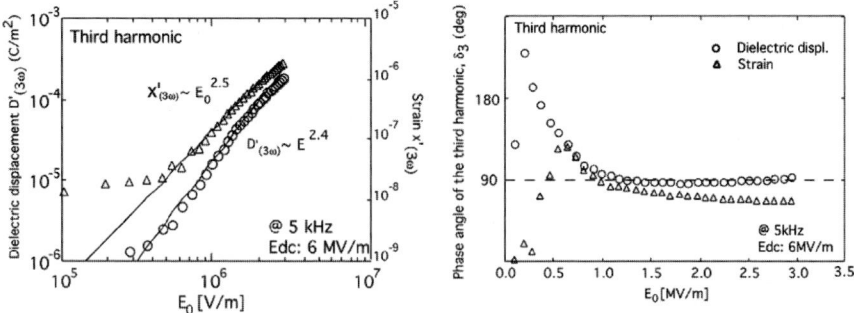

Figure 13.5: Amplitude and phase angle of the third harmonic of the field induced strain and dielectric displacement of a 60/40 PZT thin film.

In summary, we can conclude that the optical interferometry can be a useful technique not only for routine piezoelectric measurements but that it can reveal crucial information on different mechanisms operating in the material.

13.4 Investigation of the piezoelectric relaxation in ferroelectric ceramics using dynamic press

13.4.1 Maxwell-Wagner piezoelectric relaxation and clockwise hysteresis

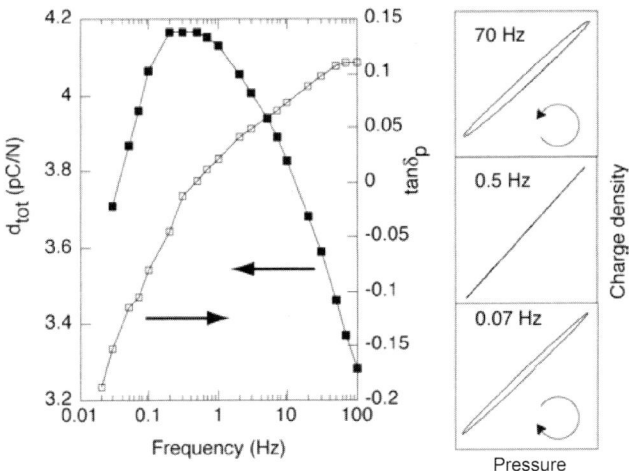

Figure 13.6: Piezoelectric coefficient d and $\tan\delta_p$ of a single phase $Bi_4Ti_3O_{12}$ ferroelectric ceramic with highly anisotropic grains, at room temperature. On the right are shown charge-pressure hysteresis loops at selected frequencies, with a clockwise hysteresis at 0.07 Hz and counter-clockwise hysteresis at 70 Hz. See [17] for details.

One of the interesting feature of the coupled electro-mechanical effects is that they can exhibit clockwise response-field hysteresis [15, 16]. In this section we present evidence of such behavior in a ferroelectric ceramic. The measurements were made using the dynamic press. Figure 13.6 shows the real and imaginary part of the piezoelectric d_{33} coefficient for the single-phase $Bi_4Ti_3O_{12}$ ceramic as a function of the driving pressure frequency, at ambient temperature. The piezoelectric coefficient exhibits a transition from the relaxation regime at low frequencies to the retardation regime at high frequencies, indicating a presence of at least two dispersion processes, one (relaxation) dominant at low and another (retardation) dominant at high frequencies. Charge-pressure hysteresis are shown in the same figure for 0.07, 5 and 70 Hz. The rotation direction is indicated for each hysteresis. The hysteresis evolution from clockwise to counterclockwise is clearly visible. These results directly demonstrates the presence of a clockwise piezoelectric hysteresis and a negative piezoelectric phase angle in a heterogeneous ferroelectric material.

The origin of this relaxation is in heterogeneity of the ceramic, in which anisotropically shaped grains exhibit strong variation in their piezoelectric and dielectric properties in different directions. As discussed in [17], in such heterogeneous materials Maxwell-Wagner like processes may lead to a behavior shown in Figure 13.6.

13.4.2 Piezoelectric relaxation and Kramers-Kronig relations in a modified lead titanate composition

An exceptionally strong piezoelectric relaxation is observed in lead titanate ceramics modified by Sm, as illustrated in Figure 13.7. The temperature dependence of the relaxation clearly suggests a thermally activated process and the activation energy of the process can be estimated from the temperature dependence of the peak in the imaginary component of d_{33}. As in relaxation of dielectric permittivity and elastic coefficients, the real and imaginary components of the piezoelectric coefficient should be related by Kramers-Kronig relations [18]. However, the possibility of having a clockwise hysteresis in the piezoelectric response (see previous section) makes question of validity of Kramers-Kronig relations for the piezoelectric effect nontrivial.

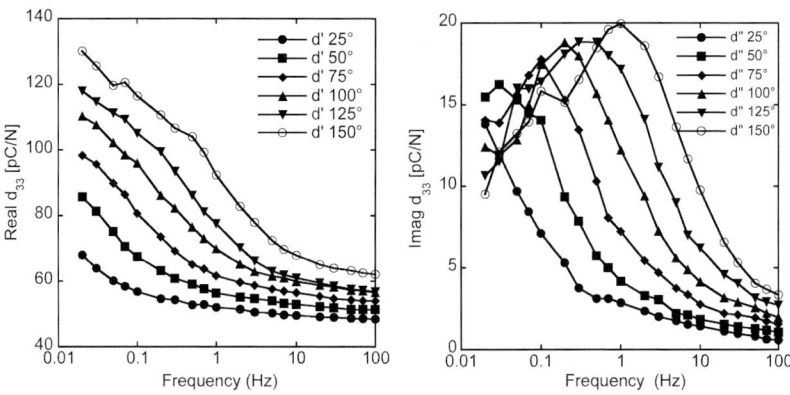

Figure 13.7: Relaxation of the longitudinal piezoelectric coefficient in Sm-modified lead titanate ceramics at different temperatures.

We show in Figure 13.8 that in the case of a well-behaved piezoelectric relaxation (counterclockwise hysteresis) presented in Figure 13.7, the Kramers-Kronig relations are indeed fulfilled. Closer inspection of the data show that the relaxation curves can be best described by a distribution of relaxation times and empirical Havriliak-Negami equations [19]. It is worth mentioning that over a wide range of driving field amplitudes the piezoelectric properties of modified lead titanate are linear. Details of this study will be presented elsewhere.

13.4 Investigation of the piezoelectric relaxation in ferroelectric ceramics using dynamic press

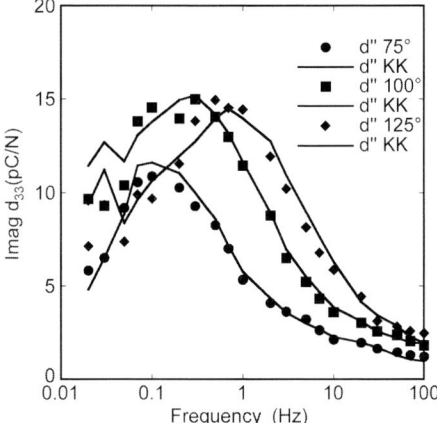

Figure 13.8: Ilustration of the validity of the Kramers-Kronig relations for the piezoelectric relaxation in Sm-modified lead titanate ceramics. The imaginary component is calculated from the real using numerical method and Kramers-Kronig relations and compared with experimentally determined data.

13.4.3 Evidence of creep-like piezoelectric response in soft PZT ceramics

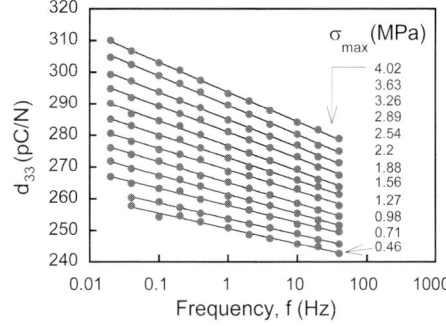

Figure 13.9: Frequency dependence of the total piezoelectric coefficient in soft PZT ceramics at different amplitudes of the driving field.

Frequency dependence of the piezoelectric response in soft PZT is very different from the piezoelectric relaxation described in the previous section, Figure 13.9. The data presented in Figure 13.9 suggest logarithmic frequency dependence [20] of the piezoelectric coefficient. However, analysis of its imaginary component reveals possibility that we are in this case in a presence of a creep-like process, as suggested by Kleemann *et al.* [21]. Indeed, data in Figure 13.10 and Figure 13.11 suggest that, if the nonlinearity is properly taken into account, both real and imaginary parts follow power-law behavior $1/\omega^\beta$ with β being very small (< 0.08) and field dependent. Details of this investigation will be published elsewhere.

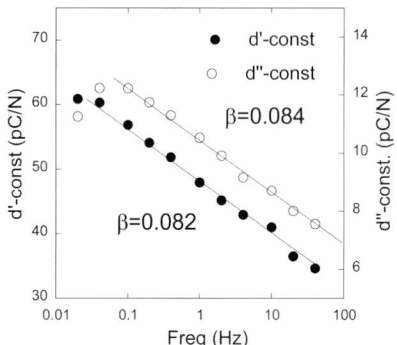

Figure 13.10: Real and imaginary parts of the piezoelectric response in a soft PZT indicating power law frequency dependence with a very small parameter β. The data are shown are for $\sigma = 4$ MPa.

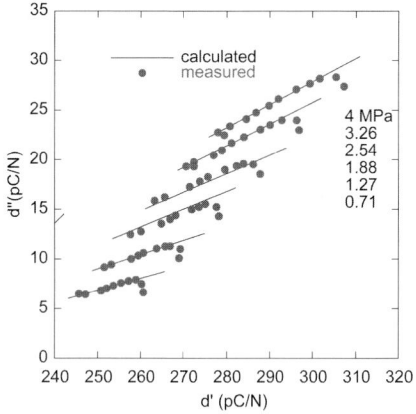

Figure 13.11: Imaginary vs real part of the piezoelectric coefficient at different dynamic stress levels with frequency taken as a parameter.

Bibliography

[1] D. V. Taylor and D. Damjanovic, *Appl. Phys. Lett.* **76**, 1615 (2000).

[2] I. D. Mayergoyz, *Mathematical Models of Hysteresis*, Springer Verlag, New York, 1991.

[3] D. V. Taylor, *Ph. D. Thesis No. 1949*, Swiss Federal Institute of Technology-EPFL, Lausanne, 1999.

[4] D. Damjanovic and G. Robert, in *Piezoelectric materials for the end user*, N. Setter ed., 2002.

[5] T. Kwaaitaal, B. J. Luymes, and G. A. van der Pijll, *J. Phys. D: Appl. Phys.* **13**, 1005 (1980).

[6] W. Y. Pan and L. E. Cross, *Rev. Sci. Instrum.* **60**, 2701 (1998).

[7] Q. M. Zhang, W. Y. Pan, and L. E. Cross, *J. Appl. Phys.* **63**, 2492 (1988).

[8] V. E. Bottom, *J. Appl. Phys.* **41**, 3941 (1970).

[9] C. Z. Tan and J. Arndt, *Physica* B **225**, 202 (1996).

[10] A. L. Kholkin, C. Wütchrich, D. V. Taylor, and N. Setter, *Rev. Sci. Instrum.* **67**, 1935 (1996).

[11] W. Y. Pan, H. Wang, L. E. Cross, and B. R. Li, *Ferroelectrics* **120**, 231 (1991).

[12] K. Yao and F. E. H. Tay, IEEE ULTRASON FERR **50**, 113 (2003).

[13] M. Sulc, L. Burianova, and J. Nosek, *Ann. Chim.-Sci. Mat.* **26**, 43 (2001).

[14] D. V. Taylor and D. Damjanovic, *Appl. Phys. Lett.* **73**, 2045 (1998).

[15] A. S. Nowick and W. R. Heller, *Adv. Phys.* **14**, 101 (1965).

[16] R. Holland, IEEE *Trans. Sonics Ultrason.* SU **14**, 18 (1967).

[17] D. Damjanovic, M. Demartin Maeder, C. Voisard, and N. Setter, *J. Appl. Phys.* **90**, 5708 (2001).

[18] R. Hayakawa and Y. Wada, in *Advances in polymer science* **11**, H. J.e.a. Cantow ed., Springer Verlag Berlin, 1973.

[19] S. Havriliak and S. Negami, *J. Polym. Sci.* C **14**, 99 (1966).

[20] D. Damjanovic, *Phys. Rev.* B **55**, R649 (1997).

[21] W. Kleemann, J. Dec, S. Miga, T. Woike, and R. Pankrath, *Phys. Rev.* B **65**, 220101 (2002).

14 Chaotic Behavior near the Ferroelectric Phase Transition

H. Beige, M. Diestelhorst and R. Habel

Martin-Luther-University Halle-Wittenberg, Department of Physics, Halle, Germany

14.1 Introduction

In many fields of nature, science and engineering one can find nonlinear dynamical systems (e.g. [1–3]). It is well known that the structural phase transitions in ferroelectric materials are connected with strong properties. So in a nonlinear system, that contains a ferroelectric material one can expect many features of nonlinear dynamical systems such as period-doubling cascades and chaotic behavior. In our case the experimentally investigated nonlinear dynamical system is a series-resonance circuit, consisting of a linear inductance and a nonlinear capacitance. Different ferroelectric materials have been used as a nonlinear capacitance. It is shown, that the methods of the nonlinear dynamics are a useful tool for the study of structural phase transitions and for the characterization of the large signal behavior of ferroelectric materials. Further the controlling of chaos in a ferroelectric system is presented. That means the chaotic system can be stabilized on an instable periodic orbit by a small control perturbation.

14.2 Dielectric nonlinear series-resonance circuit

The investigated resonance circuit consists of a linear inductance L_0 and the dielectric nonlinear capacitance C_{NL} (see Figure 14.1). Let us consider the case, that a ferroelectric triglycine sulfate crystal (TGS) is used as nonlinear capacitance.

R_s describes the loss of the resonance circuit. The resonance circuit is driven by a sinusoidal voltage $U_0 \cos \omega_e t$. The following differential equation describes the behavior of the circuit:

$$\ddot{D}_2 + \frac{R_s}{L_o} \dot{D}_2 + \frac{a}{bl L_0} E_{NL} = \frac{1}{bl L_0} U_0 \cos \omega_e t \ . \tag{14.1}$$

Here D_2 is the dielectric displacement along the ferroelectric axis, a the thickness, l the length and b the width of the sample. The electric field strength E_{NL} at the crystal along the ferroelectric b-axis is a nonlinear function of the dielectric displacement.

Polar Oxides: Properties, Characterization, and Imaging
Edited by R. Waser, U. Böttger, and S. Tiedke
Copyright © 2005 WILEY-VCH Verlag GmbH & Co. KGaA, Weinheim
ISBN: 3-527-40532-1

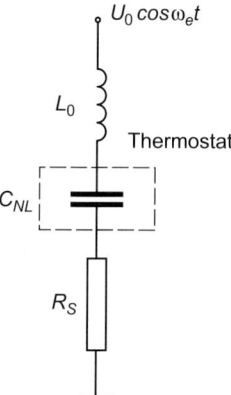

Figure 14.1: Dielectric nonlinear series-resonance circuit

14.3 Nonlinear nature of the resonant system

The ferroelectric phase transition of second-order in TGS at $\vartheta_c = 49°C$ can be described in the framework of the Landau-theory (e.g. [4]) by the thermodynamical potential

$$G = G_0 + \frac{A}{2}D_2^2 + \frac{B}{4}D_2^4 . \tag{14.2}$$

The only temperature-dependent coefficient is $A = A_0(\vartheta - \vartheta_c)$. The electric field at the sample is calculated from Equation 14.2 as

$$E_{NL} = \frac{\partial G}{\partial D_2} = AD_2 + BD_2^3 . \tag{14.3}$$

Inserting 14.3 into 14.1 provides the so-called Duffing equation (e.g. [1])

$$\ddot{D}_2 + \frac{R_s}{L_0}\dot{D}_2 + \frac{aA}{bl L_0}D_2 + \frac{aB}{bl L_0}D_2^3 = \frac{U_0}{bl L_0}\cos\omega_e t . \tag{14.4}$$

At temperatures above ϑ_c an oscillation of D_2 in a potential with one minimum may be observed. Below ϑ_c the potential becomes a double-well one.

14.4 Tools of the nonlinear dynamics

Dynamical systems may be conveniently analyzed by means of a multidimensional phase space, in which to any state of the system corresponds a point. Therefore, to any motion of a system corresponds an orbit or trajectory. The trajectory represents the history of the dynamic system. For one-dimensional linear systems, as in the case of the harmonic series-resonance circuit, described by the differential equation

$$\ddot{D}_2 + \omega_0^2 D_2 = 0 \tag{14.5}$$

14.5 Experimental representation of phase portraits

one obtains a two-dimensional phase space, a so-called phase portrait in form of an ellipse

$$\frac{D_2^2}{F^2} + \frac{\dot{D}_2^2}{F^2\omega_0^2} = 1. \tag{14.6}$$

Here F is a constant of the integration and ω_0^2 the eigenfrequency of the oscillator. Figure 14.2 shows the simple trajectory of the harmonic oscillator.

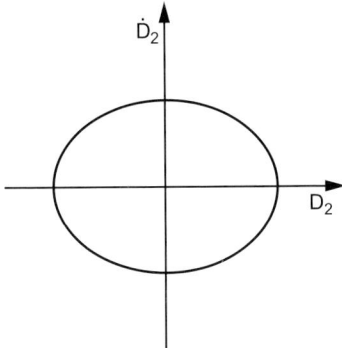

Figure 14.2: Phase portrait of a harmonic oscillator

The perpendicular slice through the phase portrait provides the stroboscopic phase portrait or Poincaré section (e.g. [3]). This is in the case of the harmonic oscillator one point in the phase portrait. A further powerful method for the analysis of nonlinear dynamical systems is the determination of the Fourier spectrum of the response function D_2.

14.5 Experimental representation of phase portraits

A computer controlled measuring system with very short measuring time for a quantitative recording of the phase portrait was developed. Figure 14.3 represents the block diagram of the measuring system.

Across the linear capacitance C_0 a signal proportional to the dielectric displacement D_2 is recorded. The temperature of the nonlinear capacitance C_{NL} can be stabilized. Further the amplitude and frequency of the driving voltage can be changed. The measuring system is based on a fully programmable true dual-channel digitizer (Sony/Tektronix RTD 710). The digitizer provides 10-bit resolution at a 200 MHz maximum sampling rate (single-channel mode). It is possible to record the excitation $U(t)$, the response function $D_2(t)$ and the phase portrait $\dot{D}_2 = \dot{D}_2(D_2)$, where $\dot{D}_2$ is calculated by the computer. By removing the inductance L_0 the hysteresis loop may be recorded at the same frequency as the phase portrait. Furthermore, the calculated Fourier spectra of the excitation and of the response function D_2 can be obtained.

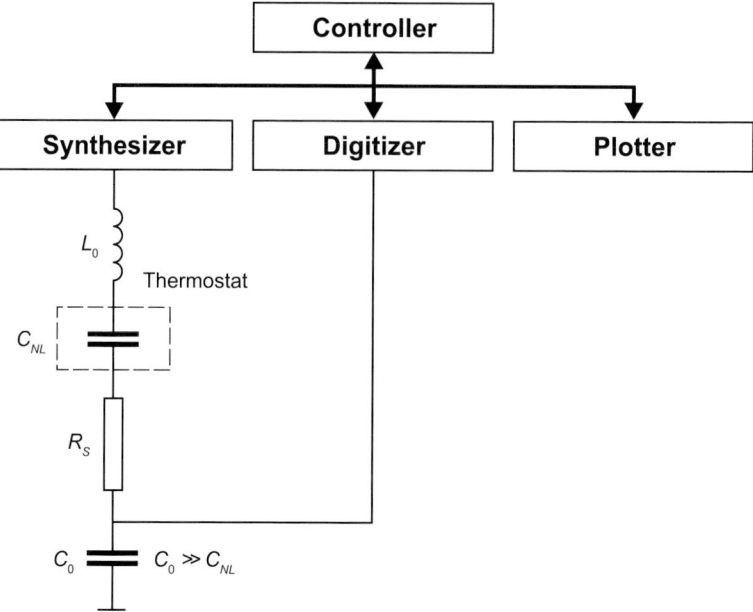

Figure 14.3: Block diagram of the measuring system for recording the phase portrait

14.6 Comparison of calculated and experimentally observed phase portraits

In order to compare calculated and experimentally observed phase portraits it is necessary to know very exactly all the coefficients of the describing nonlinear differential Equation 14.3. Therefore, different methods of determination of the nonlinear coefficient in the Duffing equation have been compared. In the paraelectric phase the value of the nonlinear dielectric coefficient B is determined by measuring the shift of the resonance frequency in dependence on the amplitude of the excitation ([1], [5]). In the ferroelectric phase three different methods are used in order to determine B. Firstly, the coefficient B is calculated in the framework of the Landau theory from the coefficient of the high temperature phase (e.g. [4]). This means $B = const.$ and B has the same values above and below the phase transition. Secondly, the shift of the resonance frequency of the resonator in the ferroelectric phase as a function of the driving field is used in order to determine the coefficient B. The amplitude of the exciting field is smaller than the coercive field and does not produce polarization reversal during the measurements of the shift of the resonance frequency. In the third method the coefficient B was determined by the values of the spontaneous polarization

$$D_{2sp} = \pm\sqrt{\frac{A}{B}} \tag{14.7}$$

14.6 Comparison of calculated and experimentally observed phase portraits

and of the coercive field

$$E_{coer} = \pm \frac{2}{3} A \sqrt{\frac{A}{3B}}. \qquad (14.8)$$

These values are obtained by recording the hysteresis loop of the TGS-crystal. In this case the field strength is so high, that a domain reorientation occurs in the sample. The results of the computer simulation and the experimentally observed phase portraits in the paraelectric phase are quantitatively in good agreement. The investigations in the ferroelectric phase can be summarized in the form, that the calculated and experimentally observed phase portraits provide only a good agreement with the computer simulation derived from values of the hysteresis loop. By measuring the hysteresis loop and the phase portrait the influence of the domain reorientation on the nonlinear dielectric properties is observed. Figure 14.4 shows the experimentally recorded phase portrait (dotted line) and the calculated phase portrait (solid line) of a TGS-crystal. The temperature is $\vartheta = 39,1°C$ and the amplitude of the excitation 14.6 V. Further the Fourier spectra of the response function are experimentally determined by a spectrum- and network-analyzer and by a computer simulation.

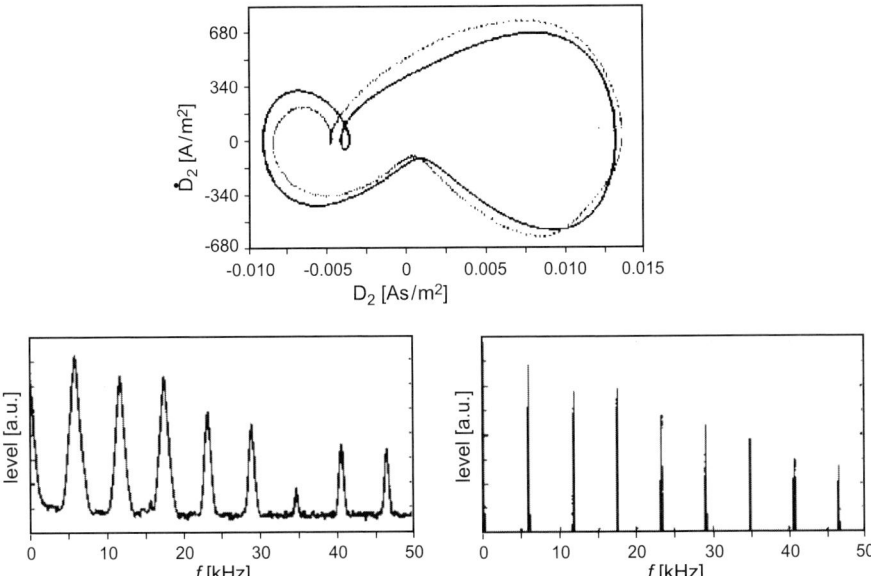

Figure 14.4: Phase portrait (dotted: experiment, solid: simulation) and experimentally determined and calculated Fourier spectra of the response function with the excitation of the amplitude 14.6 V

By increasing the amplitude of the excitation (16,4 V) a period doubling can be observed. Figure 14.5 shows the phase portrait and the Fourier spectra of the response function in the

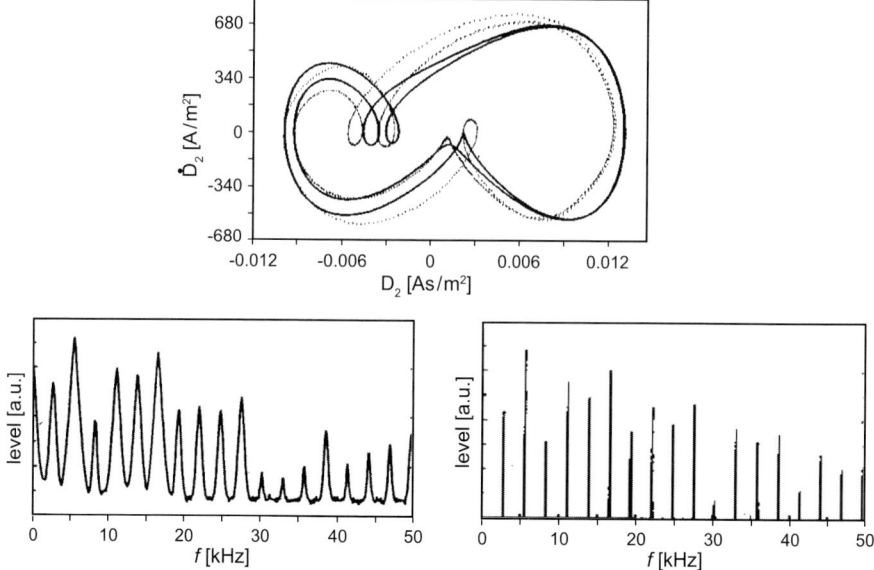

Figure 14.5: Phase portrait (dotted: experiment, solid: simulation) and experimentally determined and calculated Fourier spectra of the response function with the excitation of the amplitude 16.4 V.

case of period-2-behavior. The calculations were done with the same coefficients as in the case with the excitation of the amplitude 14.6 V.

The Duffing Equation 14.4 seems to be a model in order to describe the nonlinear behavior of the resonant system. A better agreement between experimentally recorded and calculated phase portraits can be obtained by consideration of nonlinear effects of higher order in the dielectric properties and of nonlinear losses (e.g. [6], [7]). In order to construct the effective thermodynamic potential near the structural phase transition the phase portraits were recorded at different temperatures above and below the phase transition. The coefficients in the Duffing Equation 14.4 were derived by the fitted computer simulation. Figure 14.6 shows the effective thermodynamic potential of a TGS-crystal with the transition from a one minimum potential to a double-well potential. So the tools of the nonlinear dynamics provide a new approach to the study of structural phase transitions.

The conclusion that the phase portrait gives information about the effective thermodynamic potential is supported by the high sensitivity of the phase portrait against a bias field at the sample. Figure 14.7 represents the influence of a very small bias field ($E = +2700$ V/m) on the phase portrait in the ferroelectric phase, when the field direction is changed.

The relation between the form of the phase portrait and the effective thermodynamic potential is schematically demonstrated in Figure 14.8. Figure 14.8 (a) shows the phase portrait of a TGS-crystal and the corresponding thermodynamic potential. In the L-α-alanin doped TGS-

14.7 Controlling chaos

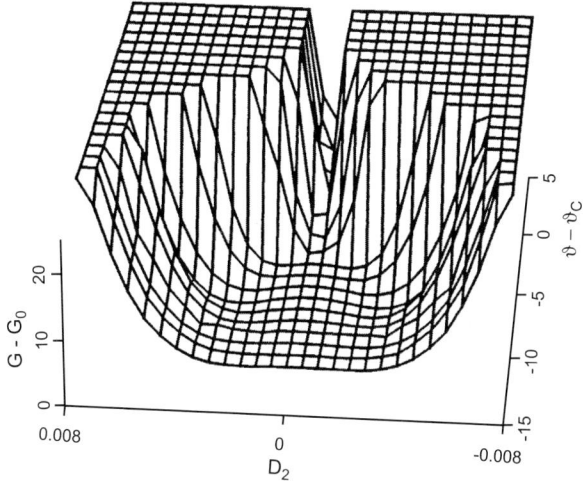

Figure 14.6: Effective thermodynamic potential of a TGS-crystal

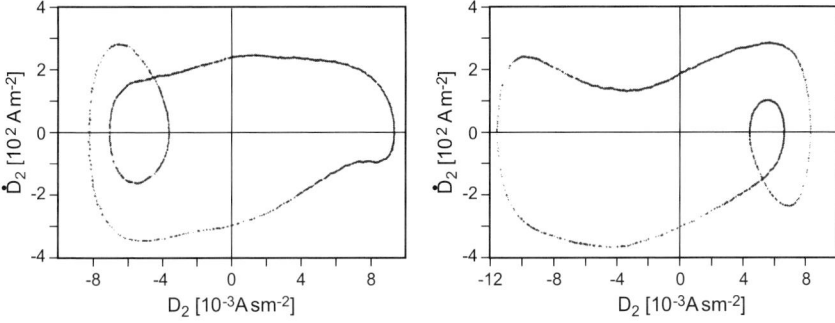

Figure 14.7: Influence of an external bias field on the phase portrait of a TGS-crystal in the ferroelectric phase

crystal the internal bias field produces a deformation of the potential (see Figure 14.8 (b)). Rochelle salt (see Figure 14.8 (c)) has not such a deep double-well potential as TGS ([8]).

14.7 Controlling chaos

It is well known that a nonlinear resonator with increasing amplitude of driving force exhibits sequences of period-doubling bifurcations leading to a chaotic motion (e.g. [2], [3]). Deterministic chaos will be used to denote the irregular or chaotic motion which is generated by nonlinear deterministic systems. Feigenbaum has shown that the route to chaos is governed

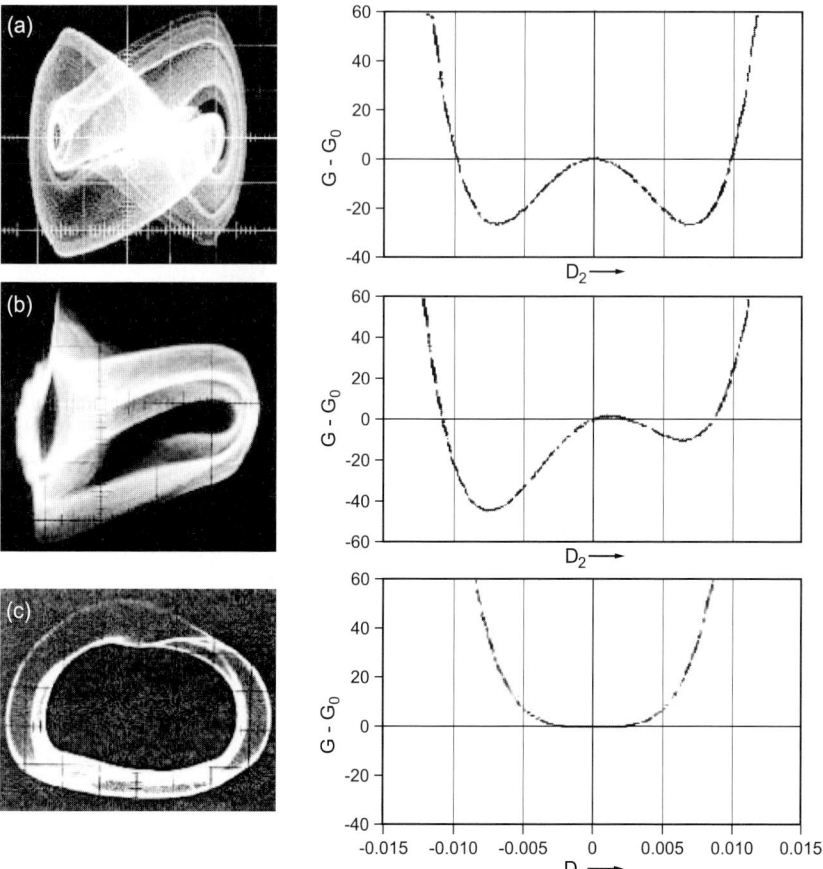

Figure 14.8: Typical phase portraits of crystals with different effective thermodynamic potentials ((a) TGS-crystal, (b) L-α-alanin doped TGS-crystal, (c) Rochelle salt)

by universal behavior ([9]). Figure 14.9 shows the experimentally recorded phase portraits of the resonance circuit with a TGS-crystal at different driving voltages below the phase transition and the corresponding Fourier spectra of the response function with the periods T, 2T, 4T and deterministic chaos. Figure 14.10 represents an experimentally recorded chaotic response function.

Two typical properties of nonlinear dynamical systems are responsible for the realization of controlling chaos. Firstly, nonlinear systems show a sensitive dependence on initial conditions. This is represented in Table 14.1 by the nonlinear equation

$$x_{n+1} = 4x_n(1-x_n) \, .$$

Small changes of the start values x_0 produce a growing separation between the neighbouring trajectories.

14.7 Controlling chaos

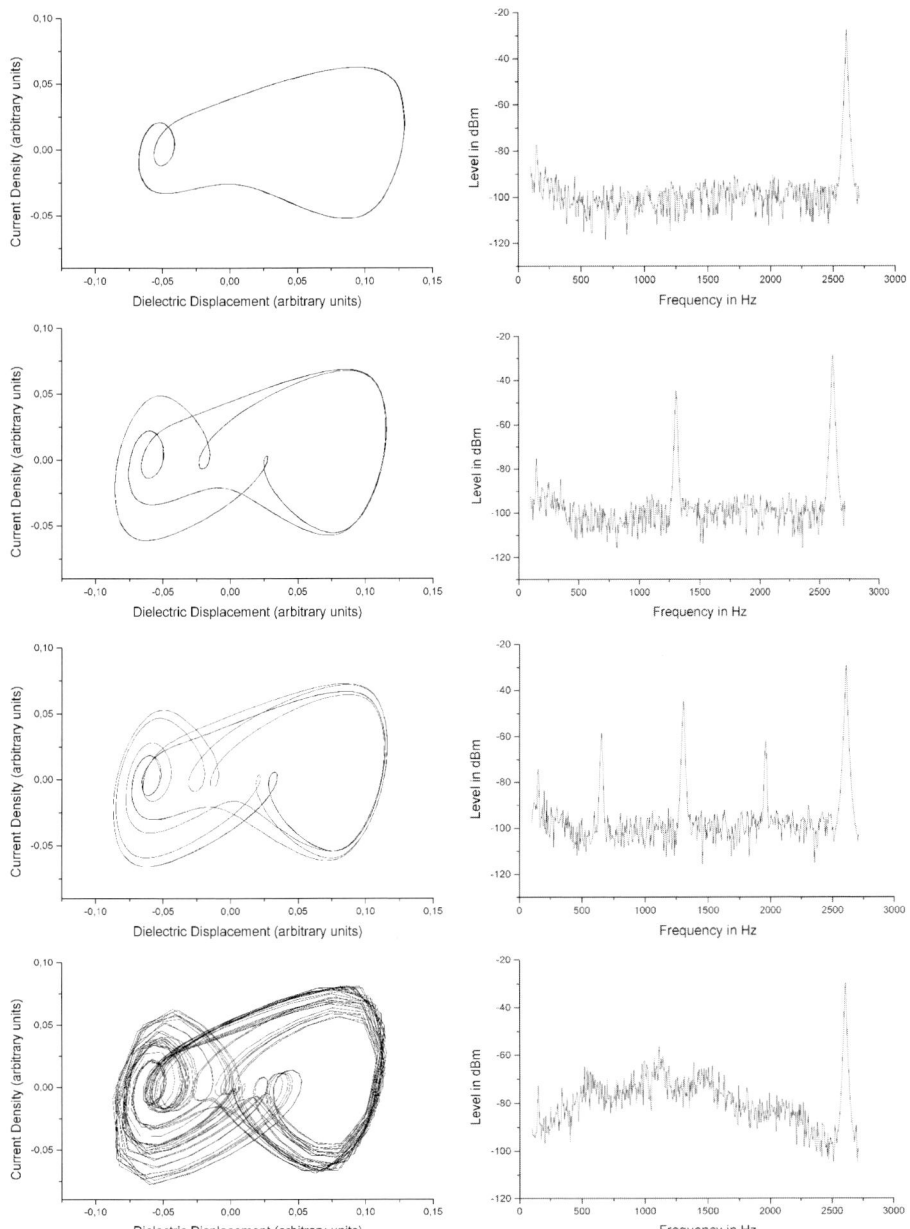

Figure 14.9: Phase portraits of the resonance circuit with a TGS-crystal at different driving voltages below the phase transition and the corresponding Fourier spectra of the response functions with the periods T, 2T, 4T and deterministic chaos

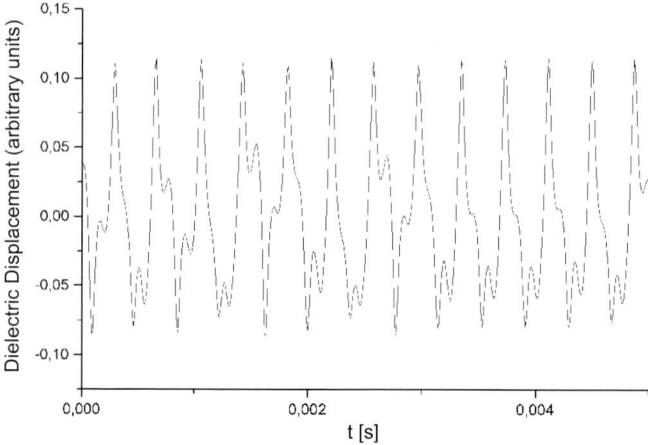

Figure 14.10: Experimentally recorded chaotic response function of the resonance circuit

number of iteration	sequence 1	sequence 2	sequence 3
0	0,3	0,3001	0,301
1	0,84	0,840159	0,8401596
2	0,5376	0,537164	0,533248
3	0,994344	0,994475	0,995578
4	0,022492	0,021977	0,017609
5	0,087945	0,085977	0,069197
6	0,320843	0,314342	0,257635
7	0,871612	0,862125	0,765038
8	0,447616	0,475461	0,719018
9	0,989024	0,997591	0,808123
10	0,043421	0,009611	0,620239
11	0,166145	0,038075	0,94217

Table 14.1: The sensitive dependence of the nonlinear equation $x_{n+1} = 4x_n(1 - x_n)$ on different initial values

Secondly, the chaotic state of a resonant system is an infinite collection of unstable periodic motions (see Figure 14.9). The basic idea of the controlling chaos is to switch a chaotic system by a small perturbation among many different periodic orbits. Ott, Grebogy and Yorke [10] developed a scheme in which a chaotic system can be forced to follow one particular unstable periodic orbit. The problem is to calculate the perturbation that will shift the system towards the desired periodic orbit. This process is similar to balancing a marble on a saddle. To keep the marble from rolling off, one needs to move the saddle quickly from side to side. And

14.8 Summary

just as the marble reacts to small movements of the saddle, the trajectory is very sensitive to changes in the parameters of the resonance system. The driving amplitude of the series-resonance circuit is selected as an external parameter of perturbation. Figure 14.11 shows the experimental setup of the control system in order to record the Poincaré section of the phase portrait. The resonance circuit is driven by a sinusoidal voltage in a frequency range between 1–20 kHz. The value of the perturbation is calculated by the controlling computer from the Poincaré section. All parameters needed for successful control can be obtained by measuring the real system without using any kind of model. An electronic unit is used in order to carry out the control as fast as possible by perturbing the nonlinear series resonance circuit with small modulation of the driving voltage [11].

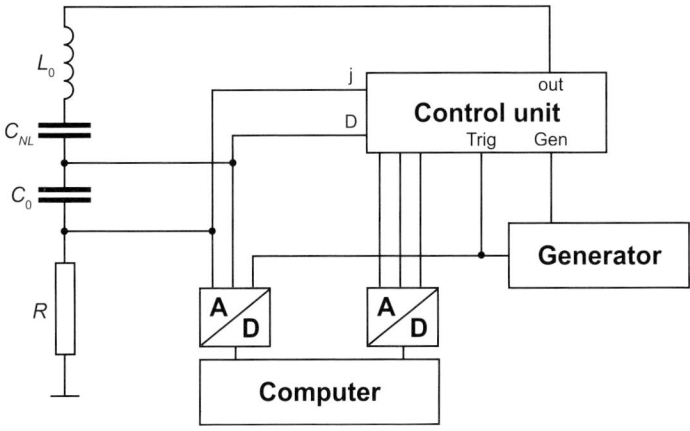

Figure 14.11: Experimental setup for the controlled nonlinear resonance circuit

The results of a control process are shown in Figure 14.12. Here the control perturbation is plotted in percent of the driving amplitude (in the upper part) and the dielectric displacement in dependence on the number of the control steps (in the lower part). At the 700th control step the control is switched on. After few steps the system moves from the chaotic state into a periodic orbit. The required perturbation decreases in time. There are two reasons for this behavior. The first is a transient state of the nonlinear series resonance circuit after varying the environmental parameter. It takes a while until a stable state is reached. The second is a thermal problem. The value of energy entering the ferroelectric crystal by loss mechanisms is different in the periodic and chaotic state of oscillations. The control unit compensates these effects by changing the driving amplitude.

14.8 Summary

The new concepts of the nonlinear dynamics are a useful tool for the study of structural phase transitions and nonlinear properties. In particular, the comparison of calculated and exper-

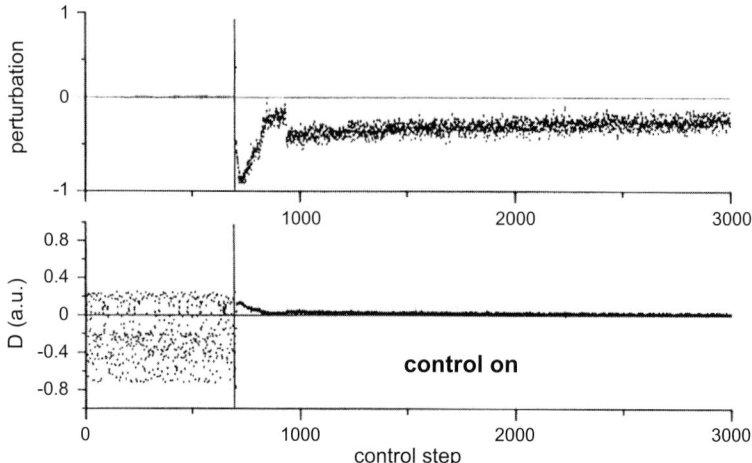

Figure 14.12: Experimental results of a control process

imentally recorded phase portraits provides additional information about the nonlinear dynamic behavior near phase transitions, the nature of the nonlinear properties and the active physical mechanisms. The recording of phase portraits provides the opportunity to investigate the nonlinear dynamics as a function of the frequency in order to study, for example, switching processes in ferroelectric and ferroelastic materials. The concept of the controlling chaos shows the high flexibility of chaotic systems and opens a new field of applications.

Bibliography

[1] A. H. Nayfeh and D. T. Mook, *Nonlinear Oscillations*, New York, 1979.
[2] H. G. Schuster, *Deterministic Chaos*, Physik Verlag, Weinheim, 1984.
[3] J. Argyris, G. Faust, and M. Haase, *Die Erforschung des Chaos*, Vieweg, Braunschweig, 1994.
[4] M. E. Lines and A. M. Glass, *Principles and Applications of Ferroelectrics and Related Materials*, Oxford, 1979.
[5] H. Beige and G. Schmidt, *Ferroelectrics* **41**, 173 (1982).
[6] R. Hegger, H. Kantz, F. Schmuser, M. Diestelhorst, R.-P. Kapsch, and H. Beige, *Chaos* **8**, 727 (1998).
[7] R.-P. Kapsch, H. Kantz, R. Hegger, and H. Diestelhorst, *Int. J. of Bifurcation and Chaos* **11**, 1019 (2001).
[8] H. Beige, M. Diestelhorst, R. Forster, and T. Krietsch, *Phase Transitions* **37**, 213 (1992).
[9] M. J. Feigenbaum, *J. Stat. Phys.* **19**, 25 (1978).
[10] E. Ott, C. Grebogy, and J. A. Yorke, *Phys. Rev. Lett.* **64**, 1196 (1990).
[11] R. Habel and H. Beige, *Int. J. Bifurcation and Chaos* **7**, 199 (1997).

15 Relaxor Ferroelectrics – from Random Field Models to Glassy Relaxation and Domain States

Wolfgang Kleemann[1], *George A. Samara*[2] *and Jan Dec*[1,3]

1. Angewandte Physik, Universität Duisburg-Essen, D-47048 Duisburg, Germany
2. Nanostructure and Advanced Materials Chemistry, Sandia National Laboratories, Albuquerque, NM 87185-1421, USA
3. Physics Department, University of Silesia, PL 40-480 Katowice, Poland

Abstract

Substitutional charge disorder giving rise to quenched electric random-fields (RFs) is probably at the origin of the peculiar behavior of *relaxor* ferroelectrics, which are primarily characterized by their strong frequency dispersion of the dielectric response and by an apparent lack of macroscopic symmetry breaking at the phase transition. Spatial fluctuations of the RFs correlate the dipolar fluctuations and give rise to polar nanoregions in the paraelectric regime. The dimension of the order parameter decides upon whether the ferroelectric phase transition is destroyed (*e.g.* in cubic $PbMg_{1/3}Nb_{2/3}O_3$, PMN) or modified towards RF Ising model behavior (*e.g.* in tetragonal $Sr_{1-x}Ba_xNb_2O_6$, SBN, $x \approx 0.4$). Frustrated interaction between the polar nanoregions in cubic *relaxors* gives rise to cluster glass states as evidenced by strong pressure dependence, typical dipolar slowing-down and theoretically treated within a spherical random bond-RF model. On the other hand, freezing into a domain state takes place in uniaxial *relaxors*. Below $T_c \approx 350$ K frozen-in nanodomains have been evidenced in SBN by piezoresponse force microscopy (PFM). RF pinning of the domain walls gives rise to non-Debye dielectric response, which is relaxation- and creep-like at radio- and very low frequencies, respectively. At T_c non-classical critical behavior with critical exponents $\gamma \approx 1.8$, $\beta \approx 0.1$ and $\alpha \approx 0$ is encountered in accordance with the RF Ising model.

15.1 Introduction

Relaxor ferroelectrics (RL) include a large group of solid solutions, mostly oxides, with a perovskite or tungsten bronze structure. In contrast to ordinary ferroelectrics (FE) whose physical properties are quite adequately described by the Landau-Ginzburg-Devonshire theory [1], RLs possess the following main features: (I) a significant frequency-dependence of the electric permittivity, (II) absence of both spontaneous polarization and structural macroscopic symmetry breaking, (III) FE-like response arising after field cooling to low temperature [2]. Figure 15.1 shows the most prominent differences between normal (FE) and *relaxor ferroelectrics* (RL):

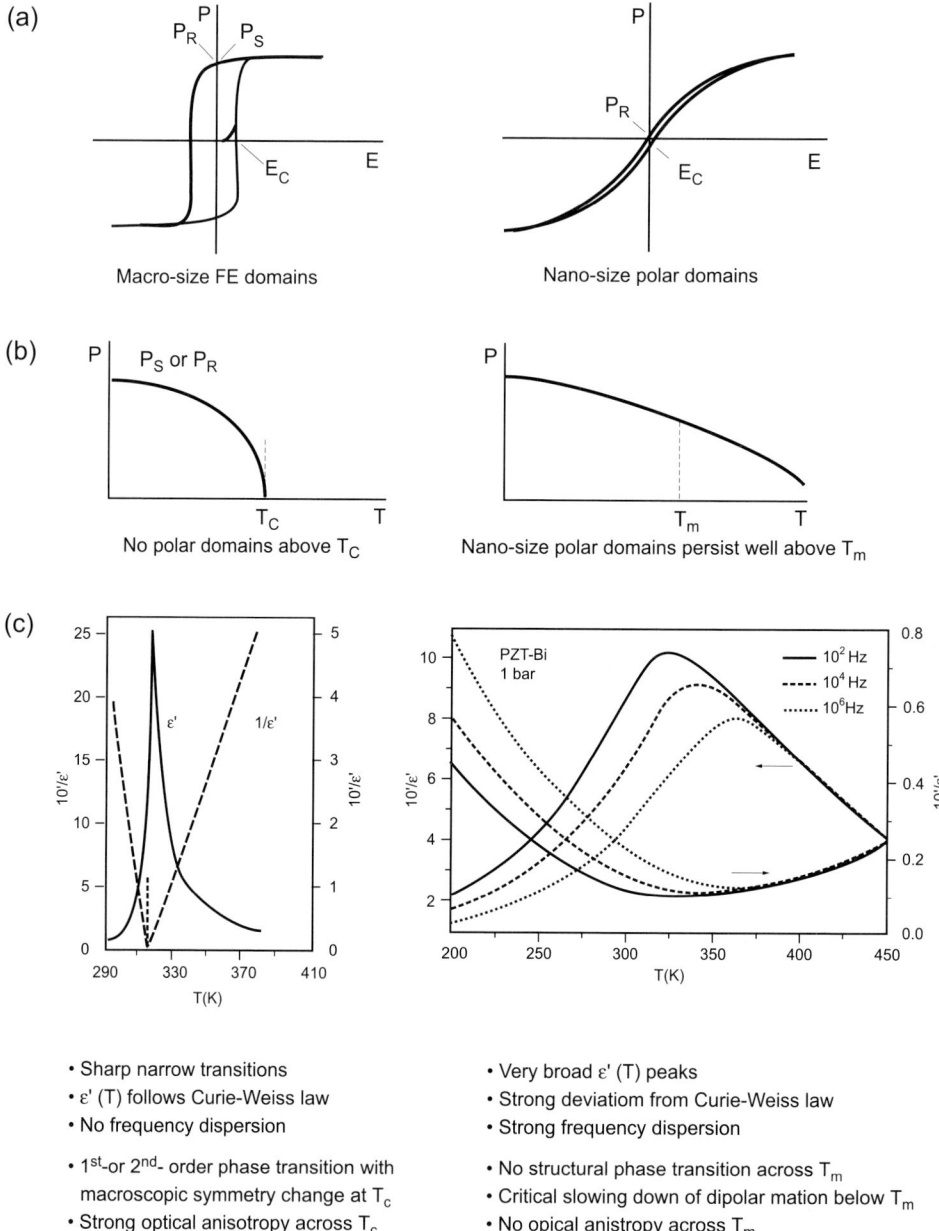

Figure 15.1: Contrast between the properties of normal and relaxor ferroelectrics (see Text) (from [14]).

15.1 Introduction

a. The FE polarization (P) versus electric Field (E) hysteresis loop is nearly of square shape and shows large remanent polarization, P_R, due to the switching of macroscopic long-range ordered domains. A RL shows a so-called slim loop, which indicates that the long-range ordering needs large fields, while in zero field the polarization decays into submicroscopic nanodomains, which re-acquire their natural random orientations, resulting in small P_R values.

b. Saturation and remanent polarization of a FE decrease with increasing temperature and vanish at the phase transition temperature T_c. No polar domains exist above T_c. By contrast, the field-induced polarization of a RL decreases smoothly through the dynamic transition temperature T_m and retains finite values up to rather high temperatures due to the fact that the nanopolar domains persist to well above T_m.

c. (I) The static dielectric susceptibility χ' or dielectric constant $\varepsilon' = 1 + \chi'$ of a FE exhibits a sharp and narrow peak at T_c (FWHM $\approx 10 - 20$ K) with frequency independence in the audio range. By contrast RLs exhibit a very broad peak and a strong frequency dispersion in both position (T_m) and height ($\varepsilon'_{\max}$) of the anomaly.

(II) The T dependence of ε' of a FE obeys a Curie-Weiss law, $\varepsilon' = C/(T - \theta)$, above T_c as shown by a linear relationship of $1/\varepsilon'$ versus T. By contrast $\varepsilon'(T)$ of a RL exhibits strong deviations from Curie-Weiss behavior well above T_m.

d. The existence of a macroscopic symmetry change for a FE at T_c, gives rise to usual phase transition features with changes of the refractive properties (occurrence or drastic change of birefringence), while in the absence of macroscopic symmetry breaking no such phenomena are observed in a RL at T_m.

Very high response coefficients and the enhanced width of the high response regime around the "ordering" temperature T_m, ("Curie range") make RLs popular systems for applications as piezoelectric/electrostrictive *actuators* and *sensors* (*e.g.* scanning probe microscopy, ink jet printer, adaptive optics, micromotors, vibration sensors/attenuators, Hubble telescope correction, ...) and as *electro-* or *elasto-optic* and *photorefractive elements* (segmented displays, modulators, image storage, holographic data storage, ...).

When reflecting on the occurrence of RL behavior in perovskites, there appear to be three essential ingredients:

I. existence of *lattice disorder*,

II. existence of *polar nanoregions* at temperatures much higher than T_m,

III. residence of these regions within the *highly polarizable host lattice* governed by a soft optic mode.

The first ingredient can be taken for granted, since RL behavior in these materials does not occur in the absence of disorder. The third ingredient is also an experimental fact in that RL behavior occurs in ABO_3 oxides with very large dielectric permittivity. The second ingredient is manifested in many experimental observations common to all perovskite RLs, as will be discussed later.

The following physical picture has emerged for RLs and has become generally accepted. Chemical substitution and lattice defects introduce dipolar entities in mixed ABO_3 perovskites. At very high temperatures, thermal fluctuations are so large that there are no well-defined dipole moments. However, on cooling, the presence of these dipolar entities manifests itself as small polar nanoregions below the so-called Burns temperature, T_d [3]. These regions grow as the correlation length, r_c, increases with decreasing T, and ultimately, one of two things happen. If the regions become large enough (macrodomains) so as to percolate (or permeate) the whole sample, then the sample will undergo a static, cooperative FE phase transition at T_c. On the other hand, if the nanoregions grow with decreasing T but do not become large enough or percolate the sample, then they will ultimately exhibit a dynamic slowing down of their fluctuations at $T \leq T_m$ leading to an isotropic RL state with random orientation of the polar regions.

A matter of dispute is still the relevance and the very origin of the Burns temperature. Very probably this is not a usual phase transition temperature. It might rather be considered as a so-called Griffiths temperature, which signifies the onset of weak singularities in a diluted ferroic system below the transition temperature of the undiluted system [4, 5]. However, the sharp onset of weak singularities is not at all confirmed in RL systems. We rather believe that the temperature regime in which the domains grow in size is a continuous one, which is determined by the correlating forces due to the underlying quenched random field (RF) distribution [6] as will be discussed in the next section.

The above describes the situation in the absence of a biasing electric field. Cooling in the presence of a biasing field, however, aligns the regions and increases their correlation length, effectively canceling the influence of the RFs. For sufficiently large E_{bias} the regions become large leading to the onset of long-range FE order. This is a field-induced nano-to-macrodomain transition. This transition occurs spontaneously in some cases in the absence of bias (e. g. for $Pb_{1-y}La_yZr_{1-x}Ti_xO_3$, PLZT). Evidence for the nano-to-macrodomain transition in RLs can be inferred, as we shall see, from the dielectric response and can often be seen in TEM images and from scattering data [7]. Its occurrence is determined by a delicate balance among E_{bias}, thermal fluctuations and the strength of dipolar correlations. Much insight about the physics of RLs has been gained from measurements in the presence of a DC bias field. Such measurements lead to distinct behavior as shown for $\varepsilon'(T)$ in Figure 15.2.

Figure 15.2 (a) is the signature of a RL in the absence of bias. The same relaxational response is observed under both zero-field cooling (ZFC) and zero-field heating (ZFH) conditions. The curve depicts the field-induced, frequency-independent FE response for sufficiently large E_{bias} which stabilizes the FE phase at low temperatures. Figures 15.2 (b) and (c) show the responses for intermediate biases. For field heating (FH) conditions after ZFC, the response has four regions (Figure 15.2 (b)). Region I is a dispersive range consisting of frozen-in, randomly oriented nanoregions. Below T_F (a freezing temperature) there is not enough thermal energy to unfreeze and align the domains with the field. However, at $T \geq T_F$ (region II) the nanoregions align and grow, forming macrodomains. This is a dispersion free FE region. On further heating, a temperature T_{F-R} is reached above which thermal fluctuations become sufficiently large so as to break the ordering tendency of the intermediate strength biasing field. As a result, the macrodomains break into randomly oriented, slowed down nanodomains, *i.e.*, a dispersive RL state (region III). Finally, above T_m the sample enters the PE state (region IV) where the nanoregions undergo rapid thermal fluctuations.

15.2 Polar nanoregions

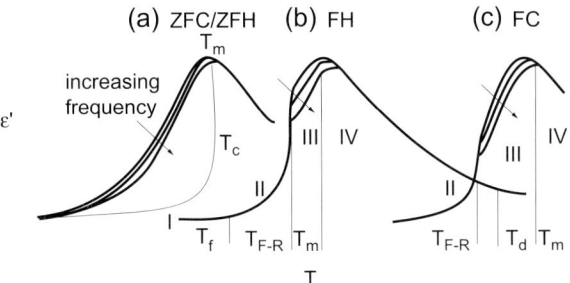

Figure 15.2: Dielectric responses of RLs both without (a) and with (c) electric field bias as discussed in the text. Response (b) defines all the various characteristic "transition" temperatures of a relaxor (from [14]).

Also shown in Figure 15.2 (b) is the fourth characteristic temperature, the Burns temperature, T_d ($> T_m$), where nanoregions first nucleate. Figure 15.2 (c) shows the response of the sample under field cooling (FC) conditions for intermediate strength fields. Regions IV and III are similar to those in Figure 15.2 (b). However, at $T \leq T_{F-R}$ the behavior is different. Here the field aligns and grows the nanoregions into macrodomains (a FE state). Once aligned, the macrodomains remain aligned and stable down to the lowest temperature in the presence of the field. Thus region I is absent in this case.

15.2 Polar nanoregions

While many researchers believe that the above mentioned three ingredients for RL behavior to appear are more or less independent, we have argued since long [6] that the primary cause of RL behavior is the lattice disorder, which is at the origin of the occurrence of polar nanodomains and their fluctuations within the highly polarizable lattice. In order to describe disordered systems and to explore their basic thermodynamic behavior simple spin models are frequently used. The model Hamiltonian

$$H = -\sum_{<ij>} J_{ij} S_i S_j - \sum_i h_i S_i \tag{15.1}$$

accounts for random interactions (or random bonds, RBs), J_{ij}, between nearest neighbor spins S_i and S_j, and for quenched random fields (RFs), h_i acting on the spins S_i. While the RBs are at the origin of spin glass behavior [8], RFs may give rise to disordered domain states provided that the order parameter has continuous symmetry [9]. This is easily shown with the help of energy arguments considering both the bulk energy decrease by fluctuations of the RFs and the energy increase due to the formation of domain walls. A remarkable exception, which does not necessarily lead to a disordered ground state, is the random-field Ising model (RFIM) system. Owing to its discontinuous spin symmetry, atomically thin domain walls are expected, which are energetically unfavorable. For this reason the three-dimensional (3-D) RFIM is expected to exhibit long-range order below the critical temperature T_c. However, as a tribute to the RFs new criticality due to a $T = 0$ fixed point [10] and strongly decelerated critical dynamics are encountered [11].

Unfortunately, the original idea [9] to realize a ferromagnetic RFIM by doping with random magnetic ions fails. Their spin dynamics always couples to that of the host system such that their dipolar fields cannot be regarded as quenched ones. Regrettably, there is no chance to dope a ferromagnet with *magnetic monopoles*, which might readily provide quenched local magnetic fields. Bearing this in mind, the situation should be much more favorable for the electric counterparts to ferromagnets, where electric charges may take the role of field-generating monopoles. Indeed, in FE systems electric charge disorder should easily give rise to quenched RFs. This was proposed previously [6] in order to understand the peculiar RL behavior of the archetypical solid solution $PbMg_{1/3}Nb_{2/3}O_3$ (PMN) [12]. Unfortunately, apart from the expected extreme slowing-down of the RFIM [11], which is closely related to the RL-typical huge polydispersivity [2, 12], no RFIM criticality was observed. This is a consequence of the high pseudo-spin dimension of the polarization order parameter, P. It has eight easy <111> directions in the cubic unit cell and thus quasi-continuous symmetry [6]. Clearly the search for an appropriate uniaxial FE (one-component order parameter $\pm P_z$, i.e. $n = 2$) with charge disorder seems advisable in order to materialize a proper ferroic 3-D RFIM system. Only recently [13] the uniaxial RL crystal $Sr_{0.61-x}Ce_xBa_{0.39}Nb_2O_6$ (SBN61:Ce, $0 \leq x < 0.02$) has been found to fulfill the conditions of a ferroic RFIM (see Section 15.6).

Evidence for the existence of polar nanoregions well above T_m has come from high resolution TEM which also showed the growth of these regions with decreasing T [14]. The evidence is also prominently reflected in certain properties of these systems. To provide the context, recall that for RLs in the absence of electrical bias there are random $+$ and $-$ fluctuations of the dipolar polarization so that $\sum P_d = 0$, i.e., there is no measurable remanent polarization. However, $\sum P_d^2 \neq 0$, and we then expect the existence of these polar regions to be manifested in properties which depend on P^2, e.g., electrostriction which is reflected in the thermal expansion and the quadratic electro-optic effect, which is reflected in the refractive index, or birefringence. Indeed, both of these properties have provided quantitative measures of this polarization for the RLs. Specifically, for a cubic perovskite it can be easily shown that the axial thermal strain, x_{11}, is given by [2, 14]

$$x_{11} = \frac{l_T - l_0}{l_0} = \alpha(T - T_0^*) + (Q_{11} + 2Q_{12})P_d^2 \qquad (15.2)$$

where l_T is the sample length at temperature T, l_0 is the reference length at the reference temperature T_0^*, α is the linear coefficient of thermal expansion and Q_{11} and Q_{12} are electrostrictive coefficients. Thus, in the presence of dipolar polarization, there are two contributions to the total thermal strain; the usual linear expansion (the first term on the right-hand side of Equation (15.2)) and the contribution due to electrostriction (the second term in Equation (15.2)). This latter contribution vanishes above T_d where $P_d \to 0$. Figure 15.3 (a) shows an example of the strain versus T for PMN [2].

The manifestation of the presence of polar nanodomains in strong RLs in terms of the electro-optic effect was first demonstrated by Burns and Dacol [3] in measurements of the T dependence of the refractive index, n. For a normal ABO_3 FE crystal, starting in the high-temperature PE phase, n decreases linearly with decreasing T down to T_c at which point n deviates from linearity. The deviation is proportional to the square of the polarization and

15.2 Polar nanoregions

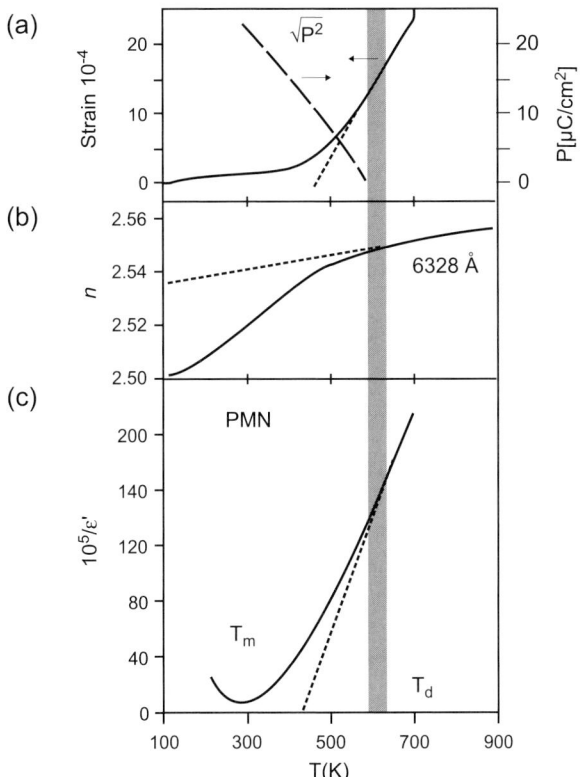

Figure 15.3: Temperature dependences of the linear thermal expansion, $\Delta l/l$ [2], refractive index, n [3] and reciprocal dielectric permittivity, $1/\chi'$ (Samara, unpublished) for PMN showing deviations from linear response at a temperature (T_d) much higher than the peak (T_m) in the dielectric susceptibility (from [14]).

increases as the polarization evolves with decreasing T. If the FE transition is first order, then there is a discontinuity in n at T_c followed by the expected deviation. This qualitative picture is representative of the behavior of many perovskite FEs. However, in the case of RLs, Burns and Dacol observed deviations from linear $n(T)$ well above T_m. In the case of PMN (Figure 15.3 (b)) these deviations became discernible on cooling at $T = 620$ K, or about 350 K above T_m, a remarkably large T range. These deviations can be quantitatively described by the relationship [3, 14]

$$\Delta n = \frac{\Delta n_{11} + \Delta n_{12}}{3} = \frac{n_0^3}{2} \frac{g_{33} + 2g_{13}}{3} P_d^2 \tag{15.3}$$

where the Δn's are the changes in the parallel and perpendicular components of n, n_o is the index in the absence of polarization (P_d), represented by the dashed straight line in Figure 15.3 (b), and the g_{ij} are the quadratic electro-optic coefficients. The temperature at the onset of the deviation from linear $n(T)$, 620 K in the case of PMN, is the Burns temperature, T_d in our notation. Above this temperature, thermal fluctuations are so large that there are no well-defined dipoles or dipolar regions. These domains nucleate at T_d by taking advantage of the statistical fluctuations of the RFs and grow on lowering T. Diffuse scattering and HRTEM results indicate that in PMN these domains grow from 2 to 3 nm in size above 400 K to ≈ 10 nm at ≈ 160 K ($T_m \approx 230$ K) [2]. Thus, they are much smaller than typical FE domains, which are orders of magnitude larger. The small size of these nanodomains explains why they cannot be detected by diffraction measurements, the bulk structures of PMN and most strong mixed ABO_3 RLs remaining cubic to both x-ray and neutron probes and to long wavelength photons down to the lowest temperatures.

Evidence for the existence of polar nanoregions well above T_m in RLs has also been deduced from the T dependence of the susceptibility. As noted earlier, it is well established that $\chi'(T)$ in the high temperature, cubic PE phase of ABO_3 FEs follows the Curie-Weiss law $\chi' = C/(T - \theta)$, where θ is the Curie-Weiss temperature, over a wide temperature range. However, $\chi'(T)$ of RLs shows large deviations from this law for $T > T_m$. This is shown in Figure 15.3 (c) for PMN. The results show that a linear $1/\chi'(T)$ response obtains only above about 625 K. An important point is that this is the same temperature, T_d, where deviations from linear $n(T)$ and $x(T)$ occur as shown (Figure 15.2). Thus, deviation from Curie-Weiss response sets in upon the nucleation of polar nanodomains, and this deviation increases with decreasing T as the size of the domains and their correlations increase. This deviation can be described by the power law expression [14]

$$\chi' - \chi'_\infty = \frac{C}{(T - \theta)^\gamma} \tag{15.4}$$

where χ'_∞ is the high frequency dielectric constant, and $\gamma > 1$. However, it has been generally observed that no single value of γ is found which uniquely describes the $\chi'(T)$ dependence for RLs.

Deviations from Curie-Weiss behavior are commonly observed in the temperature dependence of the magnetic susceptibility, χ, of spin glasses [8] above the freezing temperature of spin fluctuations, T_f (which corresponds to T_d for RLs in some sense). Also in a non-ideal superparamagnet, i.e., in interacting paramagnetic particles or clusters, $\chi(T)$ exhibits deviations from Curie-Weiss behavior. This behavior is achieved for temperatures large compared to T_f. At lower temperatures, deviations from the Curie-Weiss law are attributed to strong local magnetic correlations [8] and the onset of local (spin-glass) order below T_f. Sherrington and Kirkpatrick [15] developed a model, which relates $\chi(T)$ below T_f to the local order parameter q

$$\chi = C \frac{1 - q(T)}{T - \theta(1 - q(T))}, \tag{15.5}$$

where it is seen that q is a function of temperature. Clearly q and its temperature dependence can be evaluated from $\chi(T)$ data and the values of C and θ determined from the high-temperature $\chi(T)$ response above T_f which follows a Curie-Weiss law. In this high-temperature regime $q \to 0$ and Equation (15.5) simply reduces to a Curie-Weiss form. Equation (15.5) can be thought of as a modified Curie-Weiss law where both C and θ are functions of temperature. If we presume that the deviation from Curie-Weiss behavior in PMN and other RLs is due to correlations among local nanoregions, then we may evoke Equation (15.5) to treat the high-temperature dielectric response of RLs. Indeed this equation has been shown to provide a satisfactory description of the $\chi(T)$ response of RLs [16,17]. As expected, $q \to 0$ above, e.g., around 625 K for PMN, and it increases with decreasing temperature below T_d because of increased dipolar correlations [16]. In such a case the local order parameter due to correlations between neighboring polar domains of polarization P_i and P_j is $q =< P_i P_j >^{1/2}$.

15.3 Cubic relaxors

Figure 15.4: Random ionic distribution in PMN (Pb^{2+} = hatched, O^{2-} = small solid, Nb^{5+} = large solid, Mg^{2+} = open circles)

The archetypical perovskite-like lead-containing RL $Pb(Mg_{-1/3}Nb_{2/3})O_3$ (PMN, ABO_3 space group $Pm\bar{3}m$) system has been known for nearly fifty years [12]. As-grown PMN single crystals exhibit excellent crytalline properties with small mosaic spread ($\omega \leq 0.01°$). However, on the nanometer scale there is a significant degree of chemical and structural (displacement) disorder [18]. Figure 15.4 shows a random distribution of B site ions on an enlarged unit cell. Moreover, atoms on the B sites are occasionally short-range ordered within kinetically quenched-in chemical nanodomains with $Fm\bar{3}m$ symmetry [19]. The polar order parameter of PMN is directed along one of the eight rhombohedral <111> directions. Hence, an eight state Potts model might be applicable to this case. Figure 15.5 shows the evolution of polar regions as simulated in a two-dimensional 4-state Potts model with respective planar RFs as a function of temperature [20]. At high temperatures, $k_B T/J = 10$, it is seen that the regions image the RF distribution, while on decreasing the temperature down to $k_B T/J = 0.2$ the regions become coarse grained and merely image the local fluctuations of the RF distribution. An appropriate means to evidence the nanoregions even when being dynamic, i.e. above any transition or freezing temperature, is the optical second-harmonic generation, SHG, as evidenced for both PMN [21] and SBN [22]. In both cases the SHG intensity starts to grow well above the transition temperatures.

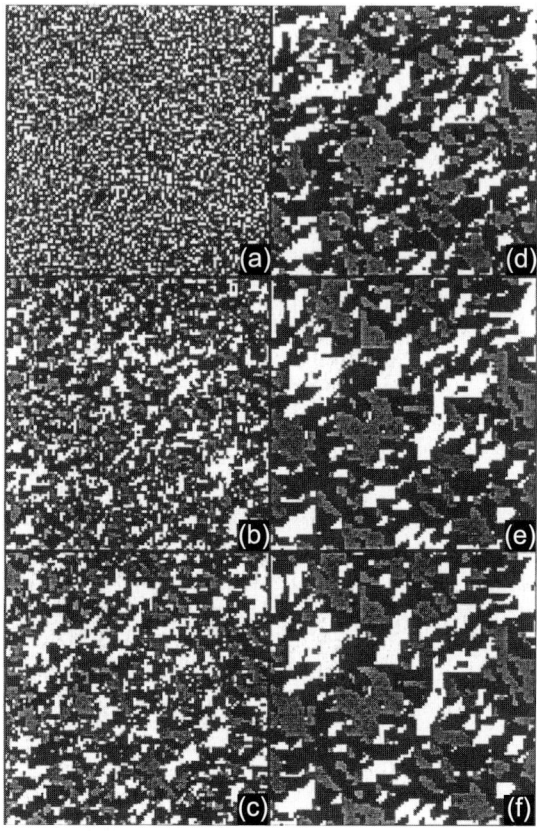

Figure 15.5: Domain distribution with polarizations $\pm P_x$ and $\pm P_y$ as indicated by different gray tones in a 4-state random field Potts model at temperatures $k_B T/J = 10$ (a), 1 (b), 0.8 (c), 0.6 (d), 0.3 (e) and 0.2 (f) (from [20]).

Since the order parameter of PMN is close to be continuous, an equilibrium phase transition into a long-range ordered FE phase is excluded [9]. However, Blinc et al. [17] developed another route towards an ordered low-T phase. Based on the existence of polar clusters and their above described correlations, $q = <P_i P_j>^{1/2}$ [16], they developed the *spherical random bond* RF (SRBRF) theory. Here polar clusters of any size fulfilling the spherical constraint are considered as randomly interacting "*superspins*" which undergo a transition into a cluster glass state. Theory has been solved for infinitely ranged interactions in mean-field approximation. Experimental tests by means of ^{93}Nb NMR reveal that the RFs are Gaussian distributed and that the Edwards-Anderson glass order parameter is finite below $T \approx 300$ K. Hence, no equilibrium static glassy freezing can be expected. Nevertheless, the preponderance of the random bonds with respect to the RFs clearly favors a glassy scenario which can be tested by measuring the (truncated) divergence of the nonlinear susceptibility, χ_3. More rigorously [23], the non-linearity parameter $a_3 = \chi_3/\chi_1^4$ has to maximize when approaching T_g on cooling, where χ_1 denotes the linear susceptibility. This has, indeed, been observed on both PMN and the related RL lead lanthanum zirconate-titanate (PLZT) [23].

A remark concerning the magnitude of the RFs seems in order. We are convinced that the primary function of the *local* RFs due to built-in charge disorder is to form energetically favored nanoregions [6]. These interact in a glass-like manner and form a spherical "*superspin*" glass [16, 17]. The glass transition, secondly, becomes smeared owing to *effective* RFs, h_i entering the *superspin* Hamiltonian, Equation (15.1). Since cluster spins containing N atomic spins experience only the fluctuations of the *local* RFs, the magnitudes of the *effective* RFs are reduced by factors $1/N^{1/2}$. This is why only weak smearing effects are observed. It should finally be remarked that the SRBRF theory is by far not yet accepted by the entire community. Still the origin of the nanoregions and their transition into a glassy state are disputed and not yet understood from a rigorous theoretical point of view. Clearly, the RF model simplifies the situation, since it neglects the possible relevance of bond disorder and the randomness of the quadrupolar degrees of freedom, which may give rise to structural glassy behavior.

15.4 Role of pressure

Compositionally-disordered ABO_3 perovskites are known to exhibit either FE or RL behavior depending on the degree and type of disorder. Hydrostatic pressure has been shown to be an excellent variable in the study of their properties [14, 24]. Results on a wide variety of these materials including PLZTs, $KTa_{1-x}Nb_xO_3$ and $[Pb(Zn_{1/3}Nb_{2/3})O_3]_{1-x}(PbTiO_3)_x$ have exhibited a pressure-induced crossover from a FE to a RL state and the continuous evolution of the dynamics of the relaxation process with increasing pressure [14]. This crossover has been shown to be a general feature of soft-mode FEs with random site dipolar impurities, or polar nanoregions, and results from a large decrease with pressure in the correlation length (λ_c) for interactions among polar nanoregions ($\propto$ size of the nanoregions) - a unique property of soft-mode, *i. e.* highly polarizable, host lattices. For those materials, like PMN, that are RLs at ambient pressure, the application of pressure simply strengthens the glass-like RL character by reducing the correlation length. Thus, unambiguously, the RL state is the ground state of these materials at high pressure or reduced volume [14].

Figure 15.6 shows the influence of pressure on the dielectric response of PMN [14]. The broad frequency-dependent peaks and the frequency dispersion in both ε' and ε'' are the characteristic signatures of the RL state. It is seen that pressure shifts the peaks to lower temperatures and suppresses ε' and ε'' in the high temperature phase. The suppression of $\varepsilon''(T)$ is mostly due to the suppression of $\varepsilon'(T)$ as $\varepsilon''(T) = \varepsilon'(T)\tan\delta(T)$, where $\tan\delta$ represents the dielectric loss. These pressure effects are characteristic of ABO_3 RLs and are well understood [14].

Taking the $\varepsilon'(T)$ peak temperature (T_m) to be representative of the dynamic glass transition temperature of PMN, the inset in Figure 15.6 shows the pressure dependence of T_m. The

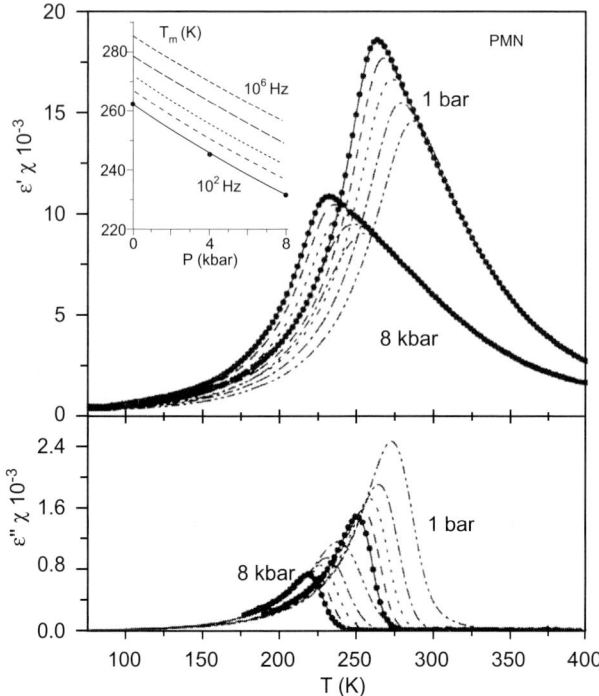

Figure 15.6: The relaxor dielectric response of PMN at 1 and 8×10^3 bar. The frequencies from left-to-right are 10^2, 10^3, 10^4 and 10^6 Hz (from [14]).

initial slope is $-(4.2 \pm 0.2)$ K/kbar – a value comparable to that for other high temperature ABO$_3$ RLs [14], and is essentially independent of frequency over the range of the data.

The effects of pressure on the properties of perovskite FEs and RLs are manifestations of the influence of pressure on the soft FE mode frequency of the host lattice [14, 24]. This frequency is determined by a delicate balance between short-range and long-range forces, and these forces exhibit markedly different dependences on interatomic separation, or pressure. Specifically, pressure increases the soft-mode frequency at constant temperature, which reduces the polarizability of the host lattice, thereby reducing λ_c. The result is a shift of the transition temperature, T_c (or T_m), to lower temperatures and a suppression of the $\varepsilon'(T)$ response in the high temperature paraelectric phase [14, 24].

Recent neutron scattering measurements [25] have revealed a well-defined FE (*TO*) soft mode in PMN at high temperatures that becomes overdamped by the polar nanoregions below the Burns temperature, T_d. Thus, while the soft mode below T_d is not a well-defined excitation in the spectrum, the large value of ε' and its strong temperature and pressure dependences between T_d and T_m clearly implicate low-lying optic mode excitations.

The $\varepsilon'(T)$ results in the high temperature paraelectric phase in Figure 15.6 provide a measure of λ_c in PMN. This follows from the fact that λ_c is determined by the polarizability of

15.4 Role of pressure

the crystal and is $\propto \sqrt{\varepsilon'}$ [14]. Figure 15.7 shows the temperature dependence of λ_c of PMN at 1 bar (solid line) and 8 kbar (dashed line). The strong enhancement of λ_c with decreasing T on approaching T_m, and the substantial decrease of λ_c with pressure are evident. The corresponding decrease in the correlation volume ($\approx$ size of the nanoregion) is, of course, quite substantial. Thus, e.g., between 1 bar and 8 kbar (i.e. a modest pressure) λ_c at 275 K decreases by 30 %. We should note here that deducing λ_c from dielectric data gives relative, and not absolute, values of λ_c. Thus, λ_c in Figure 15.7 is given in arbitrary units. However, λ_c and its temperature dependence have been determined from analysis of critical quasielastic neutron scattering data at 1 bar [26]. The results are given by the solid circles in Figure 15.7. They allow us to scale the $\lambda_c(T)$ results deduced from the dielectric data for comparison. This scaling was done by equating the arbitrary dielectrically-deduced value of λ_c at 275 K – a temperature near T_m at 1 bar, to Vakhrushev et al.'s [26] value (13.5 nm) at the same temperature. It is clearly seen that the two sets of results agree remarkably well over their common temperature range at 1 bar.

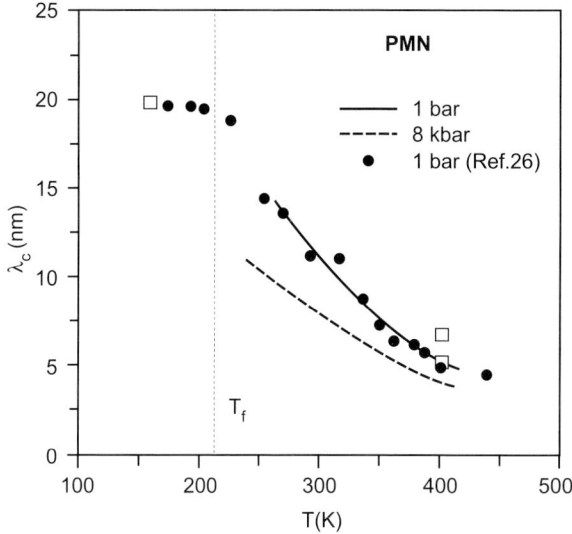

Figure 15.7: Temperature dependence of the correlation length λ_c in PMN deduced from various sources as described in the text.

The correlation length λ_c may be expected to be comparable to the size of the polar nanodomains in PMN. High resolution TEM and diffraction measurements have indicated that the domain size (diameter) in PMN increases from $\sim$ 2–3 nm at $\sim$ 400 K to 10 nm at $\sim$ 160 K and remains essentially constant down to 5 K [14, 27]. Interestingly these values are one half of the values of λ_c in Figure 15.7 at the corresponding temperatures. Doubling these diameters yields the open squares in Figure 15.7. Thus the relative values from the three sets of data in Figure 15.7 are in excellent agreement; however, the absolute values of $\lambda_c(T)$ remain in question.

15.5 Dynamics of the dipolar slowing-down process

Complex dielectric susceptibility data such as those in Figure 15.6 provide a detailed view of the dynamics of polar nanodomains in RLs. They define relaxation frequencies, f, corresponding to the $\varepsilon'(T)$ peak temperatures T_m, characteristic relaxation times, $\tau = 1/\omega$ (where $\omega = 2\pi f$ is the angular frequency), and a measure of the interaction among nanodomains as represented by the deviation of the relaxation process from a Debye relaxation. Analysis of data on PMN and other RLs clearly shows that their dipolar relaxations cannot be described by a single relaxation time represented by the Debye expression

$$\chi(\omega) \propto (1 + i\omega\tau)^{-1}. \tag{15.6}$$

Rather, a distribution of relaxation times is involved. It has become customary to assume a distribution function and fit it to the data. Alternatively, borrowing from the literature of dipolar and spin glasses, the distribution function $G(\tau, T)$ can be calculated from the susceptibility using the equation

$$\chi''(\tau, T) = \chi'_0(T) G(\tau, T), \tag{15.7}$$

where $\chi'_0(T)$ is the low frequency limit of $\chi'(\tau, T)$. Constant T cross-sections of $G(\tau)$ as a function of τ are easily calculated from the $\chi(\tau, T)$ data. Results on PMN show that $G(\tau)$ is flat near the Vogel-Fulcher temperature ($\propto$ dipolar freezing temperature), sharpens and develops a peak for $T \approx T_m$, and this peak shifts to higher frequencies at still higher temperatures [16].

On the basis of such results on PMN and other RLs, Viehland et al. [16] proposed the qualitative representation of the temperature-dependent relaxation time spectrum shown in Figure 15.8. It can be seen that the isothermal width of the spectrum is very broad at $T = T_f$ with the mean value of τ, τ_{avg}, becoming very long (macroscopic times). Below T_f, a significant fraction of the relaxations occur at short times. The shortest τ, τ_{min}, approaches macroscopic time only at $T \ll T_f$. The isothermal width of the spectrum sharpens with increasing T above T_f, with τ_{avg} and τ_{min} becoming short (microscopic times). τ_{min} approaches the Debye relaxation time, τ_D, at the "transition" temperature ($\sim T_m$), and τ_{avg} and τ_{max} begin to approach τ_D as $T \to T_d$.

Given the broad distribution of τ's in RLs, it is not surprising that the temperature dependence of the relaxation time corresponding to T_m is found to be non-Arrhenius. This is clearly shown for PMN in Figure 15.9, which gives results at 1 bar and 4–8 kbar. The departure from Arrhenius behavior can be generally satisfactorily described in a variety of ways many of which can be expressed [14, 28] in the form of the Vogel-Fulcher (V-F) equation

$$\tau^{-1} = \omega = \omega_o exp \frac{-E}{k_B(T_m - T_f)} \tag{15.8}$$

which is found to be applicable to many relaxational phenomena. The parameters in this equation have been given the following physical interpretations: ω_o is the attempt frequency related to the cut-off frequency of the distribution of relaxation times. E is the energy barrier between equivalent dipolar orientations, and T_f, the Vogel-Fulcher temperature, is a reference temperature where all relaxation times diverge (and where the distribution of τ's becomes infinitely broad). T_f can be viewed as the "static" dipolar freezing temperature for the relaxation process.

15.5 Dynamics of the dipolar slowing-down process

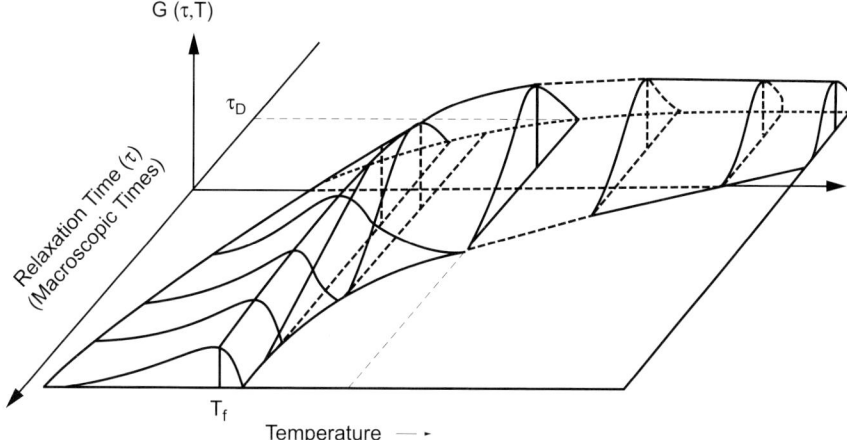

Figure 15.8: A model of temperature-dependent relaxation time spectrum for PMN. $G(\tau, T)$ is the number of polar regions having a relaxation time τ, T_f is the freezing temperature, and τ_D is the inverse Debye frequency (from [16]).

Accurate evaluation of the parameters in the V-F equation requires data over a broad range of frequencies. The PMN results in Figure 15.9 extend over only about five orders in frequency, and it was difficult to obtain unique fits of these data to Equation (15.8). A reasonable fit yielded $E = 118$ meV, $T_f = 215$ K and $w_o = 10^{14}$ Hz at 1 bar [29]. All parameters decreased with pressure. The decrease in T_f is comparable to that of T_m and E decreased by $\sim$30 % in 8 kbar, reflecting the decrease in the size of the polar nanodomains. Interestingly λ_c also decreases by about 30 % in 8 kbar. Simply stated, smaller nanodomains are easier to reorient than larger ones; thus, a lower energy barrier.

Although Equation (15.8) is widely used to analyze the dielectric relaxational response of RLs, the physical origin and general validity of this expression have been subjects of much discussion [14, 30, 31]. According to Equation (15.8), $\tau = 1/\omega \to \infty$ as $T \to T_f$ implying freezing of the relaxation time spectrum at T_f. However, arguments have been made that Equation (15.8) does not necessarily imply the freezing of the spectrum [30–32]. Thus, e.g., it has been indicated [30] that Equation (15.8) is characteristic of a dielectric in which the static dielectric permittivity, $\varepsilon'_s(T)$ has a maximum at T_f, but the spectrum exhibits only a gradual broadening upon cooling without $\tau \to \infty$ as $T \to T_f$. The question then is: Is there freezing of the τ spectrum at T_f? And is there another criterion, other than the use of $\omega(T_m)$ via Equation (15.8), to decide if freezing exists and if T_f is a true freezing temperature?

Glazounov and Tagantsev [33] have defined such a criterion. Specifically, the onset of frequency dispersion in $\chi(\omega, T)$ on cooling gives the T-dependence of the maximum relaxation time, τ_{max}, of the spectrum. It is then argued that in the case of real freezing, $\tau_{max}(T)$ should exhibit behavior similar to $\omega(T_{max})$, i.e., it should diverge at T_f according to

$$\tau_{max} = \tau_o \exp\frac{E}{k_B(T - T_f)}. \tag{15.9}$$

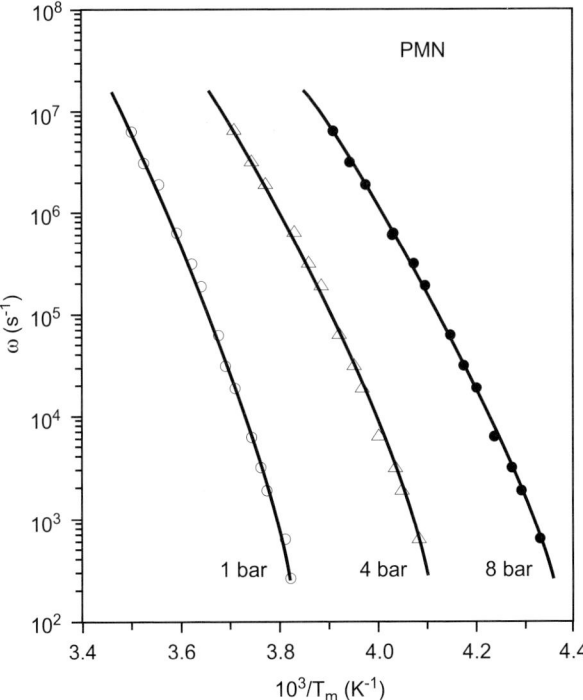

Figure 15.9: Arrhenius plots of the relaxation frequency corresponding to T_m for PMN at different pressures.

In the absence of freezing, no anomaly in $\tau_{max}(T)$ should be present at T_f. Relaxation data extending up to 9 decades on PMN and PbSc$_{0.5}$Ta$_{0.5}$O$_3$ (PST) were found to yield excellent fits to Equation (15.8) and (15.9) using the same fitting parameters (τ_o, E and T_f) within errors. The coincidence of T_f obtained from the two fits was taken as strong evidence for freezing at T_f.

Levstik et al. [34] reached similar conclusions on PMN from experimental data on $\chi(\omega, T)$ and its third harmonic component. Specifically, they demonstrated that τ_{max} diverges according to the V-F expression, while the bulk of the distribution of τ's remains finite below the freezing temperature, T_f. Furthermore, the $\chi'_3(\omega, T)$ behavior, where χ'_3 is the 3rd harmonic susceptibility, was similar to that of $\chi(\omega, T)$ indicating that both susceptibilities are governed by the same underlying relaxation spectrum responsible for the slowing-down mechanism. Additional evidence for the glass-like nature of the response of PMN below T_m was reflected in the splitting between electric field cooled and zero-field cooled $\chi'(t)$ data, that are qualitatively similar to the behavior of spin glasses with typical glassy nonergodic behavior [34]. These features have been qualitatively reproduced in the framework of the spherical random-bond/random-field model [17, 35].

Given that there had been questions about the validity of Equation (15.8), Tsurumi et al. [36] analyzed their PMN dielectric dispersion data in terms of a theory due to Ngai and White [37]. The starting point for this theory is the universal dielectric function

$$\chi(\omega) \propto \omega^{-\beta} \text{ with } 0 < \beta < 1 \tag{15.10}$$

first proposed by Jonscher [38] as applicable to many classes of materials. Ngai and White [37] explained the universality of this equation using a concept of low-energy excitation of "correlated states". Their results yielded an expression for the dielectric response function $\psi(t)$ similar to the empirical Kohlrausch, Williams and Watts [39] equation

$$\psi(t) = \frac{d}{dt} exp\left(\frac{-t}{\tau}\right)^{\beta}, \tag{15.11}$$

where the constant $\beta(T)$ represents the degree of correlations between fluctuating polar nano-regions. The Fourier transform of Equation (15.11) yields

$$\chi(\omega) = \chi(0) \int_o^\infty \psi(t) exp(i\omega t) dt \tag{15.12}$$

for the frequency-dependent susceptibility. Ngai and White's theory [37] yields the modified Arrhenius form

$$\omega = \omega_o exp \frac{-E^*}{k_B T} = \omega_o exp \frac{-E}{\beta k_B T}, \tag{15.13}$$

for the temperature dependence of the relaxation frequency. Here $E^* = E/\beta$. From a plot of their results as $\ln(\omega)$ vs. $1/\beta T$, Tsurumi et al. [36] obtained $E = 48\,\text{meV}$, a value about half that deduced from Equation (15.8), thereby raising questions about this approach.

15.6 Uniaxial relaxors

In contrast to the cubic family related to PMN the polarization of the strontium-barium niobate family, $Sr_x Ba_{1-x} Nb_2 O_6$, (SBN), is a single component vector directed along the tetragonal c direction, which drives the symmetry point group from paraelectric parent $4/mmm$ to polar $4mm$ at the phase transition into the low-T long-range-ordered polar phase as determined by X-ray diffraction [40]. Since SBN is tetragonal on the average, it belongs to the Ising model universality class rather than to the Heisenberg one as proposed for PMN-like system [6]. Assuming the presence of quenched random fields (RFs), available theory predicts the existence of a phase transition into long-range order within the RF Ising model (RFIM) universality class [9] preceded by giant critical slowing-down above T_c [11]. Only recently the SBN system has been found to fulfill the above necessary condition [13] and the ferroic RFIM seems to be materialized at last [41].

When explaining the unusual RL behavior, again the appearance of fluctuating polar precursor clusters at temperatures $T > T_c$ has to be considered as the primary signature of the polar RFIM [2,3,6]. Acting as precursors of the spontaneous polarization, which occurs below

T_c (Figure 15.10), they have been evidenced in various zero-field cooling (ZFC) experiments comprising linear birefringence [42], linear susceptibility [43], dynamic light scattering [44] and Brillouin scattering [45]. After freezing into a metastable domain state at $T < T_c$ the clusters were also directly observed with the help of high resolution piezoresponse force microscopy, PFM [46] (Figure 15.11).

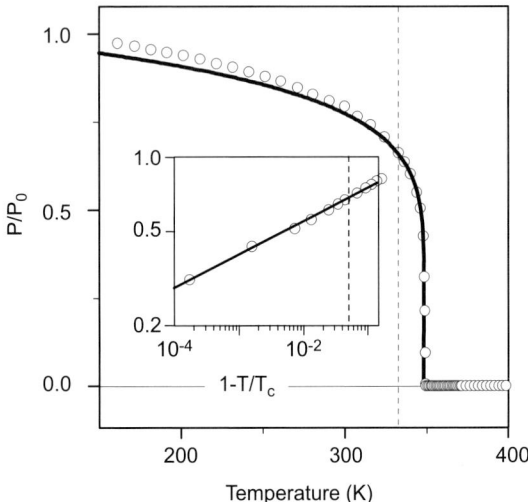

Figure 15.10: Order parameter of SBN as measured by NMR techniques displaying criticality with an exponent $\beta = 0.16$ (from [47]).

Maybe the major achievement provided by the discovery of the FE RFIM is the possibility to study the complete set of critical exponents on a ferroic system for the first time after their prediction [9, 10]. Table 15.1 shows the results as compared with predictions from theory and simulations. Most remarkably, the order parameter exponent β (Figure 15.10) clearly deviates from the prediction $\beta \approx 0$ and achieves a value which comes close to that observed recently on the standard RFIM system, the dilute uniaxial antiferromagnet $Fe_{1-x}Zn_xF_2$, $x = 0.15$, in an external magnetic field [50]. Further, the most disputed value, namely the specific heat exponent α [48] (Figure 15.12) clearly describes the same logarithmic divergence as that found on $Fe_{1-x}Zn_xF_2$, $\alpha \approx 0$ [10], which still lacks theoretical confirmation.

15.7 Domain dynamics in uniaxial relaxors

Domains in FE crystals are well-known to have a considerable influence on the value of their complex dielectric susceptibility, $\chi^* = \chi' - i\chi''$; and related quantities [51]. Owing to its mesoscopic character the domain wall susceptibility strongly reflects the structural properties of the crystal lattice. This is most spectacular in crystals with inherent disorder, where the

15.7 Domain dynamics in uniaxial relaxors

Critical Exponent	Experimental (SBN)	Theoretical (RFIM)
α (specific heat)	0.01 [48]	0.5 [49]
β (order parameter)	0.16 [47]	0.02 [49]
γ (susceptibility)	1.89 [43]	1.89 [49]
ν (correlation length)	1.0 [42]	1.37 [49]

Table 15.1: Critical exponents.

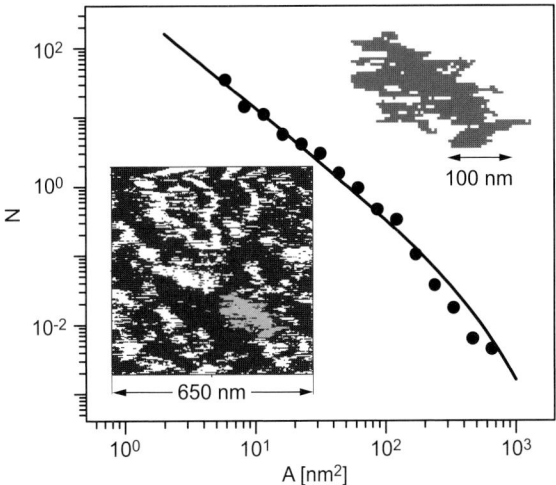

Figure 15.11: Spatial distribution of the ZFC surface polarization of SBN61:Ce ($x = 0.01$) (left-hand inset). Black and white areas refer to $\pm P_z$, respectively. One domain (highlighted) is shown in the right-hand inset. The distribution function of domain areas A (solid circles) fits to the power law $N(A) = N_0 A^{-\delta} \exp(-A/A_\infty)$ with exponential cutoff and $\delta = 1.5$ (solid line) (from [46]).

domain walls are subject of stochastic pinning forces and χ^* is highly polydispersive due to a wide distribution $g(\tau)$ of Debye-type response spectra [52, 53],

$$\chi^*(\omega) = \chi_\infty + (\chi_s - \chi_\infty) \sum_j g_j (1 + i\omega\tau_j)^{-1}, \tag{15.14}$$

where τ and ω are the relaxation time and angular frequency, respectively.

More generally, the dynamic behavior of domain walls in random media under the influence of a periodic external field gives rise to hysteresis cycles of different shape depending on various external parameters. According to a recent theory of Nattermann et al. [54] on disordered ferroic (ferromagnetic or FE) materials, the polarization, P, is expected to display a number of different features as a function of T, frequency, $f = \omega/2\pi$, and probing ac field amplitude, E_0. They are described by a series of dynamical phase transitions, whose order parameter $Q = (\omega/2\pi) \oint P dt$ reflects the shape of the P vs. E loop. When increasing the ac

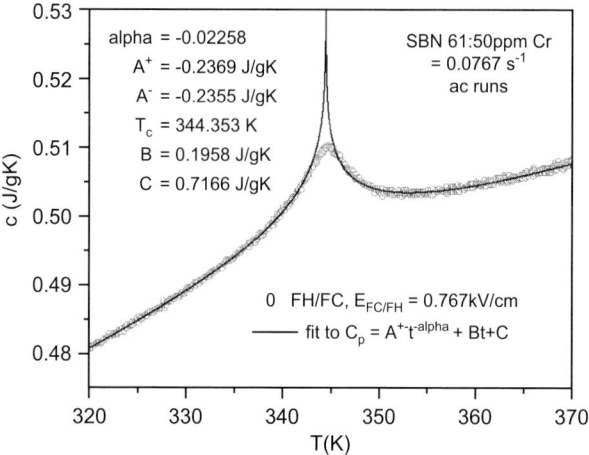

Figure 15.12: Specific heat of SBN as measured with pulsed heating techniques displaying criticality with an exponent $\alpha \approx 0$ [48].

amplitude E_0 the polarization displays four regimes. First, at very low fields, $E_0 < E_\omega$, only "*relaxation*", but no macroscopic motion of the walls should occur at finite frequencies, $f > 0$. Second, within the range $E_\omega < E_0 < E_{t1}$, a thermally activated drift motion ("*creep*") is expected, while above the depinning threshold E_{t1} the "*sliding*" regime is encountered within $E_{t1} < E_0 < E_{t2}$. Finally, for $E_0 > E_{t2}$ a complete reversal of the polarization ("*switching*") occurs in the whole sample in each half of the period, $\tau = 1/f$. It should be noticed that all transition fields, E_ω, E_{t1} and E_{t2}, are expected to depend strongly on both T and f [54].

It has been shown [55] that two different non-Debye responses referring to the field regions $E_0 < E_\omega$ ("*relaxation*") and $E_\omega < E_0 < E_{t1}$ ("*creep*") occur in the low-f dispersion of the uniaxial RL crystal $Sr_{0.61-x}Ce_xBa_{0.39}Nb_2O_6$ (SBN:Ce, $x = 0.0066$) in the vicinity of its FE transition temperature, $T_c = 320$ K. It shows both characteristics in adjacent frequency regimes. While the well-known relaxational $ln(1/f)$ characteristic of relaxing domain wall segments in a weak random field [52, 53] applies to "high" frequencies, $f > 100$ Hz, an alternative $1/f^\beta$ dependence is observed in the "low"-f regime, $f < 1$ Hz. In order to understand the latter behavior, we introduce polydispersivity via a broad distribution of wall mobilities, μ_w, which describe the viscous motion of the walls in the creep regime, where they overcome a large number of potential walls due to a high density of pinning defects. As a characteristic of irreversibility the walls stop when switching off the field. Within this concept the rapid individual Debye-type relaxation processes are averaged out on the long-time scale of a creep experiment. $(1/f)^\beta$ behavior at low frequencies has recently been reported on the RL-type single crystal $PbFe_{1/2}Nb_{1/2}O_3$ [56].

Dielectric response data were taken on a Czochralski-grown very pure crystals of SBN:Ce (size $0.5 \times 5 \times 5$ mm^3) with probing electric-field amplitudes of 200 V/m applied along the polar c axis. A wide frequency range, $10^{-5} < f < 10^6$ Hz, was supplied by a Solartron 1260 impedance analyzer with a 1296 dielectric interface. Different temperatures were chosen both

15.7 Domain dynamics in uniaxial relaxors

below and above T_c and stabilized to within ± 0.01 K. Figure 15.13 shows representative data of χ' (curve 1) and χ'' vs. f (curve 2) taken at $T = 294$ K which illustrate the main features of the dielectric dispersion of zero-field-cooled (ZFC) SBN:Ce: (I) the dielectric response strongly increases below $f_{min} \approx 25$ Hz (marked by the dotted line); (II) neither saturation of χ' nor a peak of χ'' are observed in the infra-low-frequency limit, where (III) the magnitude of χ'' exceeds that of χ' by one order of magnitude; (IV) a Cole-Cole-type plot of χ'' vs. χ' is characterized by a positive curvature at frequencies $f < f_{min}$ (Figure 15.13; inset), which is opposite to the conventional Debye-type one; (V) at higher frequencies, $f > f_{min}$, χ'' increases again in a power-law-like fashion (straight line in a log-log presentation), while χ' changes its curvature and gently bends down.

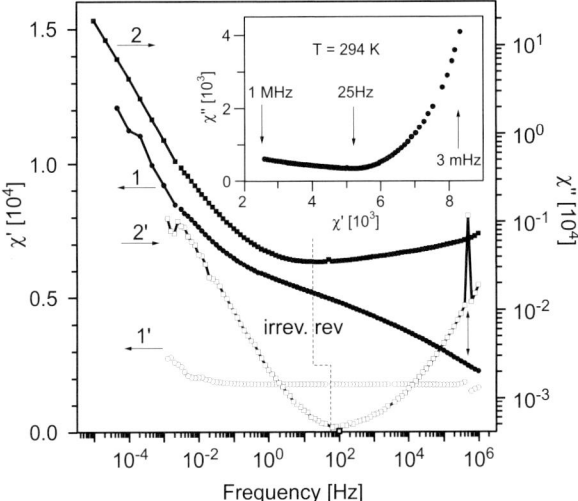

Figure 15.13: Dielectric spectra of χ' and χ'' vs. f of unpoled (curves 1 and 2) and poled (curves 1' and 2') SBN:Ce taken at $T = 294$ K. Solid lines are guides to the eye and the vertical dotted line separates different response regimes. A piezoelectric anomaly at $f = 0.5$ MHz is marked by a double arrow. The inset shows χ'' vs. χ' (from [55]).

The dominating domain-wall nature of the response is evidenced by its drastic reduction when poling the sample with $E = 350$ kV/m from above T_c into a near-single domain state as shown by the curves 1' and 2' in Figure 15.13. Despite its decrease by two orders of magnitude χ'' reveals, again, a symmetric increase on both sides of $f_{min} \approx 65$ Hz (dotted line), which becomes power-law-like in the asymptotic low- and high-f regimes, respectively. This applies also to χ' (curve 1') after subtracting a background corresponding to the minimum at $f = 65$ Hz as shown in Figure 15.13. Interestingly, a sharp piezoelectric resonance of both χ' and χ'' is observed at $f_{min} \approx 0.5$ MHz after poling. This is typical of the near-single domain state, which activates a piezoelectric resonance.

The high-f response of both the ZFC and the FC states confirms many of the characteristics predicted by Equation (15.14). Inspection shows that χ' decreases linearly on a linear-log scale prior to the steeper decrease at $f > 10^4$ Hz, while χ'' obeys linearity on a log-log scale. Clearly, the ω prefactor strongly suppresses χ'' close to f_{min} when compared with χ'. Upon increasing f the same factor determines the positive curvature of χ'' despite the competing $\ln(l/f)$ contribution (curve 2' in Figure 15.13). Simultaneously, χ' is bent down in a dispersion step-like fashion.

While the high-frequency dispersion regime is attributed to polarization processes due to the reversible motion of domain-wall segments experiencing restoring forces, viz., relaxation, the low-frequency response is due to the irreversible viscous motion of domain-walls. They experience memory-creasing friction by averaging over numerous pinning centers in a creep process. The latter type of motion becomes possible for at least two reasons: screening of depolarization fields by free charges in the bulk or at the surface and/or pinning of the domain-walls at quenched random fields, which is believed to be due to quenched charge disorder in the special case of SBN:Ce [13, 41]. Dielectric domain response under the action of an external electric field is readily modeled by considering the average polarization, $P(t) = (2P_s/D)x(t)$, of a regular stripe domain pattern of up and down polarized regions carrying spontaneous polarization, $\pm P_s$, and having an average width D. It arises from a sideways motion of walls perpendicular to the field direction by a distance x. Starting with $P(0) = 0$ at $x(0) = 0$, the favorable domains enhance their total width by an amount $2x$ until reaching (in principle) the limit $P = P_s$ for $x \to D/2$. By assuming viscous motion of the walls one obtains the rate equation

$$\dot{P}(t) = \frac{2P_s}{D}\mu_w E(t), \tag{15.15}$$

where the wall velocity $\dot{x}(t) = \mu_w E(t)$ involves the wall mobility μ_w and the driving field $E(t)$. Assuming constant mobility at sufficiently weak fields and disregarding the depinning threshold one finds

$$P(t) = \left(\frac{2\mu_w P_s}{i\omega\varepsilon_0 D} + \chi_\infty\right)\varepsilon_0 E_0 \exp(i\omega t) \tag{15.16}$$

under a harmonic field, $E(t) = E_0 exp(i\omega t)$. In Equation (15.16) the second term refers to "instantaneous" response processes due to reversible domain-wall rearrangements occurring on shorter-time scales (see above). The above relations are expected to hold in the limit of small displacements x, before the walls are stopped either by depolarizing fields (in conventional FEs) or by new domain conformations under the constraint of strong random fields (in disordered FEs). Weak periodic fields thus probe a linear ac susceptibility

$$\chi_w^*(\omega) = \chi_\infty\left(1 + \frac{1}{i\omega\tau_w}\right), \tag{15.17}$$

with $\chi_\infty/\tau_w = (2\mu_w P_s/\varepsilon_0 D)$. The "relaxation" time τ_w denotes the time in which the interface contribution to the polarization equals that achieved instantaneously, $\Delta P = \varepsilon_0\chi_\infty E$.

Since the electric fields used in our experiments ($E_0 = 200$ V/m) are well below the coercive field, $E_c \approx 150$ kV/m, we have to account for the nonlinearity of v vs E in the

15.7 Domain dynamics in uniaxial relaxors

creep regime, where thermal excitation enables viscous motion below the depinning threshold $E_{crit} \approx E_c$. Approximating this regime roughly by a power law $v \propto E^\delta$, $\delta > 2$, Equation (15.17) may be modified phenomenologically by introducing a Cole-Davidson-type exponent $\beta < 1$,

$$\chi_w^*(\omega) = \chi_\infty \left(1 + \frac{1}{(i\omega\tau_{eff})^\beta}\right), \tag{15.18}$$

similarly as used in the case of polydispersive Debye-type relaxation [57]. Here τ_{eff} denotes an effective relaxation time. Decomposition of Equation (15.18) yields

$$\chi'(\omega) = \chi_\infty \left(1 + \frac{\cos\left(\frac{\beta\pi}{2}\right)}{(\omega\tau_{eff})^\beta}\right) \tag{15.19}$$

and

$$\chi''(\omega) = \frac{\chi_\infty \sin\left(\frac{\beta\pi}{2}\right)}{(\omega\tau_{eff})^\beta} \tag{15.20}$$

such that

$$\frac{\chi''}{\chi' - \chi_\infty} = \tan\left(\frac{\beta\pi}{2}\right). \tag{15.21}$$

The power law-type spectral dependencies of χ' and χ'' are well supported by our experiments. While the unpoled sample exhibits an exponent $\beta \approx 0.2$ (not shown), i. e. large polydispersivity, the poled sample yields $\beta \approx 0.67$ for both components of χ^* (Figure 15.14 [58]). Obviously the polydispersivity is largely suppressed at low domain wall densities. This seems to show that polydispersivity is less affected by the nonlinearity in the creep regime, $v \propto E^\delta$, than by the mutual wall interactions in the nanodomain regime [46]. Very satisfactorily, also the Cole-Cole plot, Equation (15.21), which is another independent test of the ansatz, Equation (15.18), reveals a very similar exponent, $\beta \approx 0.69$ (Figure 15.15 [58]).

It should be noticed that the monodispersive relation (15.17) satisfies the Kramers-Kronig relationships, since $\chi'' \propto 1/\omega$ is a purely conductive contribution due to ohmic-like domain wall sliding and $\chi' = \chi'_\infty$ is constant. This is, however, no longer satisfied for $\beta < 1$, Equation (15.18). Hence, the spectral features displayed in Figure 15.15 must necessarily change at very low frequencies. Here we conjecture - in accordance with the theory of dynamic phase transitions in random media [54] – that monodispersivity, i.e. the sliding regime, should be attained asymptotically when approaching the static limit. This has been confirmed recently on the quantum-ferroelectric RL SrTi^{18}O$_3$ in its domain state below $T_c = 25$ K [59].

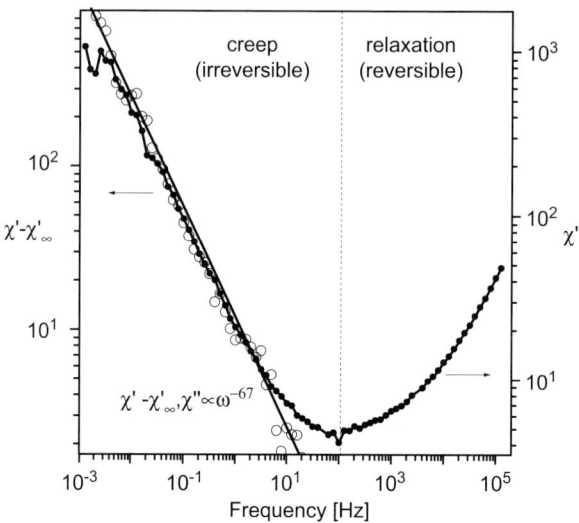

Figure 15.14: Dielectric spectra of $\chi' - \chi'_\infty$, where $\chi'_\infty = 1820$ (open circles), and χ'' (solid circles) vs. f of poled SBN:Ce taken at $T = 294$ K. The solid line is a best fit to Equation (15.20) with $\beta = 0.67$ (from [58]).

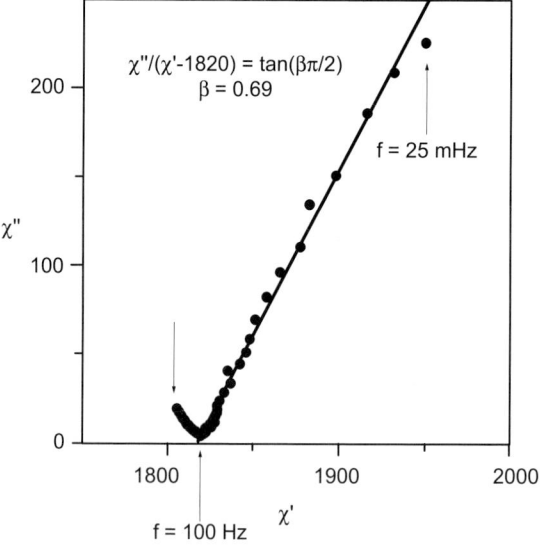

Figure 15.15: Cole-Cole plot of χ'' vs $\chi' - \chi'_\infty$, where $\chi'_\infty = 1820$, of poled SBN:Ce at $T = 294$ K (Figure 15.14). The solid line is a best fit to Equation (15.21) with $\beta = 0.69$ (from [58]).

Acknowledgement

The work at Sandia was supported by the Division of Materials Sciences and Engineering, Office of Basic Energy Sciences, U.S. Department of Energy under Contract No. DE-AC04-AL85000. The work at the Duisburg-Essen University was supported by Deutsche Forschungsgemeinschaft within Forschungsschwerpunkt "Strukturgradienten in Kristallen".

Bibliography

[1] M. E. Lines and A. M. Glass, *Principles and Applications of Ferroelectrics and Related Materials*, Clarendon, Oxford, 1977.
[2] L. E. Cross, *Ferroelectrics* **76**, 241 (1987).
[3] G. Burns and F. H. Dacol, *Solid State Commun.* **48**, 853 (1983); *Phase Trans.* **5**, 261 (1985).
[4] R. B. Griffiths, *Phys. Rev. Lett.* **23**, 69 (1968).
[5] Ch. Binek and W. Kleemann, *Phys. Rev. B* **51**, 12888 (1995).
[6] V. Westphal, W. Kleemann, and M. D. Glinchuk, *Phys. Rev. Lett.* **68**, 847 (1992).
[7] S. B. Vakhrushev, B. E. Kvyatkovsky, A. A. Naberezhov, N. M. Okuneva, and B. B. Toperverg, *Ferroelectrics* **90**, 173 (1989); G Schmidt, H. Arndt, G. Borchardt, J. V. Cieminski, T. Petzsche, K . Bormann, A. Sternberg, A. Zirnite, and A. V. Isupov, *Phys. Status Solidi* A **63**, 501 (1981).
[8] K. Binder and A. P. Young, *Rev. Mod. Phys.* **58**, 801 (1986).
[9] I. Imry and S. K. Ma, *Phys. Rev. Lett.* **35**, 1399 (1975).
[10] D. P. Belanger and A. P. Young, *J. Magn. Mater.* **100**, 272 (1991).
[11] D. S. Fisher, *Phys. Rev. Lett.* 56 (1986) 416; A. T. Ogielski and D. A. Huse, *Phys. Rev. Lett.* **56**, 1298 (1986).
[12] G. A. Smolenski and V. A. Isupov, *Dokl. Acad. Nauk* SSSR **97**, 653 (1954).
[13] W. Kleemann, J. Dec, P. Lehnen, Th. Woike, and R. Pankrath, in: *Fundamental Physics of Ferroelectrics* 2000, R. E. Cohen ed., *AIP Conf. Proc.* **535**, 26 (2000).
[14] G. A. Samara, *Solid State Physics* 56 (2001), H. Ehrenreich and F. Spaepen ed., Academic Press, New York, p. 240 and references therein; *J. Phys.: Cond. Matter* **15**, R367 (2003).
[15] D. Sherrington and S. Kirkpatrick, *Phys. Rev. Lett.* **35**, 1972 (1975).
[16] D. Viehland, M. Wuttig, and L. E. Cross, *Ferroelectrics* **120**, 71 (1991); *Phys. Rev. B* **46**, 8003 (1993).
[17] R. Blinc, J. Dolinsek, A. Gregorovic, B. Zalar, C. Filipic, Z. Kutnjak, A. Levstik, and R. Pirc, *Phys. Rev. Lett.* **83**, 424 (1999).
[18] P. Bonneau *et al.*, *Solid State Chem.* **91**, 350 (1991); L. E. Cross, *Ferroelectrics* **151**, 305 (1994); A. D. Hilton *et al.*, *J. Mater. Sci.* **25**, 3461 (1990); T. Egami *et al.*, *Ferroelectrics* **199**, 103 (1997); B. Dkhil *et al.*, *Phys. Rev.* B **65**, 4104 (2002).
[19] E. Husson, M. Chubb and A. Morell, *Mat. Res. Bull.* **23**, 357 (1988).
[20] H. Qian and L. A. Bursill, *Int. J. Mod. Phys.* B **10**, 2027 (1996).

[21] Y. Uesu, H. Tazawa, and K. Fujishiro, *J. Kor. Phys. Soc.* **29**, S703 (1998).
[22] P. Lehnen, J. Dec, W. Kleemann, Th. Woike, and R. Pankrath, *Ferroelectrics* **268**, 113 (2002).
[23] Z. Kutnjak, C. Filipic, R. Pirc, A. Levstik, R. Farhi, and M. El Marssi, *Phys. Rev. B* **59**, 294 (1999).
[24] G. A. Samara and P. S. Peercy in *Solid State Physics*, H. Ehrenreich, F. Seitz and D. Turnbull eds., Academic Press, New York, **36**, p.1 and references therein, 1981.
[25] S. Wakimoto, C. Stock, Z.-G. Ye, W. Chen, P. M. Gehring, and G. Shirane, *Phys. Rev. B* **65**, 172015 (2002).
[26] S. Vakhrushev, B. Kvyatkovsky, A. Naberezhnov, W. Okuneva, and B. Toperverg, *Ferroelectrics* **90**, 173 (1989).
[27] M. de Mathan, E. Husson, G. Calvarin, J. R. Gavarri, A. W. Hewat, and A. Morell, *J. Phys.: Condens. Matter* **3**, 8159 (1991).
[28] G. A. Samara, *Ferroelectrics* **117**, 347 (1991) and references therein.
[29] The PMN sample used for the data in Figures 15.6 and 15.9 contained 0.6 mol% Cu. This may be the reason why the deduced $E = 118$ meV is higher than what has been reported [16] for pure PMN (~ 80 meV). Additionally ω_o ($\sim 10^{14}$ Hz) is a bit high possibly due to inaccuracy of the fit over the limited range of frequencies.
[30] A. K. Tagantsev, *Phys. Rev. Lett.* **72**, 1100 (1994).
[31] J. Hemberger, H. Ries, A. Loidl, and R. Böhmer, *Phys. Rev. Lett.* **76**, 2330 (1996).
[32] Z. Kutnjak, C. Filipic, A. Levstik, and R. Pirc, *Phys. Rev. Lett.* **70**, 4015 (1993).
[33] A. E. Glazounov and A. K. Tagantsev, *Appl. Phys. Lett.* **73**, 856 (1998).
[34] A. Levstik, Z. Kutnjak, C. Filipic, and R. Pirc, *Phys. Rev. B* **57**, 204 (1998).
[35] R. Pirc, R. Blinc and V. Bobnar, *Phys. Rev. B* **63**, 054201 (2001).
[36] T. Tsurumi, K. Soejima, T. Kamiga, and M. Daimon, *Jpn. J. Appl. Phys.* **33**, 1959 (1994).
[37] K. L. Ngai and C. T. White, *Phys. Rev. B* **20**, 2475 (1979).
[38] A. K. Jonscher, *Nature* **267**, 673 (1977).
[39] N. G. McCrum, B. E. Read, and G. Williams, *Anelastic and Dielectric Effects in Polymeric Solids* (John Wiley and Sons, New York 1967).
[40] J. R. Oliver, R. R. Neurgaonkar, and L. E. Cross, *J. Appl. Phys.* **64**, 37 (1988).
[41] W. Kleemann, J. Dec, P. Lehnen, R. Blinc, B. Zalar, and R. Pankrath, *Europhys. Lett.* **57**, 14 (2002).
[42] P. Lehnen, W. Kleemann, Th. Woike and R. Pankrath, *Eur. Phys. J. B* **14**, 633 (2000).
[43] J. Dec, W. Kleemann, V. Bobnar, Z. Kutnjak, A. Levstik, R. Pirc, and R. Pankrath, *Europhys. Lett.* **55**, 781 (2001).
[44] W. Kleemann, P. Licinio, Th. Woike, and R. Pankrath, *Phys. Rev. Lett.* **86**, 6014 (2001).
[45] F. M. Jiang and S. Kojima, *Phys. Rev. B* **62**, 8572 (2000).
[46] P. Lehnen, W. Kleemann, Th. Woike, and R. Pankrath, *Phys. Rev. B* **64**, 224109 (2001).
[47] R. Blinc. A. Gregorovic, B. Zalar, R. Pirc, J. Seliger, W. Kleemann, S. G. Lushnikov, and R. Pankrath, *Phys. Rev. B* **64**, 134109 (2001).
[48] Z. Kutnjak, W. Kleemann, and R. Pankrath, *Phys. Rev. B* (submitted).
[49] A. A. Middleton and D. S. Fisher, *Phys. Rev. B* **65**, 134411 (2002).

[50] F. Ye, L. Zhou, S. Larochelle, L. Lu, D. P. Belanger, M. Greven, and D. Lederman, *Phys. Rev. Lett.* **89**, 157202 (2002).

[51] J. Fousek and V. Janovec, *Phys. Stat. Sol.* **13**, 105 (1966).

[52] L. B. Ioffe and V. M. Vinokur, *J. Phys. C* **20**, 6149 (1987).

[53] T. Nattermann, Y. Shapir, and I. Vilfan, *Phys. Rev. B* **42**, 8577 (1990).

[54] T. Nattermann, V. Pokrovsky, and V. M. Vinokur, *Phys. Rev. Lett.* **87**, 197005 (2001).

[55] W. Kleemann, J. Dec, S. Miga, Th. Woike, and R. Pankrath, *Phys. Rev. B* **65**, 220101R (2002).

[56] Y. Park, *Solid State Commun.* **113**, 379 (2000).

[57] A. K. Jonscher, *Dielectric relaxation in solids*, Chelsea Dielectric Press, London, 1983.

[58] W. Kleemann, J. Dec, and R. Pankrath, *Ferroelectrics* **286**, 21 (2003).

[59] J. Dec, W. Kleemann, and M. Itoh, *Ferroelectrics* **298**, 163 (2004).

16 Scanning Nonlinear Dielectric Microscope

Yasuo Cho

Research Institute of Electrical Communication, Tohoku University, Japan

Abstract

A sub-nanometer resolution scanning nonlinear dielectric microscope (SNDM) was developed for observing ferroelectric polarization. We also demonstrate that the resolution of SNDM is higher than that of a conventional piezo-response imaging. Secondly, we report new SNDM technique detecting higher nonlinear dielectric constants ϵ_{3333} and ϵ_{33333}. Higher order nonlinear dielectric imaging provides higher lateral and depth resolution. Thirdly, a new type of scanning nonlinear dielectric microscope probe, called the ϵ_{311}-type probe, and a system to measure the ferroelectric polarization component parallel to the surface were developed. Finally, the formation of artificial small inverted domain is reported to demonstrate that SNDM system is very useful as a nano-domain engineering tool. The nano-size domain dots were successfully formed in $LiTaO_3$ single crystal. This means that we can obtain a very high density ferroelectric data storage with the density above 1 Tbits/inch^2.

16.1 Introduction

Recently, ferroelectric materials, especially in thin film form, have attracted the attention of many researchers. Their large dielectric constants make them suitable as dielectric layers of microcapacitors in microelectronics. They are also investigated for application in nonvolatile memory using the switchable dielectric polarization of ferroelectric material. To characterize such ferroelectric materials, a high-resolution tool is required for observing the microscopic distribution of remanent (or spontaneous) polarization of ferroelectric materials.

With this background, we have proposed and developed a new purely electrical method for imaging the state of the polarizations in ferroelectric and piezoelectric material and their crystal anisotropy. It involves the measurement of point-to-point variations of the nonlinear dielectric constant of a specimen and is termed "scanning nonlinear dielectric microscopy (SNDM)" [1–7]. This is the first successful purely electrical method for observing the ferroelectric polarization distribution without the influence of the screening effect from free charges. To date, the resolution of this microscope has been improved down to the sub-nanometer order.

Here we describe the theory for detecting polarization and the technique for nonlinear dielectric response and report the results of the imaging of the ferroelectric domains in single crystals and thin films using SNDM. Especially in a measurement of PZT thin film, it was confirmed that the resolution was sub-nanometer order. We also describe the theoretical resolution of SNDM. Moreover, we demonstrate that the resolution of SNDM is higher than that of a conventional piezo-response imaging by using scanning force microscopy (SFM) technique [8, 9].

Next, we report new SNDM technique. In the above conventional SNDM technique, we measure the lowest order nonlinear dielectric constant ϵ_{333}, which is a 3rd rank tensor. To improve the performance and resolution of SNDM, we have modified the technique such that higher nonlinear dielectric constants ϵ_{3333} (4th rank tensor), ϵ_{33333} (5th rank tensor) are detected. It is expected that higher order nonlinear dielectric imaging will provide higher lateral and depth resolution. We confirmed this improvement over conventional SNDM imaging experimentally, and used the technique to observe the growth of a surficial paraelectric layer on periodically poled LiNbO$_3$ [10–12].

In addition to this technique, a new type of scanning nonlinear dielectric microscope probe, called the ϵ_{311}-type probe, and a system to measure the ferroelectric polarization component parallel to the surface have been developed. This is achieved by measuring the ferroelectric material's nonlinear dielectric constant ϵ_{311} instead of ϵ_{333}, which is measured in conventional SNDM. Experimental results show that the probe can satisfactorily detect the direction of the polarization parallel to the surface [13]. Finally, the formation of artificial small inverted domain is reported to demonstrate that SNDM system is very useful as a nano-domain engineering tool. The nano-size domain dots were successfully formed in LiTaO$_3$ single crystal. This means that we can obtain a very high density ferroelectric data storage with the density above Tbits/inch^2.

16.2 Nonlinear dielectric imaging with sub- nanometer resolution

First, we briefly describe the theory for detecting polarization. Precise descriptions of the principle of the microscope have been reported elsewhere (see [3, 4]). We also report the results of the imaging of the ferroelectric domains in single crystals and in thin films using the SNDM. Especially in the PZT thin film measurement, we succeeded to obtain a domain image with a sub-nanometer resolution.

16.2.1 Principle and theory for SNDM

Figure 16.1 shows the system setup of the SNDM using the LC lumped constant resonator probe [4]. In the figure, $C_s(t)$ denotes the capacitance of the specimen under the center conductor (the tip) of the probe. $C_s(t)$ is a function of time because of the nonlinear dielectric response under an applied alternating electric field E_{p3} (= $E_p \cos(\omega_p t)$, $f_p = 5 - -100$ kHz). The ratio of the alternating variation of capacitance $\Delta C_s(t)$ to the static value of capacitance C_{s0} without time dependence is given as [3]

$$\frac{\Delta C_s}{C_{s0}} = \frac{\epsilon_{333}}{\epsilon_{33}} E_p \cos(\omega_p t) + \frac{\epsilon_{3333}}{4\epsilon_{33}} E_p^2 \cos(2\omega_p t) \qquad (16.1)$$

where ϵ_{33} is a linear dielectric constant and ϵ_{333} and ϵ_{3333} are nonlinear dielectric constants. The even rank tensor, including the linear dielectric constant ϵ_{33}, does not change with 180°

16.2 Nonlinear dielectric imaging with sub- nanometer resolution

rotation of the polarization. On the other hand, the lowest order of the nonlinear dielectric constant ϵ_{333} is a third-rank tensor, similar to the piezoelectric constant, so that there is no ϵ_{333} in a material with a center of symmetry, and the sign of ϵ_{333} changes in accordance with the inversion of the spontaneous polarization.

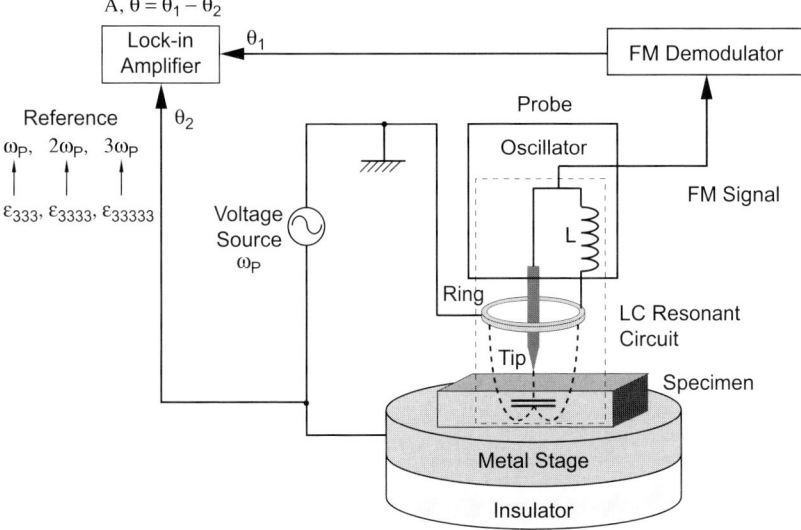

Figure 16.1: Schematic diagram of SNDM

This LC resonator is connected to the oscillator tuned to the resonance frequency of the resonator. The above mentioned electrical parts (i.e. tip, ring, inductance and oscillator) are assembled into a small probe for the SNDM. The oscillating frequency of the probe (or oscillator) (around 1.3 GHz) is modulated by the change of capacitance $\Delta C_s(t)$ due to the nonlinear dielectric response under the applied electric field. As a result, the probe (oscillator) produces a frequency modulated (FM) signal. By detecting this FM signal using the FM demodulator and lock-in amplifier, we obtain a voltage signal proportional to the capacitance variation. Each signal corresponding to ϵ_{333} and ϵ_{3333} was obtained by setting the reference signal of the lock-in amplifier at the frequency $2\omega_p$ of the applied electric field and at the doubled frequency $2\omega_p$, respectively. Thus we can detect the nonlinear dielectric constant just under the needle and can obtain the fine resolution determined by the diameter of the pointed end of the tip and the linear dielectric constant of specimens. The capacitance variation caused by the nonlinear dielectric response is quite small. ($\Delta C_s(t)/C_{s0}$ is in the range from 10^{-3} to 10^{-8}.) Therefore the sensitivity of SNDM probe must be very high. The measured value of the sensitivity of an above mentioned lumped constant probe is 10^{-22} F.

16.2.2 Nonlinear dielectric imaging

The tip of the lumped constant resonator probe was fabricated using electrolytic polishing of a tungsten wire or a metal coated conductive cantilever. The radius of curvature of the chip was 1 μm–25 nm. To check the performance of the new SNDM, first, we measured the macroscopic domains in a multidomain BTO single crystal. Figure 16.2 shows the two-dimensional image of the so called 90° a-c domain which is obtained by a coarse scanning over a large area. The sign of the nonlinear dielectric constant ϵ_{333} of the +c-domain is negative, whereas it is positive in the -c-domain.

Moreover the magnitude of $\epsilon_{111} = \epsilon_{222}$ is zero in the a-domain, because BTO belongs to tetragonal system at room temperature. Thus, we can easily distinguish the type of the domains.

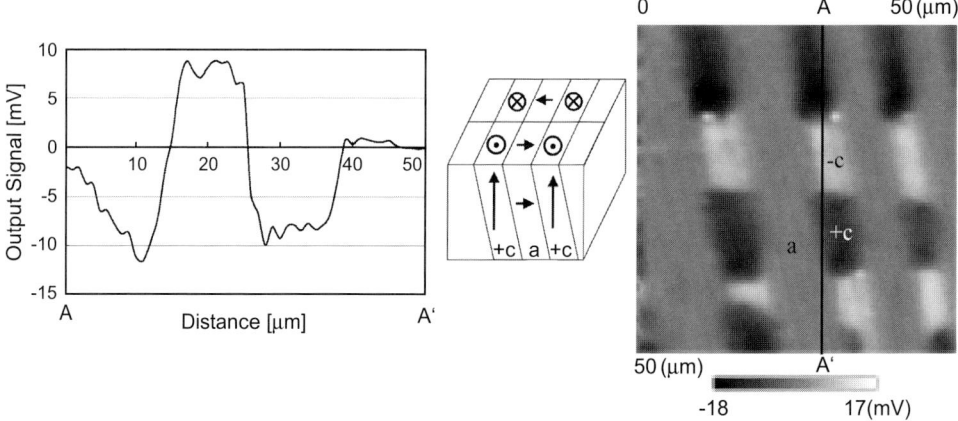

Figure 16.2: A two-dimensional image of the 90° a-c domain in a BTO single crystal and the cross-sectional (one- dimensional) image along the line A-A'.

To demonstrate that this microscopy is also useful for the domain measurement of thin ferroelectric films, we measured a PZT thin film. Figure 16.3 shows the SNDM (a) and AFM (b) images taken from a same location of PZT thin film deposited on a SrTiO$_3$ (STO) substrate using metal organic chemical vapor deposition.

From the figure, it is apparent that the film is polycrystalline (from Figure 16.3 (b)) and that each grain in the film is composed of several domains (from Figure 16.3 (a)). From X-ray diffraction analysis, this PZT film belongs to the tetragonal phase and the diffraction peaks, corresponding to both the c-axis and a-axis, were observed. Moreover, in Figure 16.3 (a), the observed signals were partially of zero amplitude, and partially positive. Thus, the images show that we succeeded in observing 90° a-c domain distributions in a single grain of the film.

These images of the film were taken from a relatively large area. Therefore, we also tried to observe very small domains in the same PZT film on STO substrate. The results are shown

16.2 Nonlinear dielectric imaging with sub-nanometer resolution

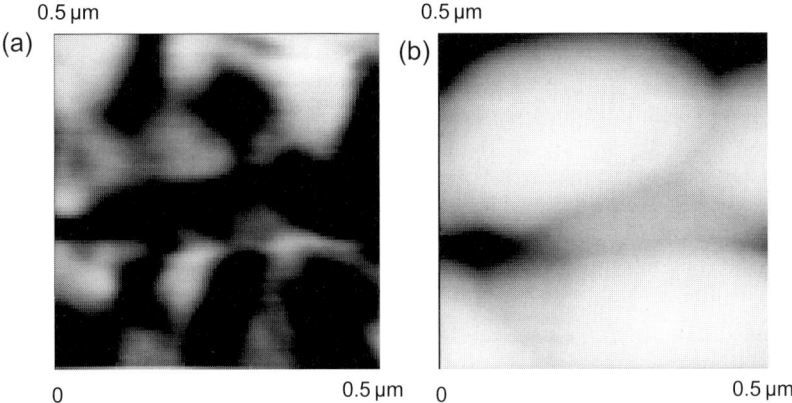

Figure 16.3: Images of a PZT film on a SrTiO$_3$ substrate. (a) Domain patterns by SNDM, (b) surface morphology by AFM.

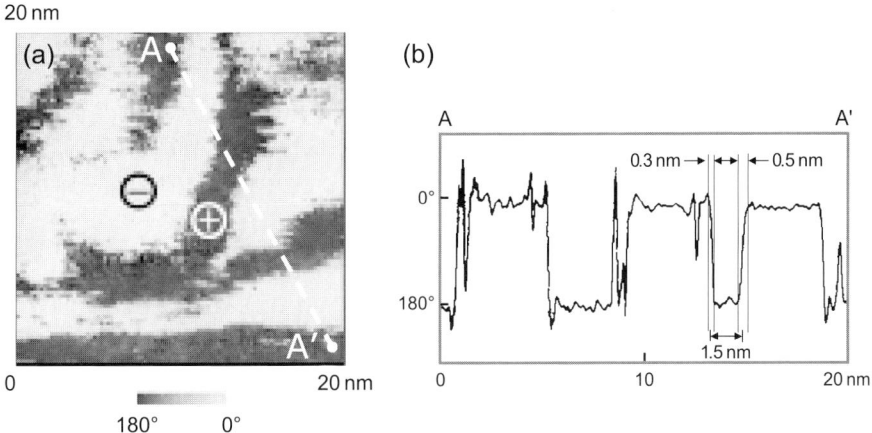

Figure 16.4: Nano-scale ferroelectric domain on PZT thin film, (a) domain image, (b) Cross sectional image of nano-scale 180° c-c domain, (one-dimensional) image of phase signal along the line A-A'.

in Figures 16.4 (a) and (b). The bright area and the dark area correspond to the negative polarization and the positive polarization, respectively. It shows that we can successfully observe a nano-scale 180° c-c domain structure. Figure 16.4 (b) shows a cross sectional image taken along line A-A' in Figure 16.4 (a). As shown in this figure, we measured the c-c domain with the width of 1.5 nm. Moreover we find that the resolution of the microscope is less than 0.5 nm.

However, as the above mentioned data shown in Figure 16.4 was phase images, some readers may think that the sub-nanometer resolution of SNDM is not convincingly proven in the references because phase profiles are invariably abrupt and can not be considered as the definition of the resolution and amplitude signals show more realistic resolution. Therefore, here, we show the amplitude images in Figure 16.5 to demonstrate the resolution of SNDM is really sub-nanometer order. These images were taken from an epitaxial PZT thin (4000 Å) / La-Sr-Co-O / SrTiO$_3$ [14]. The macroscopic surface topography and the domain pattern of this PZT thin film are shown in Figure 16.6. Square c-domains and their surrounding a-domain strip pattern are clearly observed.

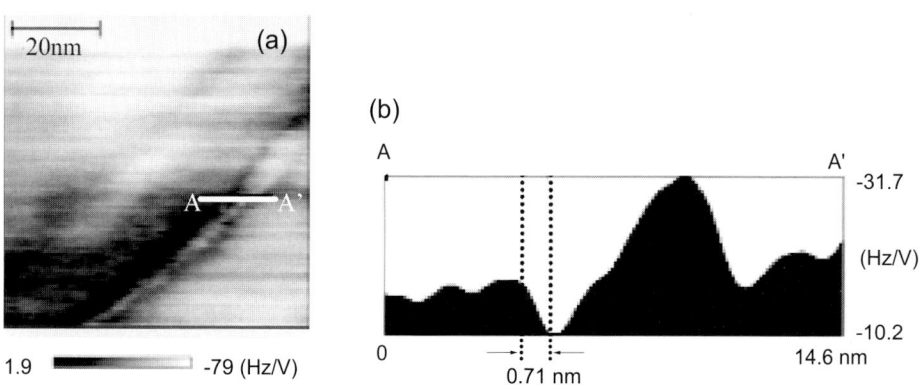

Figure 16.5: (a) Amplitude image of nanoscale ferroelectric domain on PZT thin film (b) Cross-sectional amplitude image taken along A-A'.

The strip shape domain pattern is seen in Figure 16.5. Figure 16.5 (b) is a cross sectional image taken along line A-A' in Figure 16.5 (a). From the distance between the clearly distinguishable structures in the image, it is apparent that SNDM has sub-nanometer resolution.

To clarify the reason why such high resolution can be easily obtained, even if a relatively thick needle is used for the probe, we show the calculated results of the one dimensional image of 180° c-c domain boundary lying at $y = 0$ (We chose y direction as the scanning direction) [15, 16]. Figure 16.7 shows the calculated results where Y_0 is the tip position normalized with respect to the tip radius a. The resolution of the SNDM image is heavily dependent on the dielectric constant of the specimen. For example, for the case of $\epsilon_{33}/\epsilon_0 = 1000$ and $a = 10$ nm, an atomic scale image will be able to be taken by SNDM.

16.2.3 Comparison between SNDM imaging and piezo-response imaging

Another frequently reported high-resolution tool for observing ferroelectric domains is piezo-electric response imaging using SFM [8,9]. From the viewpoint of resolution for ferroelectric domains, SNDM will surpass the piezo-response imaging because SNDM measures the nonlinear response of a dielectric material which is proportional to the square of the electric field,

16.2 Nonlinear dielectric imaging with sub-nanometer resolution

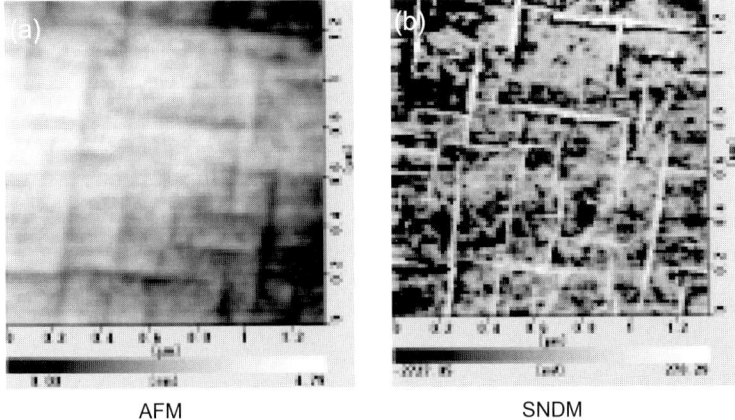

Figure 16.6: Macroscopic surface topography and domain pattern taken from an epitaxial PZT thin (4000 Å) / La-Sr-Co-O / SrTiO$_3$.

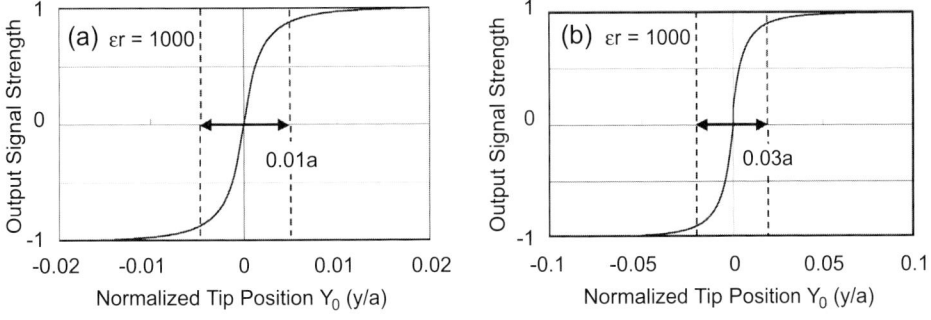

Figure 16.7: Theoretical images of the 180° c-c domain boundary.

whereas the piezoelectric response is linearly proportional to the electric field. The concentration of the distribution of the square of the electric field in the specimen underneath the tip is much higher than that of the linear electric field. Thus, SNDM can resolve smaller domains than that measured by piezo-imaging technique. To prove this fact experimentally, we also performed the simultaneous measurements of the same location of the above mentioned PZT film sample by using AFM (topography)-, SNDM- and piezo-imaging. The result is shown in Figure 16.8. These domain images were taken under identical conditions except that the applied voltage was $2V_{pp}$ in the SNDM imaging and $8V_{pp}$ in the piezo-imaging. In both SNDM image and piezo image, large negative signal was observed on the -c-domain and almost zero signal was detected on the a-domain, because there is a crystal symmetry along the depth direction on the a-domain. From the images, we can prove that SNDM can resolve greater detail than a conventional piezo-response imaging by using SFM technique.

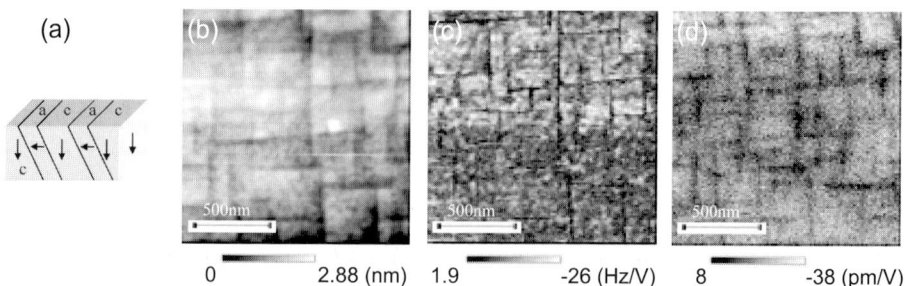

Figure 16.8: Simultaneously taken images of a PZT film. (a) Schematic domain structure (b) Topography by AFM, (c) domain patterns by SNDM, and (d) domain patterns by SFM (piezo-imaging).

16.2.4 Observation of domain walls in PZT thin film using SNDM

Several studies have examined the thickness of the domain wall in ferroelectric materials [17]. The 90° a-c domain wall thickness has been measured using a transmission electron microscope (TEM), [14] but distinguishing the positive and negative domains in 180° opposite polarization areas is difficult because these methods are used to observe the strain of arrangements of molecules. Direct clarification of the domain wall thickness is important with respect to both scientific and engineering aspects.

Therefore we observed the 90° a-c domain walls and the 180° c-c domain walls in the above mentioned same epitaxial PZT thin film grown on $SrTiO_3$ using SNDM.

The linear dielectric constant of a-domain is expected to be larger than that of c-domain based on the BTO single crystal analogy. To the authors' knowledge, no actual observation of linear dielectric constant of a- and c-domains in PZT has been reported. Because obtaining a PZT single crystal of sufficient size in order to compare the linear dielectric constant of a-domain and that of c-domain by the bulk method using a parallel plate capacitor is difficult. Therefore, at first, we measured the linear dielectric constant of a- and c-domains. The observation images are shown in Figure 16.9. In the linear dielectric constant measurements, we measure directly the frequency shifts immediately under the tip. If the high linear dielectric constant area is measured, the frequency shifts low. In Figure 16.9, we can see the frequencies in the a-domain areas are lower than those in the c-domain areas. So the linear dielectric constant of the a-domain is larger than that of the c-domain. The a-domain was first proven to have a higher linear dielectric constant than that of the c-domain just like $BaTiO_3$.

Next we observed the domain walls of 90° a-c and 180° c-c domains. Figure 16.10(a) is a two-dimensional SNDM image of a-c domain and Figure 16.10(b) is the cross-sectional image along the A-A' line in Figure 16.10(a). In Figure 16.10(b), the thickness of the boundary between a-domain and c1-domain was larger than that between a-domain and c2-domain. Moreover, according to the above-mentioned observation, a-domains have larger linear dielectric constants than c-domains. Therefore, the depth sensitivity at the a-domain becomes thinner than that at the c-domain. As a result, when the c-domain is measured, a signal of thicker area can be obtained than when the a-domain is measured. At the boundary between

16.2 Nonlinear dielectric imaging with sub-nanometer resolution

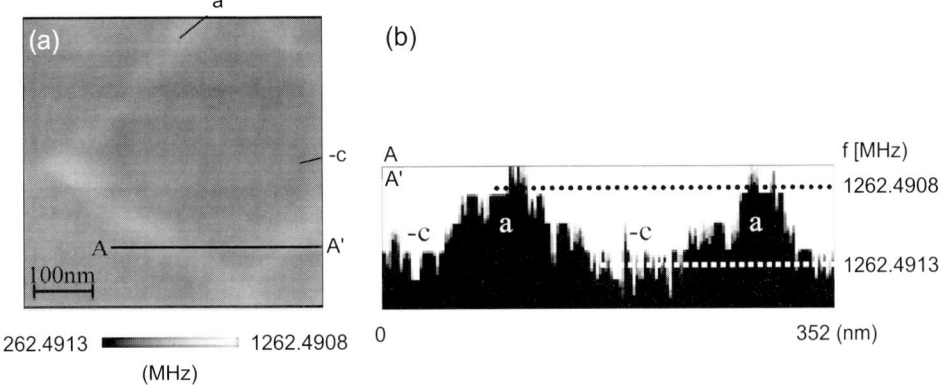

Figure 16.9: (a) Linear dielectric image of a-c domain in PZT, (b) cross-sectional image along the A-A' line.

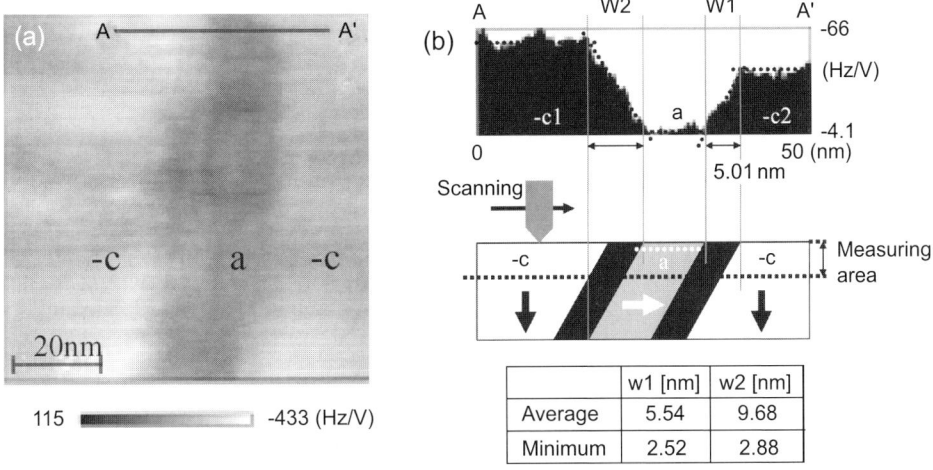

Figure 16.10: (a) Image of a-c domain wall in PZT, (b) cross-sectional image of a-c domain along the A-A' line.

the c1-domain and the a-domain, the tip senses the a-domain under the c-domain before the tip reaches the real a-c domain boundary. As a result, the signal transition distance between c1-domain and a-domain becomes larger than the thickness of the actual a-c domain boundary. On the other hand, at the boundary between a-domain and c2-domain, the signal shows the change with no influence of c-domain under the a-domain, because the depth sensitivity at the a-domain is thinner then that at the c-domain. As a result, the distance between a-domain and c2-domain can be regarded as the actual a-c domain wall thickness. This wall thickness was measured to be 5.01 nm, as shown in Figure 16.10 (b). The average thickness and minimum thickness of a-c domain wall were 5.54 nm and 2.52 nm, respectively. The minimum value seems to indicate the ideal domain wall thickness without the influences of the internal electric field or the residual stress stored in the boundary.

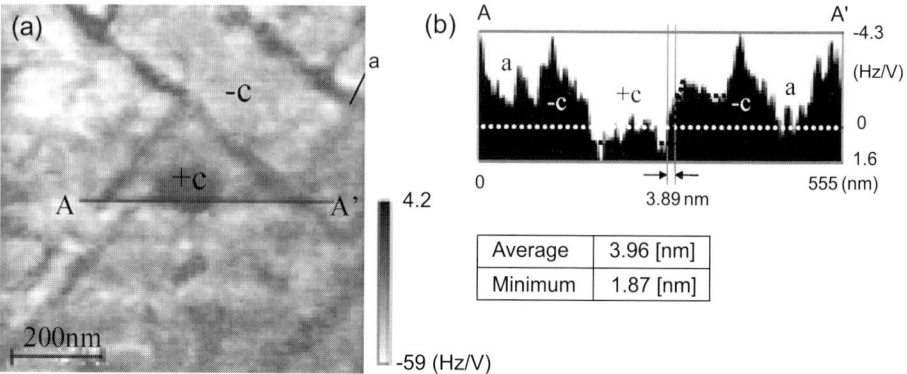

Figure 16.11: (a) Image of c-c domain wall in PZT, (b) cross-sectional image of c-c domain along the A-A' line.

In a similar way, we observed the domain wall between the 180° c-c domains. Before observation, we applied +10 V voltage to the -c domain area to make the +c domain area. Figure 16.11 (a) shows the two-dimensional SNDM image. The dark area in the center of the image is the +c domain area. Figure 16.11 (b) is a cross-sectional image along line A-A' shown in Figure 16.11 (a). The thickness of the c-c domain wall is 3.89 nm in this image. The average thickness and minimum thickness of the c-c domain wall were 3.95 nm and 1.87 nm, respectively. The c-c domain wall had a size of a few unit cells.

In the case of the a-c domain wall, the boundary of a-domain and c-domain is constructed to include lattice strain in order to match the length of the a-axis and the c-axis. On the other hand, in the case of the c-c domain wall, the lattice lengths do not need to match. Therefore, the c-c domain wall does not contain strain, which depends on lattice mismatching. As a result, the c-c domain wall appears to be smaller than the a-c domain wall. Based on the above result, we determined experimentally that the 180° c-c domain wall is thinner than the 90° a-c domain wall.

16.3 Higher order nonlinear dielectric microscopy

A higher order nonlinear dielectric microscopy technique with higher lateral and depth resolution than conventional nonlinear dielectric imaging is investigated. The technique is demonstrated to be very useful for observing surface layers of the order of unit cell thickness on ferroelectric materials.

16.3.1 Theory for higher order nonlinear dielectric microscopy

Equation (16.2) is a polynomial expansion of the electric displacement D_3 as a function of electric field E_3.

$$D_3 = P_{s3} + \epsilon_{33}E_3 + \frac{1}{2}\epsilon_{333}E_3^2 + \frac{1}{6}\epsilon_{3333}E_3^3 + \frac{1}{24}\epsilon_{33333}E_3^4 + ... \qquad (16.2)$$

Here, ϵ_{33}, ϵ_{333}, ϵ_{3333}, and ϵ_{33333} correspond to linear and nonlinear dielectric constants and are tensors of rank 2nd, 3rd, 4th and 5th, respectively. Even-ranked tensors including linear dielectric constant ϵ_{33} do not change with polarization inversion, whereas the sign of the odd-ranked tensors reverses. Therefore, information regarding polarization can be elucidated by measuring odd-ranked nonlinear dielectric constants such as ϵ_{333} and ϵ_{33333}.

Considering the effect up to E^4, the ratio of the alternating variation of capacitance ΔC_s underneath the tip to the static capacitance C_{s0} is given by

$$\frac{\Delta C_s}{C_{s0}} \approx \frac{e_{333}}{e_{33}} E_p \cos(\omega_p t) + \frac{1}{4}\frac{e_{3333}}{e_{33}} E_p^2 \cos(2\omega_p t) + \frac{1}{24}\frac{e_{33333}}{e_{33}} E_p^3 \cos(3\omega_p t) + ... \qquad (16.3)$$

This equation shows that the alternating capacitance of different frequencies corresponds to each order of the nonlinear dielectric constant. Signals corresponding to ϵ_{333}, ϵ_{3333} and ϵ_{33333} were obtained by setting the reference signal of the lock-in amplifier in Figure 16.1 to frequency ω_p, $2\omega_p$ and $3\omega_p$ of the applied electric field, respectively.

Next, we consider the resolution of SNDM. From (16.2), the resolution of SNDM is found to be a function of electric field E. We note that the electric field under the tip is more highly concentrated with the increase of ϵ_{33} [18], and the distributions of E^2, E^3 and E^4 fields underneath the tip become much more concentrated in accordance with their power than that of the E field, as shown in Figure 16.12. From this figure, we find that higher order nonlinear dielectric imaging has higher resolution than lower order nonlinear dielectric imaging.

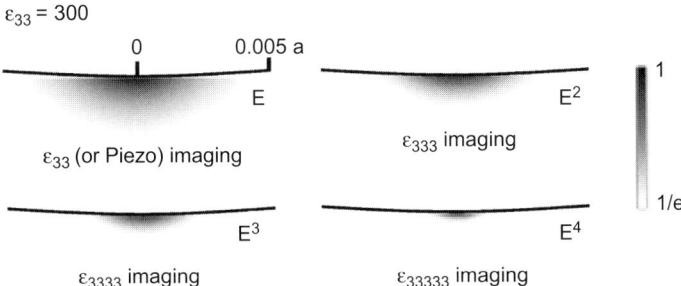

Figure 16.12: Distribution of E, E^2, E^3 and E^4 field under the needle tip. a denotes the tip radius.

16.3.2 Experimental details of higher order nonlinear dielectric microscopy

We experimentally confirmed that ϵ_{33333} imaging has higher lateral resolution than ϵ_{333} imaging using an electroconductive cantilever as a tip with a radius of 25 nm. Figures 16.13 (a) and (b) show ϵ_{333} and ϵ_{33333} images of the two-dimensional distribution of lead zirconate titanate PZT thin film. The two images can be correlated, and it is clear that the ϵ_{33333} image resolves greater detail than ϵ_{333} image due to the higher lateral and depth resolution.

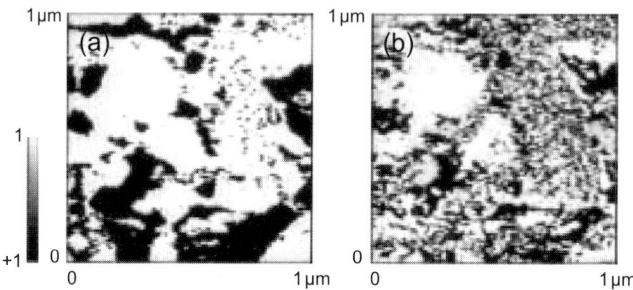

Figure 16.13: (a) ϵ_{333} and (b) ϵ_{33333} images of PZT thin film.

Next, we investigated the surface layer of periodically poled LiNbO$_3$ (PPLN) by ϵ_{333}, ϵ_{3333} and ϵ_{33333} imaging. Figure 16.14 (a) shows ϵ_{333}, ϵ_{3333} and ϵ_{33333} signals of virgin unpolished PPLN. In this figure, only ϵ_{333} imaging detects the c-c domain boundary, while ϵ_{33333} imaging does not. The ϵ_{3333} signal shows weak peaks at domain boundaries. This is because ϵ_{3333} and ϵ_{33333} imaging is affected by the surface paraelectric layer. To prove the existence of a surface paraelectric layer, we polished and measured the PPLN. Figure 16.14 (b) shows the images of it. In this figure, it is clear that ϵ_{33333} imaging can detect the c-c domain boundary after removal of the paraelectric layer. Moreover, ϵ_{3333} imaging can also detect periodic signals, in contrast to our expectation. The nonlinear dielectric signals of a positive area of PPLN are stronger than those of a negative area immediately after polishing, possibly because the negative area is more easily damaged than positive area and has already been covered by a very thin surface paraelectric layer with weak nonlinearity even immediately after polishing. One hour after polishing, we conducted the ϵ_{333}, ϵ_{3333} and ϵ_{33333} imaging again, and the results are shown in Figure 16.14 (c). In this figure, the ϵ_{33333} signal disappears and the ϵ_{3333} signal becomes flat again, whereas ϵ_{333} imaging clearly detects the c-c domain boundary (Figure 16.14 (a)). This implies that the entire surface area of PPLN is covered by the surface paraelectric layer again. From theoretical calculations, on the LiNbO$_3$ substrate, ϵ_{33333} and ϵ_{3333} imagings have sensitivities down to 0.75 nm depth and 1.25 nm, respectively, whereas ϵ_{333} imaging has sensitivity down to 2.75 nm depth when a tip of 25 nm radius is used. Thus, we conclude that the thickness of this surface paraelectric layer ranges between 0.75 nm and 2.75 nm.

From these results, we succeed in observing the growth of the surface layer and we confirm that the negative area of LiNbO$_3$ can be more easily damaged than the positive area.

16.3 Higher order nonlinear dielectric microscopy

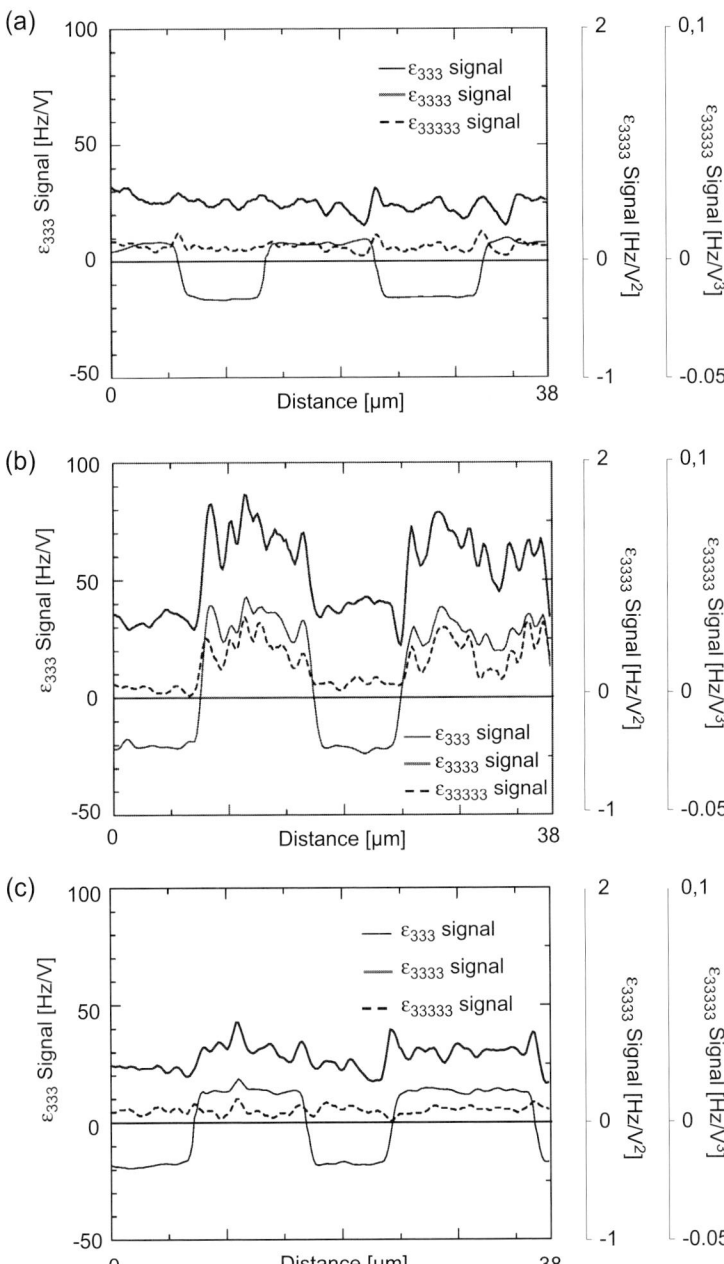

Figure 16.14: (a) ϵ_{333}, ϵ_{3333} and ϵ_{33333} images of virgin PPLN, (b) immediately after polishing, and (c) 1 h after polishing.

16.4 Three-dimensional measurement technique

A new type of scanning nonlinear dielectric microscope (SNDM) probe, named ϵ_{311} type probe, and a system to measure the ferroelectric polarization component parallel to the surface using SNDM has been developed [19]. This is achieved by measuring a nonlinear dielectric constant of ferroelectric material ϵ_{311} instead of ϵ_{333}, which is measured in conventional SNDM. Experimental results show that the probe can detect the polarization direction parallel to the surface with high spatial resolution. Moreover, we propose an advanced measurement technique using rotating electric field. This technique can be applied to measure three-dimensional polarization vectors.

16.4.1 Principle and measurement system

Figure 16.15 shows parallel plate models of nonlinear dielectric constant measurements. Since precise descriptions of the ϵ_{333} measurement have been mentioned above, we explain only the ϵ_{311} measurement. We consider the situation in which a relatively large electric field $\bar{E}_3$ with the amplitude E_p and angular frequency ω_p is applied to the capacitance C_s, producing a change of the capacitance resulting from the nonlinear dielectric response. We detect the capacitance variation ΔC_s, which is perpendicular to the polarization direction (z-axis) by a high frequency electric field with small amplitude along x-axis ($\tilde{E}_1$) as shown in Figure 16.15 (b). (In the ϵ_{333} measurement, we detect ΔC_s along the direction of spontaneous polarization P_{s3}.) That is, in the ϵ_{311} measurement, $\bar{E}$ is perpendicular to $\tilde{E}$. We call this kind of measurements, which use the crossed electric field, "ϵ_{311} type" measurement. In this case, final formula is given by

$$\frac{\Delta C_s(t)}{C_{s0}} \approx \frac{e_{311}}{e_{11}} E_p \cos(\omega_p t) + \frac{1}{4} \frac{e_{3311}}{e_{11}} E_p^2 \cos(2\omega_p t) \tag{16.4}$$

where ϵ_{11} is the linear dielectric constant, and ϵ_{311} and ϵ_{3311} are nonlinear dielectric constants. From this equation, by detecting the component of capacitance variation with the angular frequency of the applied electric field ω_p, we can detect the nonlinear dielectric constant ϵ_{311}. According to this principle, we develop a ϵ_{311} type probe for measuring the polarization direction parallel to the surface. Figure 16.16 shows a schematic diagram of the measurement system. We put 4 electrodes around the probe tip to supply the electric field $\bar{E}$, which causes the nonlinear effect. The electrode A and B supply $\bar{E}_3$, which is along the z-axis, and electrode C and D supply $\bar{E}_2$, which is along the y-axis. We apply voltages to the electrodes, as satisfying the condition that, $\bar{E}_3$ and $\bar{E}_2$ just under the tip become parallel to the surface without concentrating at the tip as shown in Figure 16.16 (b). (In Figure 16.16 (b) the component related to y-direction are omitted for simplification.) On the other hand, the electric field $\tilde{E}$ for measuring the capacitance variation concentrates at the probe tip as the conventional measurement. It is sufficient to consider only x- component of $\tilde{E}$, because we confirmed that most of $\tilde{E}$ underneath the tip is perpendicular to the surface.

Moreover, we can obtain any electric field vector $\bar{E}$ with arbitrary rotation angle by combining the amplitude of $\bar{E}_2$ and $\bar{E}_3$. Therefore, we need not to rotate the specimen for detecting a lateral polarization with an arbitrary direction.

16.4 Three-dimensional measurement technique

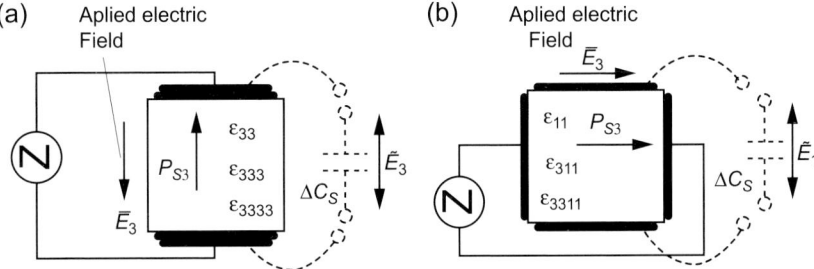

Figure 16.15: Capacitance variation with alternating electric field. (a) ϵ_{333} measurement, and (b) ϵ_{333} measurement.

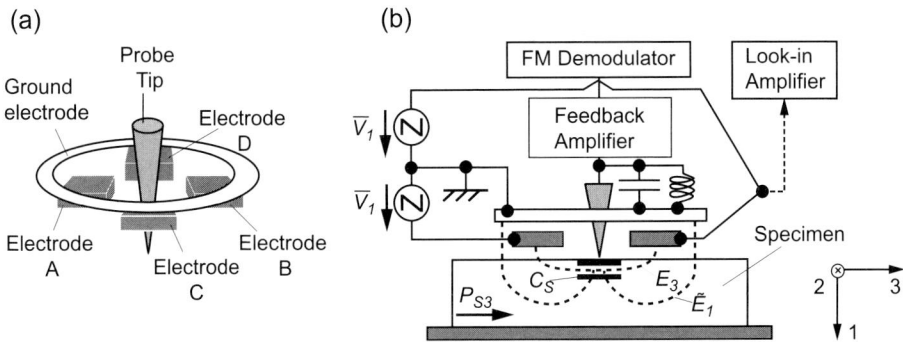

Figure 16.16: Schematic configuration of (a) new ϵ_{311} probe and (b) measurement system.

16.4.2 Experimental results

Figure 16.17 shows measurement result of PZT thin film as changing the direction of applied electric field $\bar{E}$. In Figure 16.17 (a), when $\bar{E}$ is parallel to the polarization direction, the pattern corresponding to the polarization is observed, while no pattern is observed in Figure 16.17 (b) because, in this case, $\bar{E}$ is perpendicular to the polarization direction. Figures 16.17 (c) and (d) are the cases where $\bar{E}$ is applied along the intermediate direction. The pattern can be observed from both Figures 16.17 (c) and (d), because the vector $\bar{E}$ can be divided by the component along the polarization direction. However the opposite contrast was obtained because the signs of the component along the polarization are opposite.

Moreover, the new probe can measure both ϵ_{311} and ϵ_{333} independently. Figure 16.17 (e) shows an ϵ_{333} image, which corresponds to the perpendicular component of the polarization. From the same position in Figure 16.17 (a), signals are observed. It means that this polarization has both parallel and perpendicular component, that is, the polarization tilts from the surface. Figure 16.17 (f) is a topography, which is also measured simultaneously. From these

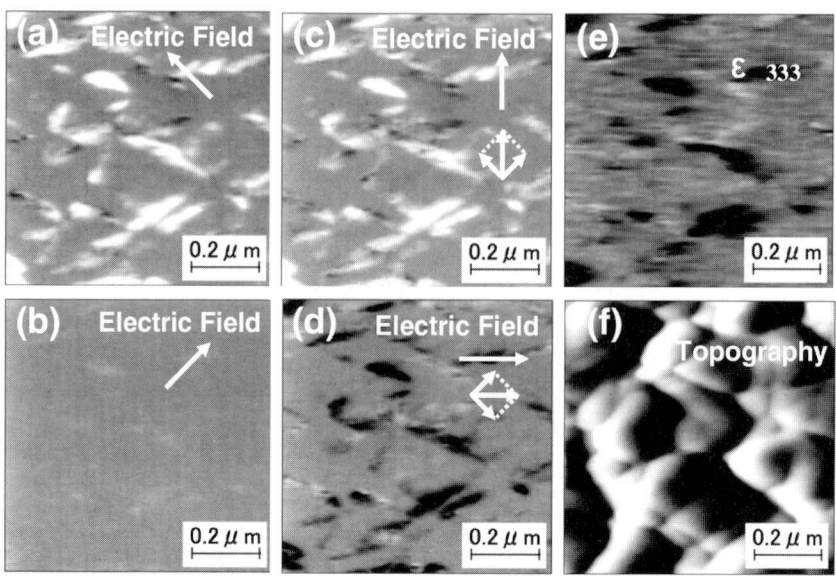

Figure 16.17: Images of PZT thin film, (a)–(d) ϵ_{311} images, (e) ϵ_{333} image, and (f) topography.

results, we confirmed that the new probe and system can be applied to the 3-dimensional polarization measurements.

Based on this technique, we have developed a more advanced method to measure the distribution of polarization directions parallel to the surface. We use a rotating electric field by applying a 90° phase shifted electric voltage between the electrodes A, B, and C, D; that is, we apply the electric fields $\bar{E}_3 = E \cos \omega_p t$ and $\bar{E}_2 = E \cos \omega_p t$ just under the probe tip of Figure 16.16. Under this condition, the electric field rotates with an angular frequency of ω_p and the amplitude of the capacitance variation is changed periodically. When the electric field is parallel to the polarization direction, the capacitance variation is at a maximum, and when the electric field is perpendicular to the polarization direction, the capacitance variation is at a minimum. Consequently, if we detect the capacitance variation by a lock-in amplifier using a reference signal of angular frequency ω_p, we can obtain the angle of the polarization direction directly from the phase output of the lock-in amplifier. Figure 16.18 shows the measurement results on a PZT thin film using a rotating electric field. Histogram of the measured data clearly shows three peaks, which correspond to 0°, 90° and 180°. In Figure 16.18, we can see these three regions. One region is white and black. Within the white and black regions, the white parts and the black parts show the polarization directions +180° and -180°, respectively. Therefore these regions can be considered to be the same domain. Another region is dark gray. The polarization direction in this region is 0°. The third region is bright gray, where the polarization direction is 90°. We suppose that the reasons why a -90° region does not exist are related to a film growth condition. The polarization direction for each region is shown by arrows. The domain structure of this figure is not the typical 90° a-c domain and 180° c-c domain which is usually seen in PZT ceramics. This is because the sample is a thin film. Since the grain size of this film is about 300 nm, Figure 16.18 shows the polarization distribution in a grain. We suppose that the domain structure in a grain depends on a film growth condition. As shown in this figure, we could successfully observe the domain structure, showing the different directions of the polarization parallel to the surface with high spatial resolution.

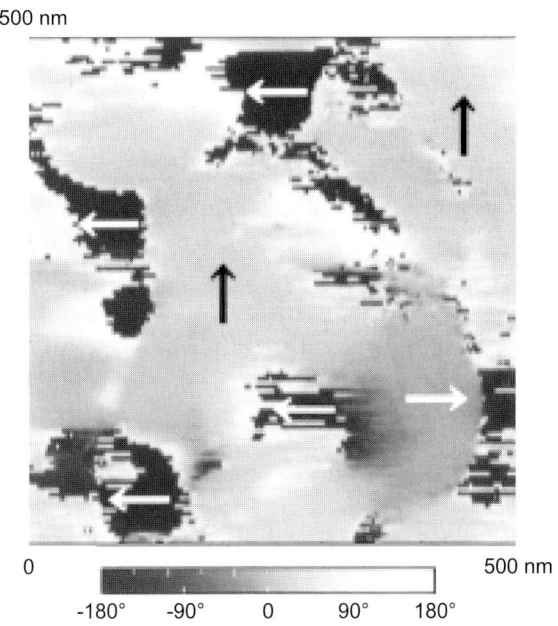

Figure 16.18: Image of a PZT thin film measured by SNDM using rotating electric field.

16.5 Ultra High-Density Ferroelectric Data Storage Using Scanning Nonlinear Dielectric Microscopy

Ferroelectrics have created considerable interest as promising storage media. Here, an investigation for ultra high-density ferroelectric data storage based on scanning nonlinear dielectric microscopy (SNDM) was carried out. For the purpose of obtaining the fundamental knowledge of high-density ferroelectric data storage, several experiments of nano-domain formation in lithium tantalate ($LiTaO_3$) single crystal were conducted. As a result, very small inverted domain with radius of 6 nm was successfully formed in stoichiometric $LiTaO_3$ (SLT), and besides, domain dot array with areal density of 1.5 Tbit/inch^2 was written in congruent $LiTaO_3$ (CLT).

16.5.1 SNDM domain engineering system

Figure 16.19 shows the schematic diagram of SNDM domain engineering system. The probe is composed of a metal-coated conductive cantilever (typical tip radius is 25 nm), an oscillator and a grounded metal ring. Polarity distinction is performed by SNDM technique. On the other hand, writing is performed by applying relatively large voltage pulse to the specimen and locally switching the polarization direction.

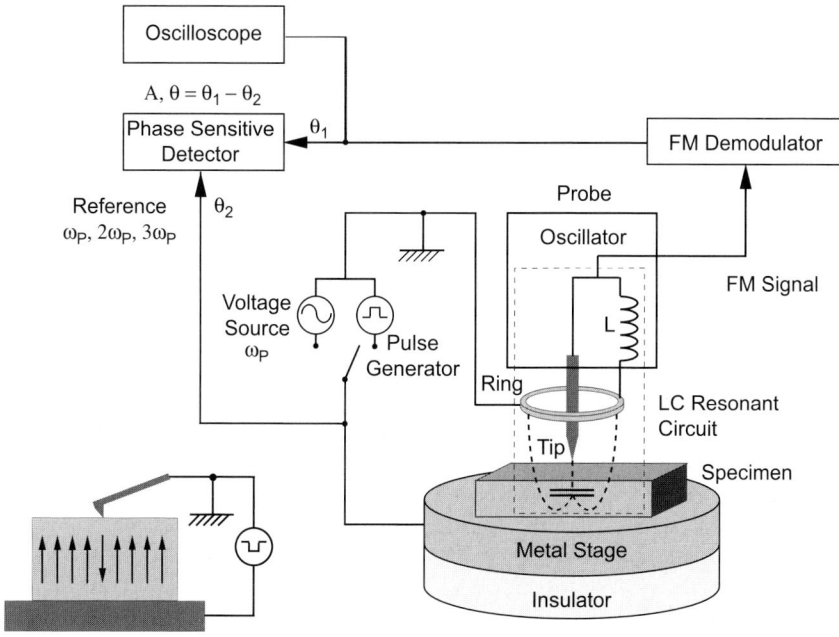

Figure 16.19: Schematic diagram of SNDM domain engineering system.

16.5.2 Nano-domain formation in LiTaO$_3$ single crystal

In this study, we selected LiTaO$_3$ single crystal as a recording medium because this material has suitable characteristics as follows:

(1) There exist only 180° c-c domains.

(2) It does not possess transition point near room temperature.

(3) High-quality and large single crystal can be fabricated at low cost.

At the present time, two types of LiTaO$_3$ single crystals are widely known; one is stoichiometric LiTaO$_3$ (SLT) [20] and the other is congruent LiTaO$_3$ (CLT). It is reported that domain inversion characteristics of these crystals are distinctly different each other [21–23]. SLT has

16.5 Ultra High-Density Ferroelectric Data Storage Using SNDM

few pinning sites of domain switching derived from Li point defects. Therefore, SLT has the characteristics that the coercive field is low and the switching time is short. This means that SLT is favorable for low-power and high-speed writing.

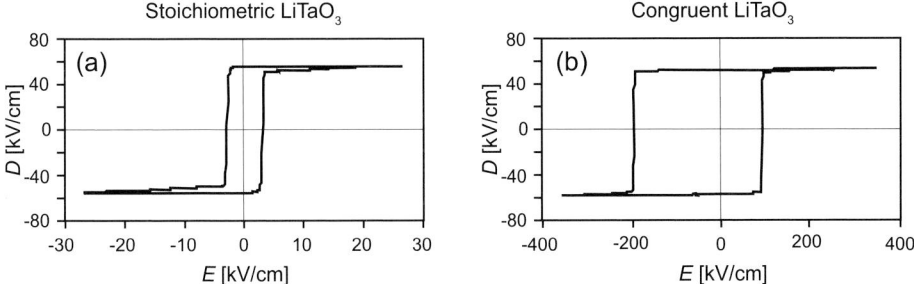

Figure 16.20: D-E hysteresis loop of LiTaO$_3$ measured by applying 10 mHz triangular wave voltage.

Figure 16.20 shows D-E hysteresis of SLT and CLT measured by applying 10 mHz triangular wave voltage, and lower coercive field of SLT compared with that of CLT can be confirmed. On the other hand, CLT has many pinning sites because it is Ta-rich crystal, and it is known that natural domain size of CLT is much smaller than that of SLT. Therefore, we expect that CLT is suitable for higher density storage with smaller domain dots. In the case of forming inverted domain by means of applying voltage to a specimen using a sharp-pointed tip, electric field is highly concentrated just under the tip. So, if the specimen is very thick as compared to the tip radius, large voltage is required for domain switching. Therefore, making thin specimens is very important. In this study, we prepared specimens with thickness of about 100 nm.

Figure 16.21 shows SNDM images of typical nano-sized inverted domains formed in a 100 nm thick SLT medium by means of applying voltage pulses with amplitude of 15 V and duration time of (a) 500 ns (b) 100 ns (c) 60 ns. From these figures, we found that the area of the domain decreases with decreasing voltage application time.

The dependence of domain size on voltage application time is derived from sidewise motion of domain wall. More detailed experimental result with regard to the relationship between the radius of the inverted domain and voltage application time is shown in Figure 16.22.

Figure 16.23 shows the smallest inverted domain at the present time. The radius of this domain is 6 nm, which corresponds to storage density of 4 Tbit/inch^2 if more than one dot can be formed in close-packed array. The result of studying the retention of small inverted domains is depicted in Figure 16.24. These images are observed (a) 50 minutes after pulse application (b) 8 hours after pulse application (c) 24 hours after pulse application. From this result, we found that small inverted domains remained stably for a long time.

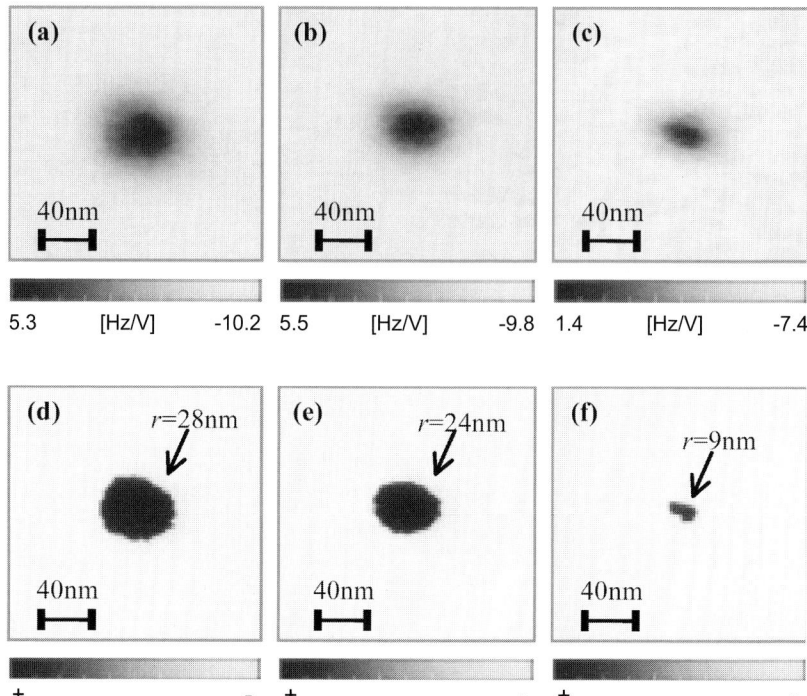

Figure 16.21: Images of typical inverted domains formed in SLT by means of applying a voltage pulse with amplitude of 15 V and duration time of (a), (d) 500 ns, (b), (e) 100 ns, (c), (f) 60 ns. (a)–(c) cos Θ images, (d)-(f) polarity images.

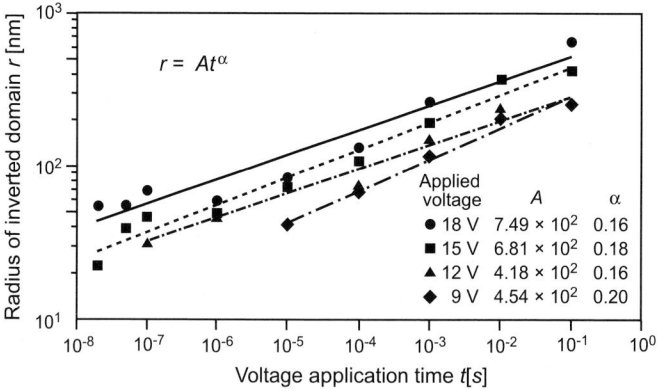

Figure 16.22: Relationship between the radius of the inverted domain and voltage application time in SLT. Sample thickness is 250 nm.

16.5 Ultra High-Density Ferroelectric Data Storage Using SNDM

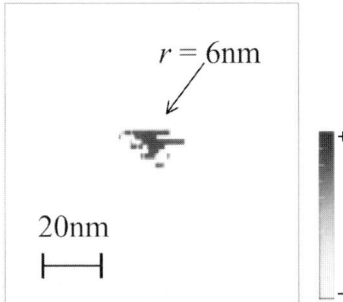

Figure 16.23: The smallest inverted domain at the present time, which was formed in 100 nm thick SLT by applying 15 V, 60 ns pulse.

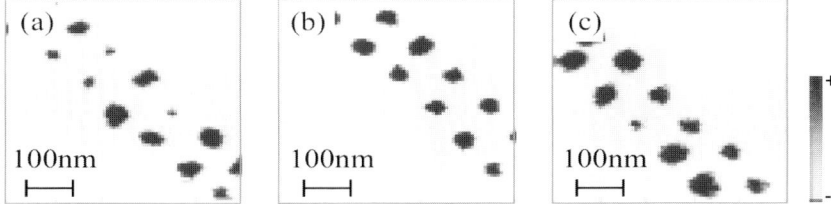

Figure 16.24: Images of inverted domain dot array in 150 nm thick SLT. These images are observed (a) 50 minutes after pulse application (b) 8 hours after pulse application (c) 24 hours after pulse application.

Figure 16.25 shows the relationship between the radius of the inverted domain and voltage application time in SLT and CLT. Voltage application time required for forming a certain size of domain in SLT and CLT are different by five to six orders of magnitude. This difference in switching time is derived from the difference in the number of domain pinning sites. Short switching time in SLT is favorable for low-power and high-speed writing. Subsequently, we conducted some experiments of forming any domain shape by means of applying voltage pulses in multipoint while controlling the probe position.

Figure 16.26 shows the domain characters "TOHOKU UNIV." written in SLT and CLT. The inverted domain in Figure 16.26 (a) was formed in 150 nm thick SLT by applying 15 V, 100 ns voltage pulses, and the inverted domain in Figure 16.26 (b) was formed in 70 nm thick CLT by applying 14 V, 10 μs voltage pulses for the left figure and 14 V, 5 μs voltage pulses for the right figure. From these figures, we found that small inverted domain pattern was successfully formed in CLT despite of applying relatively long pulses on a thinner sample. This result is vividly reflects that pinning sites derived from lithium nonstoichiometry prevent the sidewise motion of domain walls. Thereby, we verified the feasible storage density using CLT.

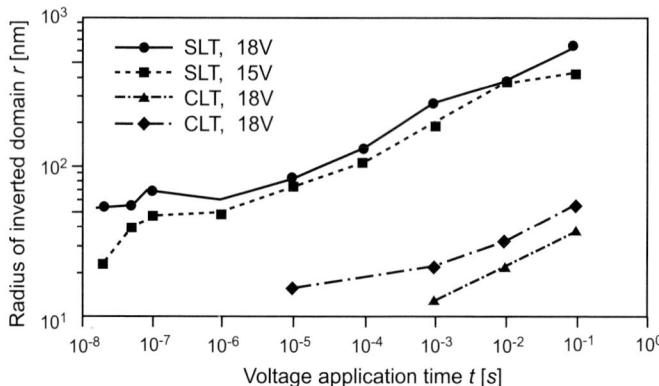

Figure 16.25: Relationship between the radius of the inverted domain and voltage application time in SLT and CLT. Sample thickness is 250 nm.

Figure 16.27 shows images of the inverted domain pattern in CLT with density of (a) 0.62 Tbit/inch^2 (b) 1.10 Tbit/inch^2 (c) 1.50 Tbit/inch^2. These domain patterns were formed by applying voltage pulses with amplitude of (a) 11 V (b),(c) 12 V and duration time of (a) 10 μs (b) 500 ns (c) 80 ns. Close-packed dot array composed of positive and negative domain can be seen in these figures.

Although the dots in the 1.50 Tbit/inch^2 array may not be resolvable with sufficient accuracy for practical data storage, this system is fully expected to become practically applicable as a storage system after further refinement. We have thus demonstrated, using a ferroelectric medium and nano-domain engineering, that rewritable bit storage at a data density of more than 1 Tbit/inch^2 is achievable. To the best of our knowledge, this is the highest density reported for rewritable data storage, and is expected to stimulated renewed interest in this approach to next-generation ultrahigh-density rewritable electric data storage systems.

16.6 Conclusions

In this paper, first, a sub-nanometer resolution scanning nonlinear dielectric microscope (SNDM) for the observation of ferroelectric polarization was described. We also demonstrated that the resolution of SNDM is higher than that of a conventional piezo-response imaging. Using SNDM, we measured the wall thickness of PZT thin film. We also described the theoretical resolution of SNDM. This theoretical result predicted that an atomic scale image can be taken by SNDM.

Next, we reported new SNDM technique detecting higher nonlinear dielectric constants ϵ_{3333} and ϵ_{33333}. It is expected that higher order nonlinear dielectric imaging will provide higher lateral and depth resolution. Using this higher order nonlinear dielectric microscopy technique, we successfully investigated the surface layer of ferroelectrics. Moreover, a new

16.6 Conclusions

(a) SLT (Sample thickness: 150 nm)

Applied pulse: 15 V, 100 nsec

(b) CLT (Sample thickness: 70 nm)

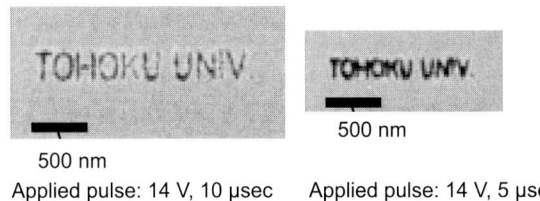

Applied pulse: 14 V, 10 μsec   Applied pulse: 14 V, 5 μsec

Figure 16.26: Nano-domain characters "TOHOKU UNIV." written in (a) SLT and (b) CLT.

type of scanning nonlinear dielectric microscope probe, called the ϵ_{311}-type probe, and a system to measure the ferroelectric polarization component parallel to the surface was developed.

Finally, the formation of artificial small inverted domain was reported to demonstrate that SNDM system is very useful as a nano-domain engineering tool. The nano-size domain dots were successfully formed in LiTaO$_3$ single crystal. This means that we can obtain a very high density ferroelectric data storage with the density above 1 Tbits/inch^2.

Therefore, we have concluded that the SNDM is very useful for observing ferroelectric nano domain and local crystal anisotropy of dielectric material with sub-nano meter resolution and also has a quite high potential as a nano-domain engineering tool.

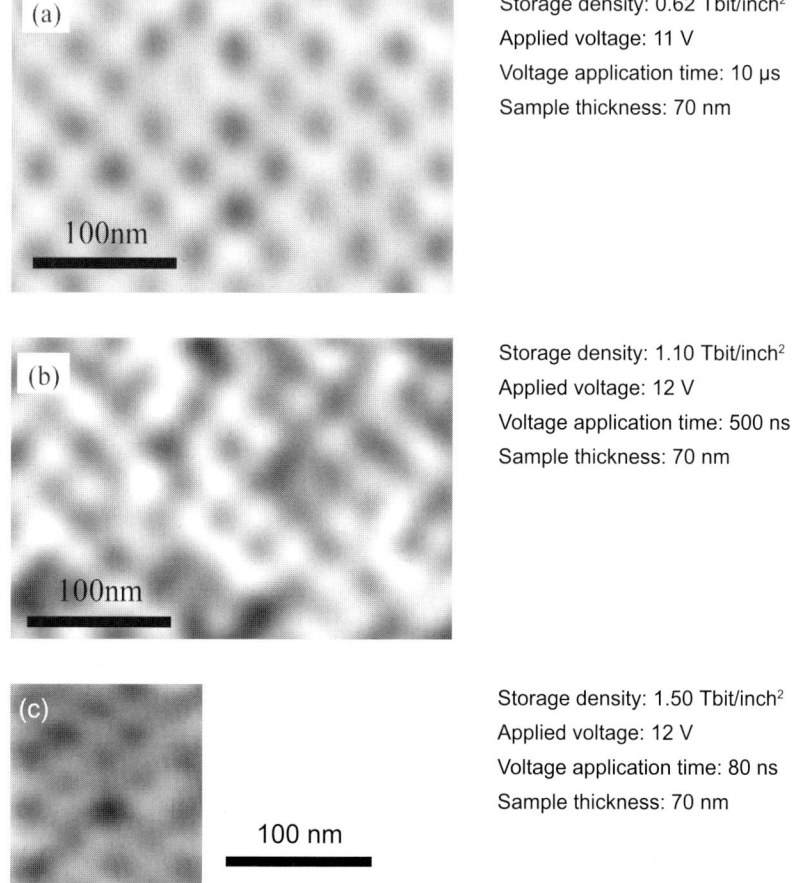

Figure 16.27: Images of the inverted domain pattern formed in CLT with density of (a) 0.62 Tbit/inch^2 (b) 1.10 Tbit/inch^2 (c) 1.50 Tbit/inch^2.

Acknowledgement

We would like to sincerely thank Prof. H. Funakubo of Tokyo Institute of Technology and Prof. R. Ramesh of University of Maryland for supplying the PZT thin film samples.

Bibliography

[1] Y. Cho, A. Kirihara, and T. Saeki, *Denshi Joho Tsushin Gakkai Ronbunshi* **J78-C-1**, 593 (1995) [in Japanese].
[2] Y. Cho, A. Kirihara and T. Saeki, *Electronics and Communication in Japan*, Part 2, **79**, Scripta Technica, Inc., 68 (1996).
[3] Y. Cho, A. Kirihara and T. Saeki, *Rev. Sci. Instrum.* **67**, 2297 (1996).
[4] Y. Cho, S. Atsumi and K. Nakamura, *Jpn. J. Appl. Phys.* **36**, 3152 (1997).
[5] Y. Cho, S. Kazuta and K. Matsuura, *Appl. Phys. Lett.* **72**, 2833 (1999).
[6] H. Odagawa and Y. Cho, *Surface Science* **463**, L621 (2000).
[7] H. Odagawa and Y. Cho, *Jpn. J. Appl. Phys.* **39**, 5719 (2000).
[8] A. Gruverman, O. Auciello, R. Ramesh, and H. Tokumoto, *Nanotechnology* **8**, A38 (1997).
[9] L. M. Eng, H.-J. Günterodt, G. A. Schneider, U. Köpke, and J. Muñoz Saldaña, *Appl. Phys. Lett.* **74**, 233 (1999).
[10] Y. Cho, K. Ohara, A. Koike, and H. Odagawa, *Jpn. J. of Appl. Phys.* **40**, 3544 (2001).
[11] K. Ohara and Y. Cho, *Jpn. J. Appl. Phys* **40**, 5833 (2001).
[12] Y. Cho and K. Ohara, *Appl. Phys. Lett.* **79**, 3842 (2001).
[13] H. Odagawa and Y. Cho, *Ferroelectrics*, to be published.
[14] C. S. Ganpule, V. Nagarajan, H. Li, A. S. Ogale, D. E. Steinhauer, S Aggarwal, E. Williams, R. Ramesh, and P. De Wolf, *Appl. Phys. Lett.* **77**, 292 (2000).
[15] Y. Cho, K. Ohara, S. Kazuta, and H. Odagawa, *J. Eur. Ceram. Soc.* **21**, 2135 (2001).
[16] H. Odagawa and Y. Cho, *Jpn. J. Appl. Phys.* **39**, 5719 (2000).
[17] Y. Ishibashi, *J. Phys. Soc. Jpn.* **62**, 1044 (1993).
[18] K. Matsuura, Y. Cho, and H. Odagawa, *Jpn. J. of Appl. Phys.* **40**, 3534 (2001).
[19] H. Odagawa and Y. Cho, *Appl. Phys. Lett.* **80**, 2159 (2002).
[20] Y. Furukawa, K. Kitamura, E. Suzuki, and K. Niwa, *J. Cryst. Growth* **197**, 889 (1999).
[21] K. Kitamura, Y. Furukawa, K. Niwa, V. Gopalan, and T. E. Mitchell, *Appl. Phys. Lett.* **73**, 3073 (1998).
[22] V. Gopalan, T. E. Mitchell, and K. E. Sicakfus, *Solid State Commun.* **109**, 111 (1999).
[23] S. Kim, V. Gopalan, K. Kitamura and Y. Furukawa, *J. Appl. Phys.* **90**, 2949 (2001).

17 Electrical Characterization of Ferroelectric Properties in the Sub-Micrometer Scale

T. Schmitz[1], S. Tiedke[1], K. Prume[1], K. Szot[2], A. Roelofs[3]

1. aixACCT Systems GmbH, Aachen, Germany
2. Research Center Juelich, Germany
3. Seagate Technologies, Pittsburgh, USA

Abstract

The progress in the development and integration of ferroelectric memories (FeRAM) leads to increasing demand for electrical characterization of sub-micron structures. This article will point out the measurement problems arising from the reduction of the ferroelectric capacitor size e.g. from memory cells or nanostorage devices. Procedures and solutions are presented to overcome these problems and to increase further the resolution and speed of ferroelectric characterization to be ahead of the technological demand.

17.1 Introduction

The measurement of the hysteresis loop of ferroelectric materials is well known since 1930, when Sawyer and Tower used a capacitive divider to record the hysteresis loop of their material [1]. Later on current to voltage converters have been used to record the $P(V)$ curve as these so called virtual ground measurements are more precise. Today this method is the standard method for this kind of measurement, when no specific requirements need to be fulfilled, like high speed or high resolution. In Figure 17.1 the $P(V)$ curve for SrBi$_2$Ta$_2$O$_9$ (SBT) capacitors at different capacitor sizes is shown. It can be seen that it is not straight forward using a standard set up to derive precisely the hysteresis loop of sub-micrometer capacitors. A method for testing these properties would allow the investigation of pad size dependence and to investigate single cell capacitors.

A scheme of a typical memory cell of an integrated 32 Mbit Chain FeRAM device is shown in Figure 17.2 [2]. Here the cell capacitor has an area of 0.49 μm^2, but sizes down to 0.09 μm^2 for integrated capacitor area used in devices. The fabrication of these devices can only be done by memory manufacturers since the whole processing has to be performed, including CMOS, ferroelectric capacitor and electrodes, and metalization to get a functional device. To perform measurements on these capacitors, the device must have the possibility to access their bottom and top electrodes, e.g. through the CMOS circuit or through additional contact pads.

Polar Oxides: Properties, Characterization, and Imaging
Edited by R. Waser, U. Böttger, and S. Tiedke
Copyright © 2005 WILEY-VCH Verlag GmbH & Co. KGaA, Weinheim
ISBN: 3-527-40532-1

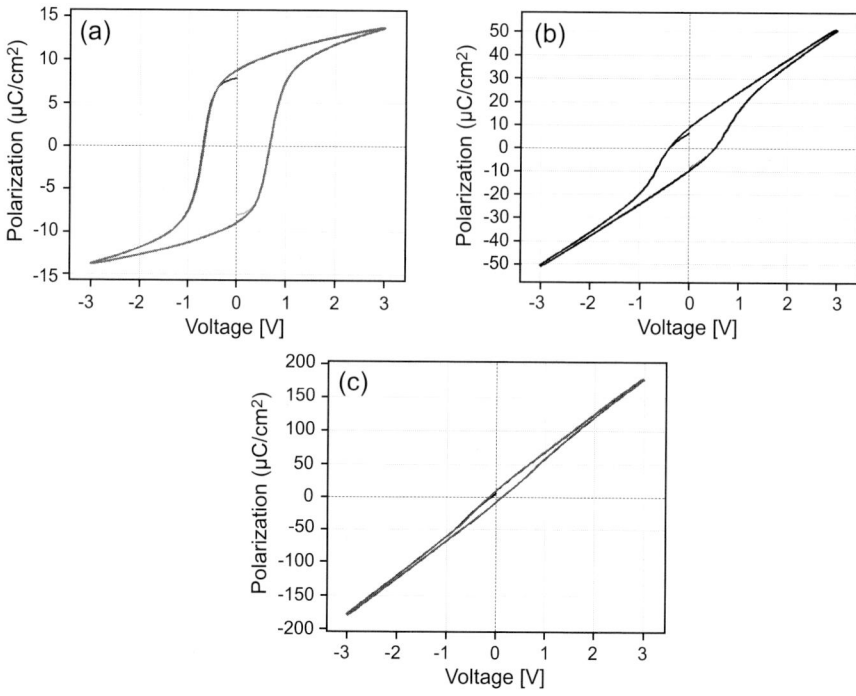

Figure 17.1: Increasing Polarization for small pad hysteresis loops

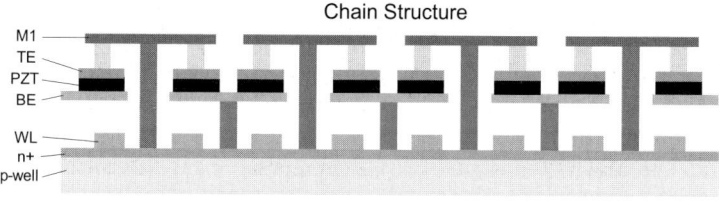

Figure 17.2: Test structure of an integrated 32 Mbit Chain FeRAM cell with contact pads

17.2 Sample preparation

However, to evaluate the properties and size constraints of only the ferroelectric capacitor and its materials there are easier approaches in terms of manufacturing. So typically stand-alone capacitors are manufactured on continuous or structured bottom electrodes with structured top electrodes. A powerful tool is FIB (focused ion beam) where a beam of highly accelerated ions is used to etch or even deposit materials. Stand-alone sub-micron ferroelectric capacitors are shown in Figure 17.3, which have been etched from a continuous thin film structure of

17.2 Sample preparation

bottom electrode, PZT, and top electrode [3]. Furthermore FIB can be used to etch and deposit electrodes to get electrical contact to an integrated capacitor as shown in Figure 17.4.

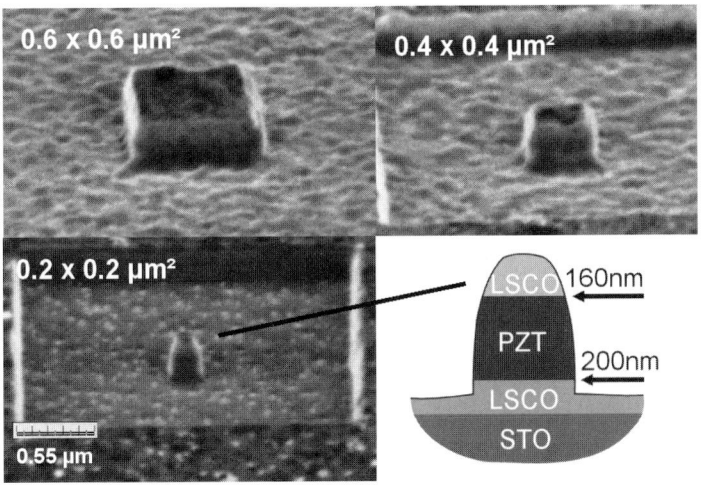

Figure 17.3: Stand alone sub-micron PZT capacitors manufactured by FIB etching

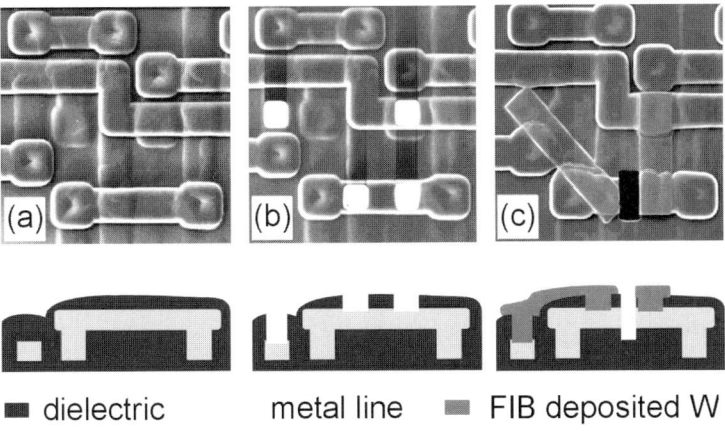

Figure 17.4: Methods of FIB deposition and etching for contacting

Even more simple is the deposition of ferroelectric material e.g. PZT using highly diluted precursors, as it is shown by Roelofs in [4]. Due to the low precursor concentration, small islands of PZT form with sizes down to few nanometers. Another way uses the defined growth based on titanium oxide seeds, which e.g. are used in nanostorage devices based on ferroelectric materials. This approach allows to grow sample sizes on the nanometer scale. However, it is difficult to contact these capacitors without top electrode, so additionally some top electrodes can be deposited on the PZT islands by using FIB again. A further approach is to deposit a continuous film of PZT on the bottom electrode and deposit a "highly diluted" top electrode by sputtering the electrode at low vapor pressure. This leads to the formation of small top electrode islands of various sub-micron sizes, as shown in Figure 17.5, however the electrical field distribution in the film becomes increasingly inhomogeneous when the electrode diameters come into the range of the film thickness.

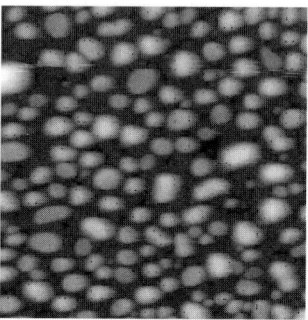

Figure 17.5: Small Pt top electrode "islands" formed by sputtering

17.3 Contact problems

From the fabrication of the capacitor we run into the problem of contacting the capacitor, which can mostly be done in two ways. The easiest way is to contact by standard metal needle probes on a probe station, however, this requires large contact areas larger than $3 \times 3\,\mu m^2$. But this also requires fully structured top and bottom electrodes as in Figure 17.6, or integrated devices, which are much more difficult to manufacture, especially for sub-micron capacitors.

For few micrometer or sub-micron sized electrodes, an AFM (atomic force microscope) is the only tool for contacting. Hereby a conductively coated cantilever of the AFM is used to contact the top electrode and also to scan the sample surface to retrieve a topographic image to find the top electrode position. But when we look at AFM measurements of sub-micron capacitors, we sometimes face hysteresis loops with strangely increased coercive voltages of e.g. 10 Volts, at a film thickness of e.g. 170 nm (Figure 17.7).

But the film thickness doesn't change, so how can the coercive voltage increase? And the same capacitor can show its true coercive voltage of e.g. 2 Volts after a few measurements. So it appears to be a contact problem, though a pure platinum electrode is contacted to a pure PtIr alloy coated cantilever, so an excellent contact should be expected. To investigate the reason of this poor contact behavior, AFM conductivity scans were performed on a pure Pt coated

17.3 Contact problems

Figure 17.6: Structured Pt top and bottom electrodes for probe station probing

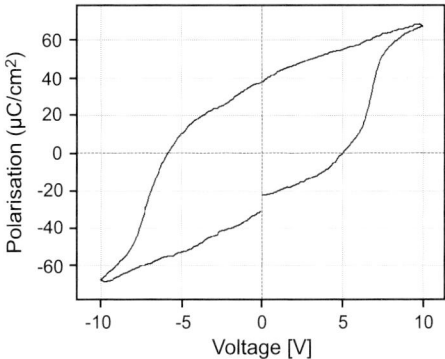

Figure 17.7: Artificially increased coercive voltage created by poor contact quality

wafer [5]. One contact was established using a simple probing needle to guarantee excellent contact due to the high pressure that can be applied to the surface, and the opposite contact was managed by the platinum coated AFM cantilever.

The initial scan shows indeed a very poor contact (Figure 17.8). At a voltage of 1 V the current is below 1 nA and the conductivity is dependent on voltage and also on the location of the contact within the scanned area. After a thermal treatment of 5 min at 200°C the conductivity scan showed excellent conductivity on the whole scan area and at 0.01 V the current amplitude rises to more than 300 nA, so the conductivity increases by more than five orders of magnitude, see Figure 17.9. The reason for this behavior can be described by surface contamination. But the analytical description of the electrical behavior of this surface layer, which is mostly related to water and carbon, is hardly possible. An analytical description of the electri-

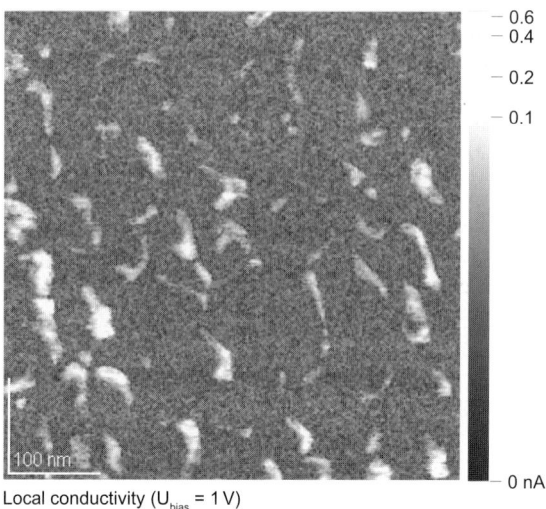

Figure 17.8: AFM conductivity scan of pure Pt surface as received

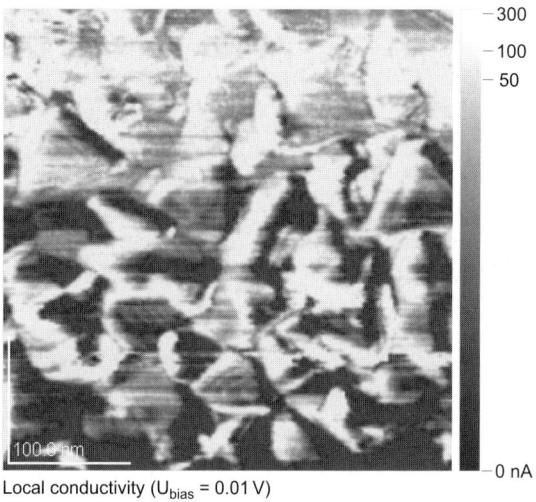

Figure 17.9: AFM conductivity scan of pure Pt surface after thermal treatment

cal behavior in two special cases can be found in [6]. What we can see from this publication is a capacitive behavior from the surface contamination if more than two monolayers of water molecules are involved. But, these surface conditions are mostly not stable and their electrical behavior is hardly predictable. This can explain why adjacent hysteresis measurement results strongly vary and especially why the switching appears at virtually higher voltages. The mon-

17.3 Contact problems

itored voltage is the externally applied voltage. Due to the capacitive divider of the surface layer on the electrode and the ferroelectric sample the effective voltage drop across the ferroelectric sample is smaller than across the surface contamination. As a result the switching appears at a higher value of the externally applied voltage, but at the same internal field in the ferroelectric film. For the Pt scan we verified by XPS that we investigated metallic platinum. A more detailed study of the surface conditions showed physisorbates of OH- and chemisorbates of CO. The physisorbates could be removed by the thermal treatment, but up to 315°C the chemisorbates were still remaining on the surface. Different cantilever materials were tested, where the rhodium coated type could get in very good contact to the top electrode. Often no thermal treatment was necessary.

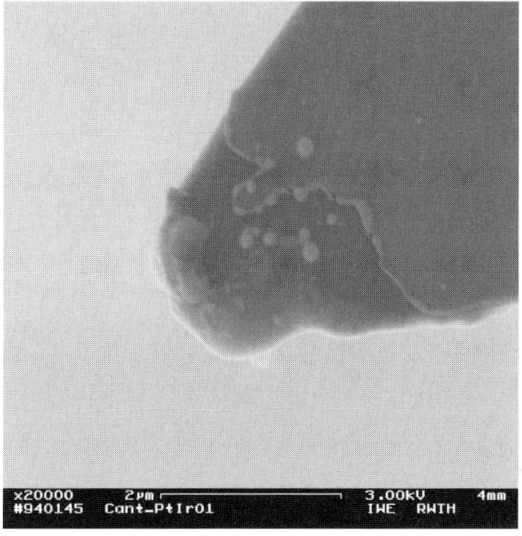

Figure 17.10: SEM picture of Pt coated AFM cantilever tip after measurement

Beside these contact problems another effect was found. Depending on the sample area and the measurement speed the current magnitude and so the current density in the cantilever tip reach very high values. This could lead to the delamination of the cantilever coating. A standard coating thickness like 20 nm is too small in this case. Thicknesses like 200 nm Rh allow the testing of larger sample sizes and at high speed without being damaged. Figure 17.10 shows the cantilever tip after a number of electrical measurements. The delamination/evaporation of the coating at the tip end is obvious.

To circumvent these difficulties in testing integrated structures whose cell capacitor size moved into the dimension of 500 nm × 500 nm and below [7], contact pads with a size of a few hundred μm^2 at the word line, bit line and plate line are used as shown above in Figure 17.2. These pads can be used for simple probing in an automated probing station.

17.4 Parasitic capacitance

Figure 17.1 has shown the change in shape of the hysteresis loop at smaller scale. If we analyze the reason for this change a linear capacitor which is much larger than the ferroelectric one can be identified. Waser and Lohse have given a nice overview of different contributions to the shape of the hysteresis in [8].

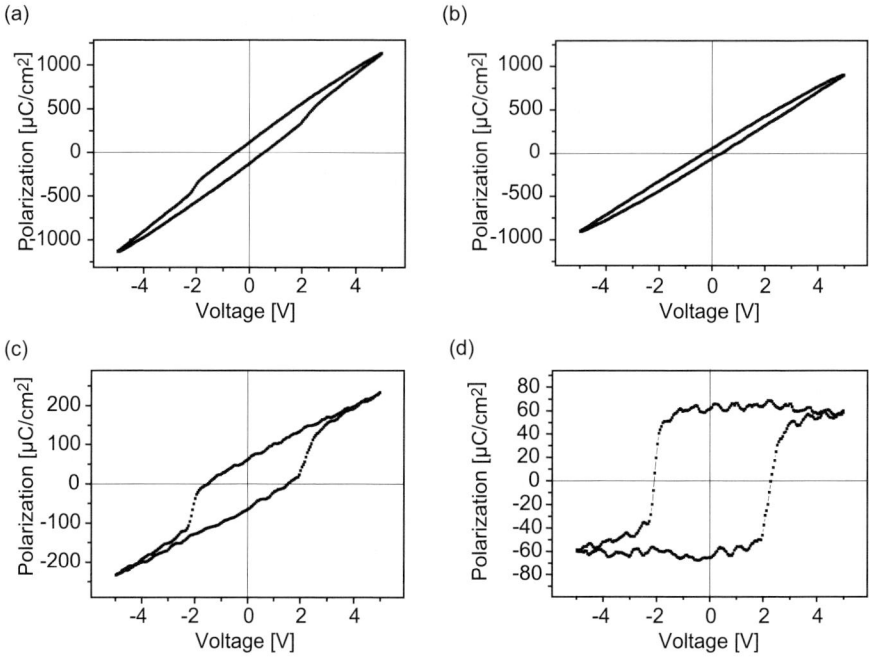

Figure 17.11: Procedure of compensation for linear capacitance by mathematical subtraction

But where does this capacitance arise from? Is it a sample property? No, because finally, the setup to contact the sample could be identified as the origin in [9]. Figure 17.15 demonstrates the effect schematically. Between the needles of the probing station, which are on different voltage potentials a capacitance is existing, which is in parallel with the capacitor of the ferroelectric thin film sample. For measurements using AFM this capacitance exists between cantilever and bottom electrode. The first attempt to eliminate the influence of this capacitor is based on an open measurement, which means recording the $P(V)$ loop without the device under test at the same set up conditions. Therefore the probe tips of the probing station or the cantilever is lifted slightly and the data is recorded. If this open measurement is subtracted from the results of the $P(V)$ curve of the sample the data looks more like a hysteresis. But still a linear capacitor of the setup is remaining, which results from the difference in set up conditions in case of the open measurements from those of the measurement on the sample.

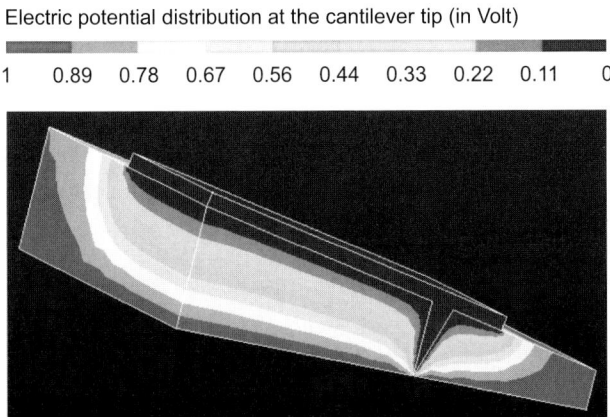

Figure 17.12: FEM simulation of cantilever potential distribution to calculate parasitic stray capacitance

If we look more closely at these changes, which was done by FEM simulations by Prume in [10], we can find a correction factor, which helps to derive the electrical properties precisely for measurements down to 1 μm^2. Figure 17.12 shows the influence of the cantilever tip and the body part of the cantilever in FEM simulation.

Nevertheless, this method does not help to derive the $P(V)$ curve for even smaller capacitors. The reason is the saturation of the recording amplifier and noise increase by the linear contribution of the set up stray capacitance. This is demonstrated in Figure 17.13 (a), where the current response and the $P(V)$ curve on a 300 nm × 300 nm capacitor is shown. The rectangular part of the current response is related to the parasitic capacitor of the set up and the slight switching peak which can hardly be estimated, correspond to the ferroelectric switching of the material. The switching current contribution is similar to the noise level of the current signal. From this measurement the hysteresis loop shown in Figure 17.13 (b) can only be derived after a high number of averages and mathematical subtraction of the parasitic capacitance as described above.

In order to reduce the contribution of the linear stray capacitor an inverse current compensation is used which directly reduces the influence of this capacitor [11]. The optimized results using this method are shown in Figures 17.13 (c) and (d). Now the ferroelectric switching can be seen clearly. The increase in resolution for the $P(V)$ curve is in the range of a factor of 50.

17.5 In-situ compensation

How can we get rid of the parasitic capacitance? The measurement principles of recording the hysteresis loop have been introduced in the basic chapter "Electrical Characterization of Ferroelectrics" within this book. If we take a look at a typical measurement set-up as shown in Figure 17.14 we find that most of the parasitic contributions can be eliminated using the

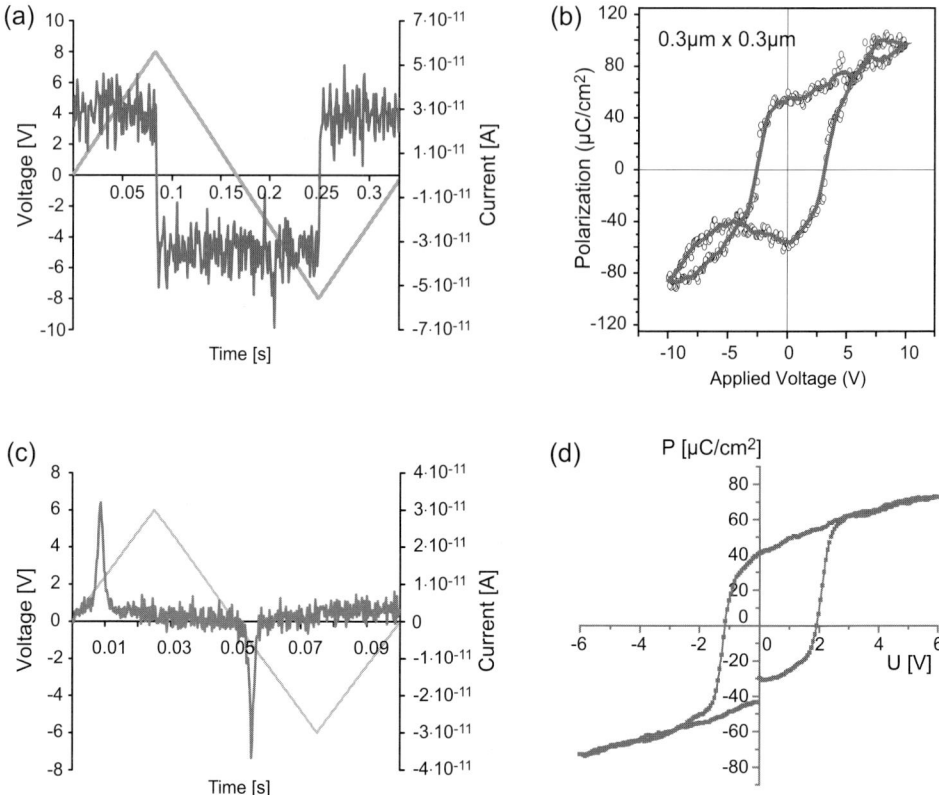

Figure 17.13: Comparison of mathematical and in-situ compensation, current response and hysteresis loop of stand-alone 300 nm × 300 nm PZT capacitor

virtual ground method. Figure 17.15 shows that most of the capacitances, especially cabling capacitance C_{pu} and C_{pi} are not effective in case of the virtual ground method, only the residual parasitic stray capacitance in parallel to the sample C_{ps} needs to be compensated as described in [11]. Beside the reduction of the influence of the parasitic capacitance also the bandwidth of the recorded signal must be taken into account. If the bandwidth of the recording amplifier is too small a widening of the hysteresis loop will be the effect. But the bandwidth requirements are tough as the bandwidth of the recording amplifier must be limited due to stability criteria.

By introduction of this method, it is possible to measure the hysteresis loop of single integrated memory cell capacitors. It is even possible to receive a single shot measurement which is important to measure the initial state of polarization of the capacitor, e.g. to evaluate its retention and imprint properties. This information is lost when the measurement is averaged. A single shot measurement of an integrated 0.49 μm^2 ferroelectric PZT memory capacitor out of the chain FeRAM structure in Figure 17.2 is shown in Figure 17.16 [2].

17.5 *In-situ compensation*

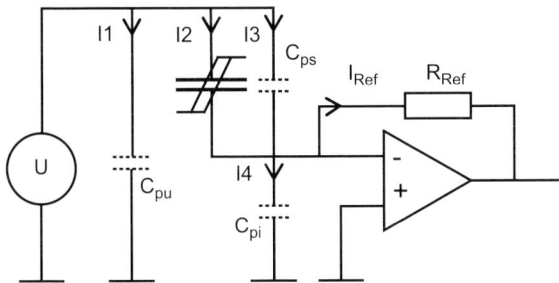

Figure 17.14: Virtual ground amplifier with parasitic capacitances

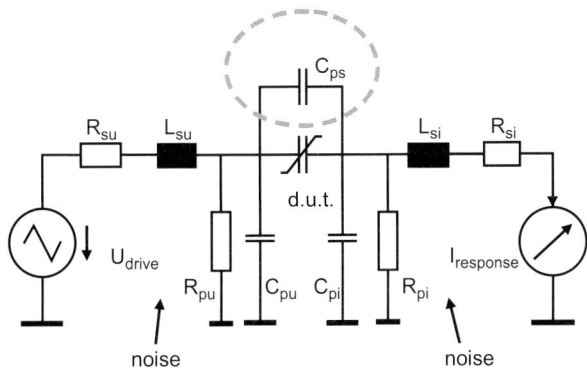

Figure 17.15: Schematic of a typical measurement set-up with parasitic elements

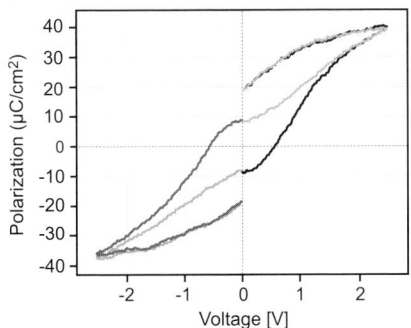

Figure 17.16: Hysteresis pulse measurement of a single FeRAM cell capacitor (0.49 μm^2 PZT)

Furthermore this method can be applied to fast pulse measurements, which are also of great interest in view of memory applications. An example is shown in Figure 17.17. This method can also be applied as small signal measurements on small structures, e.g. $C(V)$ measurements on sub-micron capacitors, and is of great interest to gain further insight in ferroelectric behavior in the nano-scale [12], [13].

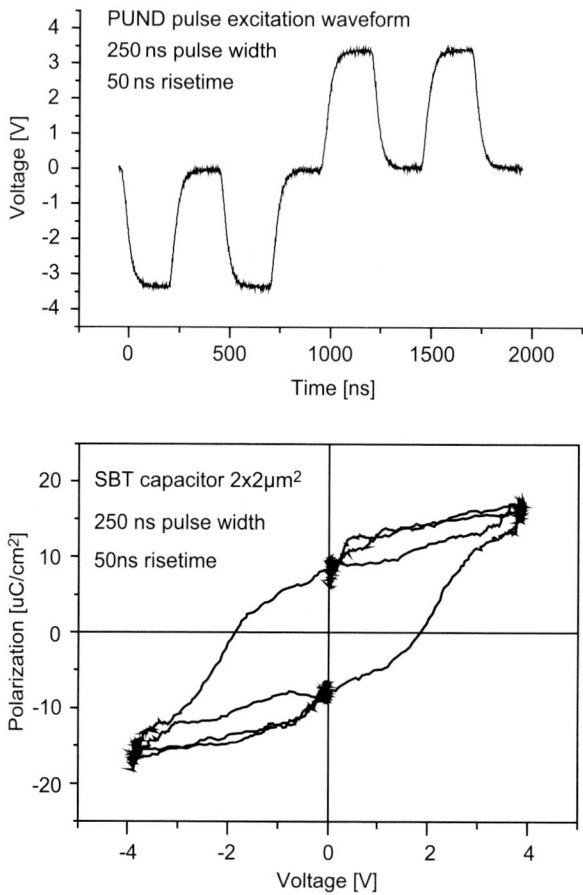

Figure 17.17: Pulse waveform and resulting hysteresis loop of small SBT capacitor

Bibliography

[1] C. B. Sawyer and C. H. Tower, *Phys. Rev.* **35**, (1930).
[2] S. Tiedke, T. Schmitz, R. Bruchhaus, M. Jacob, and I. Kunishima, ISIF *2004 Proceedings*, to be published
[3] C. S. Ganpule, A. Stanishevsky, S. Aggarwal, J. Melngailis, E. Williams, and R. Ramesh, *Appl. Phys. Lett.* **75**, 3874 (1999).
[4] A. Roelofs, *PhD thesis*, RWTH Aachen (to be published).
[5] T. Schmitz and S. Tiedke, *Polecer Meeting*, Capri, Italy, 2003.
[6] J. Halbritter *et al.*, *Appl. Phys.* A **68**, 153 (1999).
[7] S. R. Gilbert *et al.*, *Reliability and Switching Characteristics of Sub-Micron IrOx/MOCVD Pb(Zr,Ti)O_3/Ir Capacitors*, oral presentation ISIF, 2000.
[8] O. Lohse *et al.*, *Mat. Res. Soc.* **493**, 267 (1998).
[9] S. Tiedke *et al.*, *Appl. Phys. Lett.* **79**, 22 (2001).
[10] K. Prume, A. Roelofs, T. Schmitz, B. Reichenberg, S. Tiedke, and R. Waser, *Jpn. J. Appl. Phys.* **41**, 7198 (2002).
[11] T. Schmitz, K. Prume, B. Reichenberg, S. Tiedke, A. Roelofs, and R. Waser, JECS **24**, 1145 (2004).
[12] S. Tiedke, *Nanoscale Phenomena in ferroelectric thin films*, Seungbum Hong ed., Kluwer Academic Publisher, expected publication date autumn 2003.
[13] R. Waser *et al.*, *Towards the Superparaelectric Limit of ferroelectric Nanosized grains*, oral presentation, Trends in Nanotechnology, Santiago de Compostela, Spain, 2002.

18 Searching the Ferroelectric Limit by PFM

A. Roelofs[1], T. Schneller[1], U. Böttger[1], K. Szot[2], and R. Waser[2]

1. Institut für Werkstoffe der Elektrotechnik, RWTH Aachen, Germany
2. Research Center Juelich, Germany

Abstract

Ferroelectric Random Access Memory (FeRAM) devices which have been introduced into the market in recent years are non-volatile, use low voltage, and have high read/write speed. This shows the large potential of future *Gbit* scale universal non-volatile memories. The ultimate limit of this concept will depend on the ferroelectric limit, i.e. the size limit below which the ferroelectricity is quenched. While there are clear indications that 2-D ferroelectric oxide films may sustain their ferroelectric polarization below 4 nm in thickness [1], the limit is quite different for isolated 3-D nanostructures (nanograins, nanoclusters). To investigate the scaling effects of ferroelectric nanograins we studied $PbTiO_3$ (PTO) islands grown by a self-assembly chemical solution deposition method. Preparing highly diluted precursor solutions we achieved single separated ferroelectric grains with grain sizes ranging from 200 nm down to less than 10 nm. The grain size dependent domain configuration is studied using three-dimensional piezoresponse force microscopy (PFM) [2]. While using the PFM to investigate single PTO grains several measurement artifacts can appear which will be discussed in detail. An approach to prevent these measurement artifacts will be proposed. It is found that the PTO grains in a dense film contain laminar 90° domain walls whereas separated grains show more complicated domain structures. For PTO grains smaller than 20 nm no piezoresponse is measured. We suppose that this is caused by the transition from the ferroelectric to a superparaelectric phase which shows no piezoresponse due to the absence of the spontaneous polarization. PFM measurements on standard x-cut quartz were performed to demonstrate the high vertical resolution of 2 pm. Finite element simulations have shown that the expected displacement of these grains (if ferroelectric) would be detectable. Together this proves that the lack of piezoresponse for 20 nm grains is not due to insufficient resolution of the PFM but is a real size effect. Recent calculations [3] and experiments [4] showed that the ferroelectricity of fine ferroelectric particles decrease with decreasing particle size. From these experiments the extrapolated critical size of PTO particles was found to be around 4 - 20 nm, which is in conformity with our findings.

18.1 Introduction

In the growing field of applications for ferroelectric thin films it is of great interest to investigate the ferroelectric properties on a nanometer scale. A powerful tool for the monitoring and

Polar Oxides: Properties, Characterization, and Imaging
Edited by R. Waser, U. Böttger, and S. Tiedke
Copyright © 2005 WILEY-VCH Verlag GmbH & Co. KGaA, Weinheim
ISBN: 3-527-40532-1

the manipulation of the domain structure of ferroelectric thin films is the Piezoresponse Force Microscope (PFM). During scanning a ferroelectric or piezoelectric thin film an alternating voltage is applied through a conducting cantilever. The cantilever can be either highly doped silicon or metal coated (all the measurements shown here were carried out with Pt/Ir coated cantilever.). Due to the converse piezoelectric effect the thin film is locally deformed. The deformation of the ferroelectric thin film is depending on the orientation of the polarization. The PFM is applied in order to monitor ferroelectric and topological properties of individual grains along all 3 dimensions [5, 6]. Comparable experimental setups can be found elsewhere [7, 8]. In contrast to conventional macroscopic electrical characterization using extended top electrodes, PFM may be used to study ferroelectric thin films in the early growth stage having not a complete coverage at all. Whereas extended top electrodes would be shorted making macroscopical electrical characterization impossible PFM may be addressed both for inspection and manipulation of single grains in the nanometer range.

18.2 Polycrystalline ferroelectric PTO thin films on platinized silicon substrates

Polycrystalline $PbTiO_3$ thin films were deposited onto a commercially available Pt(111)/ TiO_2/ SiO_2/ Si substrate from aixACCT laboratories. The used precursor solution was an all-propionate-in-propionic acid (APP) route developed by Hasenkox et al. [9]. Exactly weighted quantities of Titan-tetra-n-butylat (TBT) and dehydrated lead propionate, with a lead excess of 16% were dissolved in propionic acid under dry nitrogen using the Schlenk apparatus. Thereafter the mixture was distilled by reflux in vacuum at 130°C removing the solvent and the by-products. The stock solution was obtained by rediluting the mixture with butoxyethanol. Adding further butoxyethanol led to the coating solution with the desired precursor concentration. The precursor solution was spin coated onto the substrate and dried for 2 minutes on a hot plate at 350°C. Crystallization was initiated by a rapid thermal annealing (RTA) process at 700°C for 30 minutes, resulting in polycrystalline films having no predominant crystallographic orientation as proven by X-ray measurements. The film consists of relatively large single grains generating a non-dense film exhibiting holes down to the bottom electrode (Figure 18.1).

Sputtering extended top-electrodes on this film, trying to build ferroelectric capacitors will lead to short circuits within the device. Nevertheless using a PFM it can be proved that the film is ferroelectric. Operation in the PFM mode allows simultaneous recording of at least three different information: sample topography, out-of-plane polarization (OPP) and in-plane-polarization (IPP) along the x-scan direction (via lever torsion). A conducting p-doped silicon tip, serving as movable top electrode (Figure 18.2), was used for domain visualization (by applying an imaging ac-voltage well below the coercive voltage of the PTO-film.

Monitoring the converse piezoelectric effect and separating both the x (in-plane) and z (out-of-plane) signals enables us to determine the projection of the polarization onto the $x - z$ plane. Please note that a polarization along the y-direction causes a small lifting of the cantilever leading to an overlap with the z signal. Nevertheless careful calibration on single crystalline standard samples allows full disentangling [5].

18.2 Polycrystalline ferroelectric PTO thin films on platinized silicon substrates

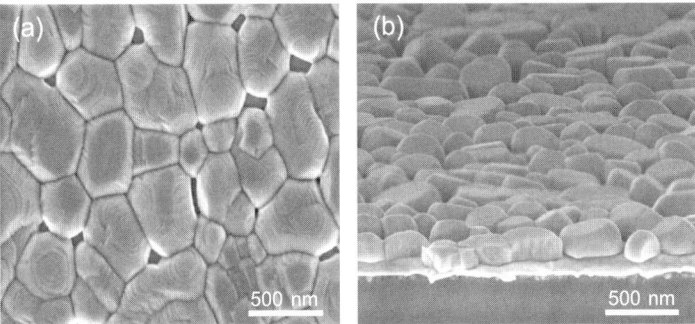

Figure 18.1: SEM of a non dense PTO film prepared by the CSD method. The obtained film is exhibiting several holes reaching down to the bottom electrode. (a) top-view and (b) side-view.

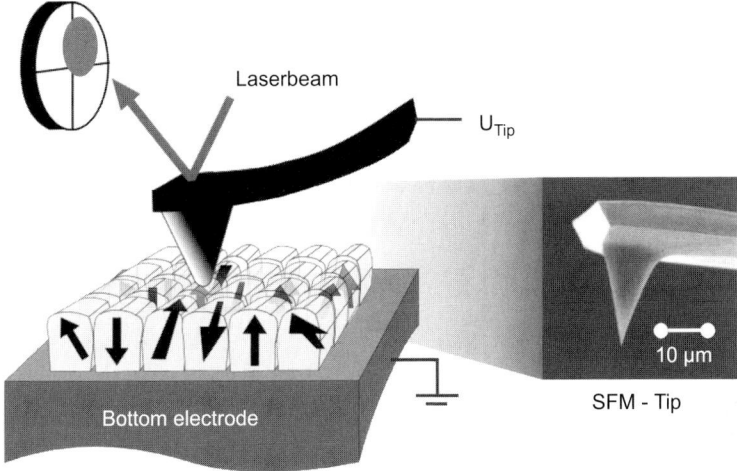

Figure 18.2: Sketch of the PFM. A conductive cantilever is scanned in contact mode over a ferroelectric sample surface.

In Figure 18.3 a measurement (topography and piezoresponse images) of a ferroelectric PTO thin film is shown. The measurement was taken in air in the repulsive force range using only a few Pico-Newton as contact force. It is noticed that not all areas exhibit a piezoelectric signal (Figure 18.3 in- and out-of-plane images). The area denoted with 1 does not show any piezoresponse, whereas the area denoted with 2 shows clearly the domain pattern of the inspected grain.

This is due to contamination of the PTO thin film with a natural water layer and physical adsorbates. This contamination leads to a reduced piezoelectric response in the PFM measurement. There are two possibilities of desorbing the contamination. The first one is to scan the

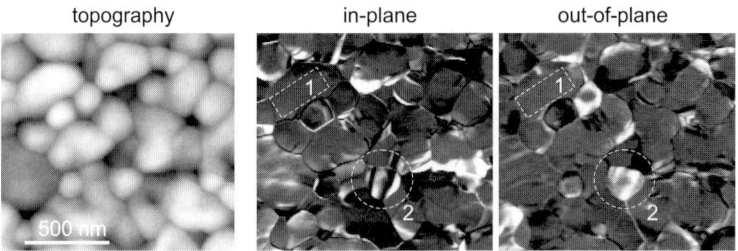

Figure 18.3: Topographical and PFM images of a PTO thin film. Due to physical adsorbates and a natural water layer on the surface of the film the piezoresponse is suppressed in several areas (see rectangle denoted with 1). In other areas a clear piezoresponse can be detected as marked by the circle denoted with 2.

area of interest several times in contact mode while applying an ac or dc voltage in the range from 2 to 3 Volts. This locally blasts the adsorbed molecules from the surface of the ferroelectric and after several scans a PFM image can be monitored (Figure 18.4) where no area of reduced or suppressed piezoresponse appears (as far as all grains are ferroelectric). Note, that this method could partly switch the ferroelectric and change the as grown domain pattern if the voltage applied for cleaning the surface is chosen higher than the coercive field of the film.

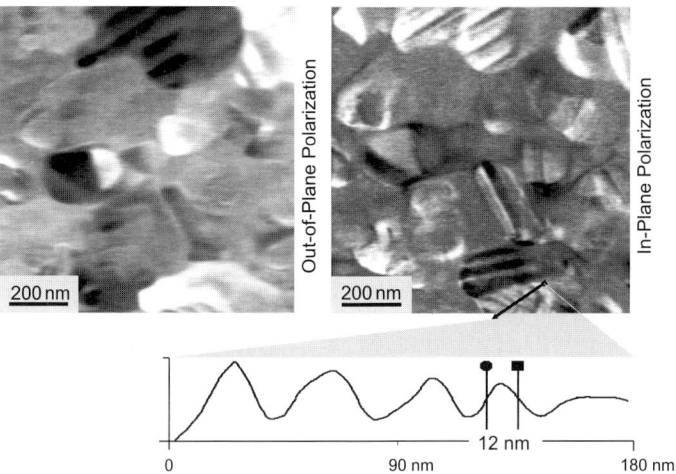

Figure 18.4: PFM image of the same PTO film shown in Figure 18.3 after scanning the area with an ac-voltage of 3 V for 5 times. The whole image area exhibits a piezoresponse. In the linescan below laminar 90° domains of 12 nm width can be observed, proving a good lateral resolution.

18.2 Polycrystalline ferroelectric PTO thin films on platinized silicon substrates

A more gentle method to get rid of most of the surface contamination (mostly the water film reduces the resolution) is to heat the sample for several minutes on a hot plate at about 150°C to 200°C. If the piezoresponse measurement is additionally taken under vacuum conditions the image measurement accuracy is highly enhanced. Figure 18.5 displays the topographical and the piezoresponse images (of the identical PTO sample shown in Figures 18.3 and 18.4) taken under vacuum conditions after the heat treatment. Especially attention should be paid to the topographical image, where now steps become visible corresponding to the 90° domain walls. These steps are easily measured in epitaxial grown ferroelectric thin films exhibiting a $c/a/c$ domain pattern because due to the higher mechanical constraints the surface steps are more pronounced [10].

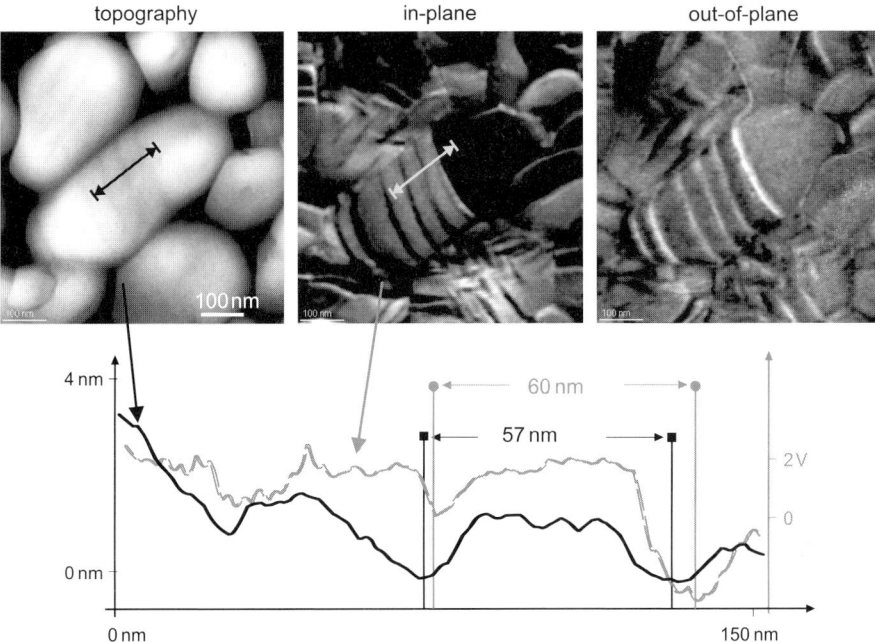

Figure 18.5: PFM measurement of the PTO sample after heating it for 10 min at 200°C. The measurement was taken under vacuum conditions. In the topographical image surface steps due to the 90° domainwalls can be found. Note the linescan, determining the domain width through the topographical measurement leads to a smaller value (51 nm) than determining it from the PFM images (56 nm). This discrepancy is explained by the inspection depth of the PFM mode (Figure 18.6).

If we determine the domain spacing from the topographical and from the piezoresponse image, we find that the PFM image leads to a larger value than the value determined from the topographic image. The difference is caused by the fact, that the surface tilt (seen in the topography) changes its sign exactly at the 90° domain boundary of the a and c-domains

(Figure 18.6), whereas the domain spacing seems to be larger in the PFM images, as the domain spacing is defined by minima (or maxima) in the piezoresponse signal. However, the minimum in the piezoresponse image appears in the center of the domain. Consequently the domain spacing defined by the topographical image and by the PFM image are in principle different. Only in the case that the domains are equal in width both the topographical and the PFM image reveal the same domain spacing. Similar results are found measuring the domain width from the topographical image and from the PFM image.

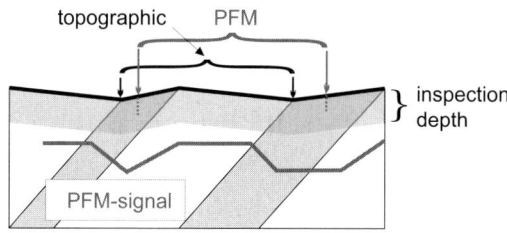

Figure 18.6: Sketch of the correlation between the domain spacing determined from the PFM image and from the topographical image. Only if the width of the two domains are identical the domain spacing measurement in the PFM and in the topographical image would lead to the same value.

18.3 Separated lead titanate nano-grains

In standard CSD processing the aim is to produce dense polycrystalline, columnar or even epitaxial thin films. The above described CSD precursor solution did not lead to satisfactory dense $PbTiO_3$ thin films. Nevertheless, the opposite of dense films, i.e. separated nano-structures are of great interest for studying intrinsic size effects. For this reason the before introduced $PbTiO_3$ precursor synthesis was modified. The flow chart in Figure 18.7 displays the main modifications of the basic CSD route.

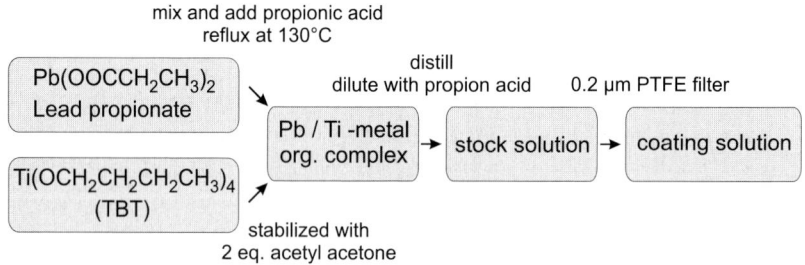

Figure 18.7: Flow chart of the precursor synthesis for PTO single grain deposition.

18.3 Separated lead titanate nano-grains

The solvent propionic acid was replaced by 2-butoxyethanol and the lead propionate was exchanged for lead acetate. However, the crucial idea was to work with highly diluted coating solutions in order to achieve separated PbTiO$_3$ grains. Spinning the precursor solution onto the substrate and performing a subsequent drying step at 350°C on the hotplate led to a *dense amorphous film*. Not until the final crystallization annealing at 700°C the single grains are formed. Figure 18.8 points out the dilution influence on the grain size and the number of grains per area. A dilution of 1:10 stock solution (1.2 molar) to butoxyethanol led to an average grain size of approximately 70 nm and a grain density of about 50 grains/μm^2 (Figure 18.8 left). An additional coating led mostly to grain growth rather than to an increase in the number of grains per area.

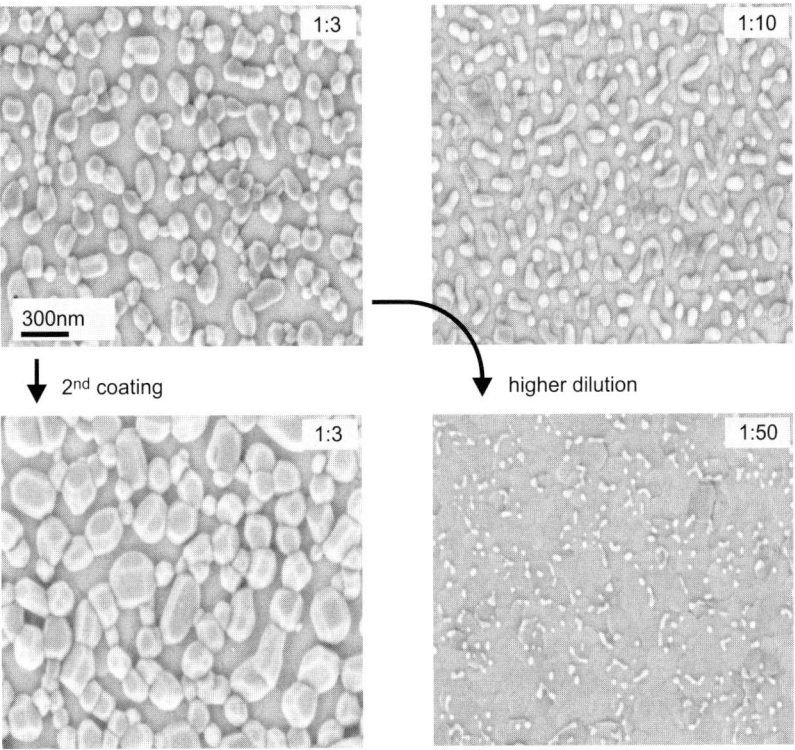

Figure 18.8: Scanning electron microscope micrograph of PTO nanograins deposited onto a Si/SiO$_2$/TiO$_2$/Pt substrate using differently diluted precursor solutions. The higher the dilution is the smaller the grain size and the less grains per area are obtained. The grain sizes range from about 100 nm down to several 10 nm.

When we look at the size and height distribution of the separated PTO grains (Figure 18.9) it is noticed that the grain height increases with increasing grain diameter.

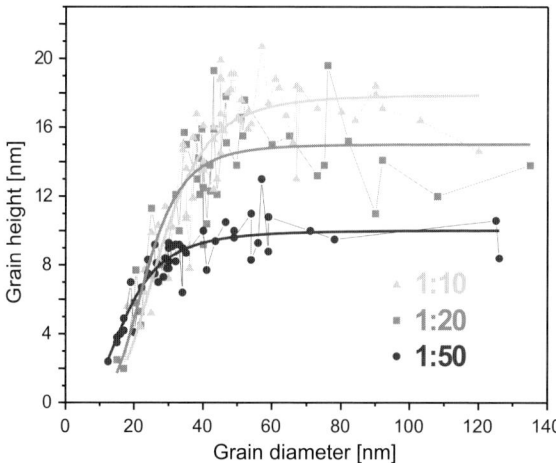

Figure 18.9: The grain height as a function of the grain diameter for three different precursor dilutions. The maximum grain height for larger grain diameters is correlated to the thickness of the amorphous film after the drying step.

Above a certain diameter (depending on the precursor concentration) the grain height saturates. This maximum grain height is correlated to the thickness of the amorphous film after the drying step. Smaller grains have a decreasing height. The smallest measured grain had a diameter of about 12 nm and the height was 2.4 nm. Since the sample consists of single PTO grains that are separated usually more than one hundred nanometers from each other the PFM tip gets into direct contact with the bottom electrode making a short circuit. In order to prevent the tip from damage, the current flow through the tip needs to be controlled, because a too high current flow will damage the metal Pt/Ir - coating. Figure 18.10 displays scanning electron micrographs of a Pt/Ir coated tip before and after current induced damage (a drastic example is shown). The inset in Figure 18.10 on the left depicts the intact metal coating.

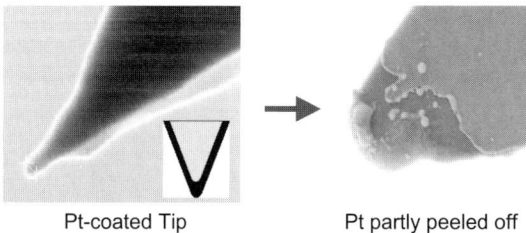

Figure 18.10: SEM micrographs of a new Pt/Ir coated tip (left) and a tip damaged by a high current density. The Pt/Ir coating is partly peeled off and even the underlying silicon is partly removed. The inset on the left is a sketch of an intact metal coating compare Figure 18.12.

18.3 Separated lead titanate nano-grains

Acquiring a PFM image with a cantilever which lost the metal coating only at the tip apex can still provide a good topographic image but leads to strange PFM results that could be misinterpreted (Figure 18.11 1st measurement). In Figure 18.11 only the grain edges lead to a piezoresponse. This could be understood in the way, that only the grain edges exhibit a spontaneous polarization caused by a size effect.

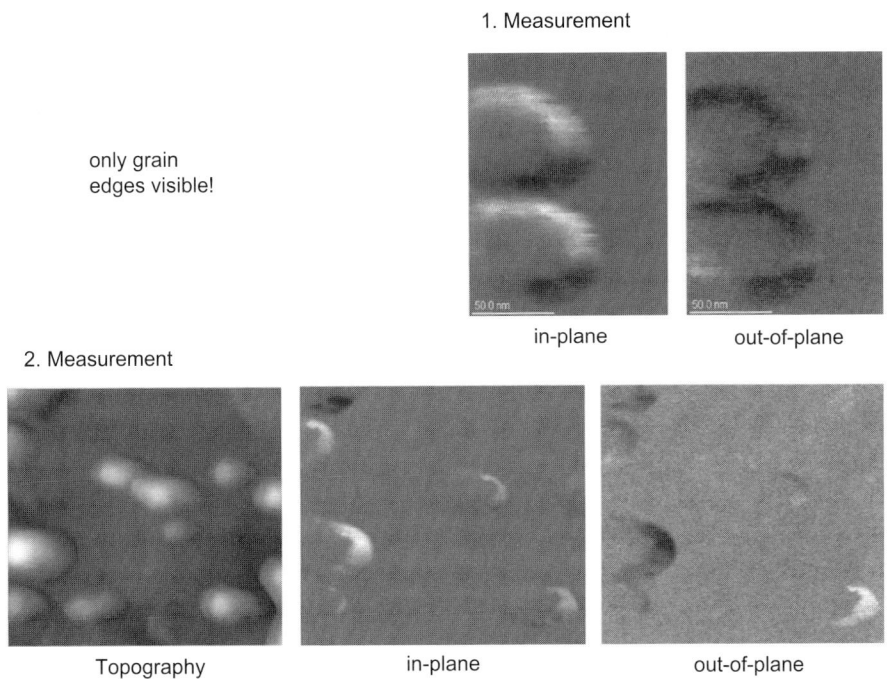

Figure 18.11: Two measurements of PTO single grains on a platinized silicon substrate. In the first measurement the metal coating on the tip apex is removed, causing only the grain edges to contribute to the PFM images. In the second measurement the tip wear increased and only the right side of the larger grains are visible in the PFM images. Note the topographical image still provides an acceptable resolution.

However, this is a measurement artifact because the cantilever lost its metalization at the tip apex. A sketch of the PFM scanning process with a damaged cantilever is shown in Figure 18.12. When the tip approaches the grain, the metal coating on the tip face makes contact to the ferroelectric grain and the piezoresponse is detected (Figure 18.12(a).1). Is the tip positioned upon the grain center the metal coating is not in contact with the grain surface (Figure 18.12(a).2) and the voltage drops mostly over the air gap between metal and grain surface. Consequently, no converse piezoelectric effect is detected. If the tip approaches the other edge of the grain electrical contact is again established (Figure 18.12(a).2). The more of the metal coating is lost the less of the grain volume contributes to the PFM image.

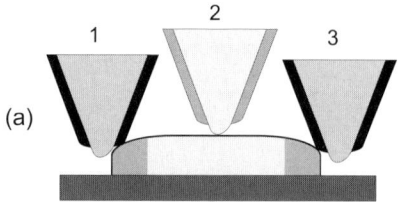 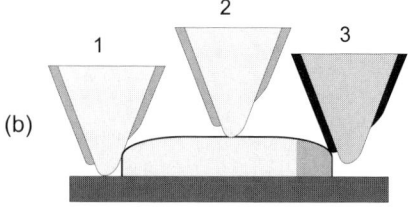

Figure 18.12: Sketch of the PFM scanning process using a metal coated tip where the metal is partly removed. (a) As a result only the grain edges contribute to the PFM image. (b) An asymmetrical tip wear leads to an asymmetrical piezoresponse image.

Is the tip damage asymmetric, obviously the piezoresponse image will exhibit the same asymmetry, see Figure 18.11 2nd measurement and Figure 18.12 (b). This is normally easier to recognize as a measurement artifact. If all grains exhibit a special symmetry either in the topographic or in the piezoresponse image one should become suspicious of the measurement and should replace cantilever.

Most fascinating is the finding represented in Figure 18.13. The topographic image clearly shows nine PTO grains of which two (marked with the white circle) do not lead to a piezoresponse signal as shown in the piezoresponse images. The grain size can be estimated from the line scan to be about 18 nm. We believe that at this grain size the transition from the ferroelectric to the superparaelectric phase could take place [11].

18.4 Conclusion

In conclusion, we presented a simple bottom up approach to achieve separated ferroelectric PTO grains down to 20 nm. It was found that the domain configuration was strongly dependent on the size of the grains. For dense PTO films the grains are dominated by 90° domain walls whereas 180° domain walls are governing the separated grains. Grains that were smaller than 20 nm did not show any piezoresponse that led us to the conclusion that this could be the limiting size for the ferroelectric phase. As ferroelectricity is viewed as a collective phenomenon with a spontaneous polarization resulting from the alignment of localized dipoles within a correlation volume, hence, it is important to shrink all three dimensions. Theoretical studies [3, 12] predict a phase transition for PTO from the tetragonal phase (ferroelectric phase) to non-ferroelectric phase, frequently called superparaelectric phase if the volume of a PTO cell becomes smaller than $1000\,\text{nm}^3 - 2000\,\text{nm}^3$. Furthermore, it was shown that the domain configuration for single ferroelectric grains become more simple with the reduction in grain size. For grains in the range of about $40 - 50$ nm we often find only two domains, but if the grain size becomes smaller than 40 nm we find only mono-domain grains.

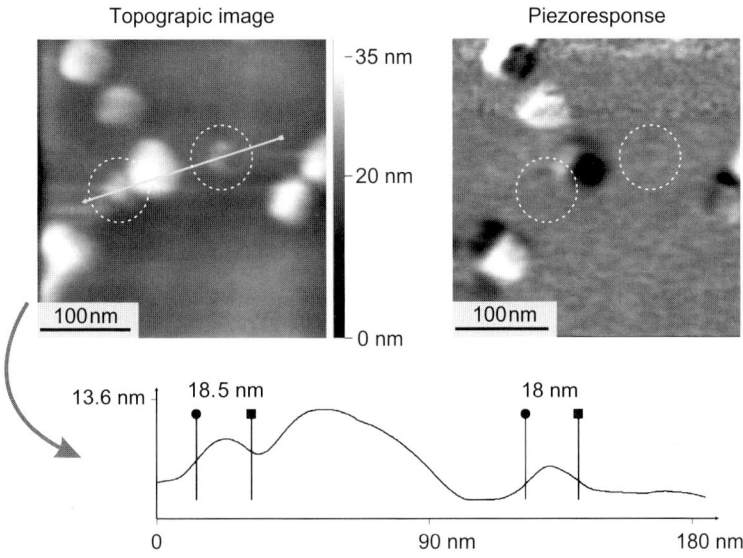

Figure 18.13: The topographic image shows nine PTO grains of sizes between 100 nm down to 18 nm. In the line-scan over the grains denoted with an arrow, shown at the bottom, the size of the grains can be determined. In the PFM image the grains of the size of 18 nm (indicated by the circles) are not visible, leading to the assumption they do not have any permanent polarization. Note, all ferroelectric grains exhibit two domains except one (entirely white), which is smaller then 40 nm in diameter.

Bibliography

[1] T. Tybell, C. H. Ahn, and J.-M. Triscone, *Appl. Phys. Lett.* **75**, 856 (1999).

[2] L. Eng, F. Schlaphof, S. Trogisch, R. Waser, and A. Roelofs, *Ferroelectrics* **251**, 11 (2001).

[3] W. L. Zhong, Y. G. Wang, P. L. Zhang, and B. D. Qu, *Phys. Rev. B* **50**, 698 (1994); C. L. Wang, and S. P. Smith, *J. Phys.: Condens. Matter* **7**, 7163 (1995); S. Li, J. A. Eastman, J. M. Vetrone, C. M. Foster, R. E. Newnham, and L. E. Cross, *Jpn. J. Appl. Phys.* **36**, 5169 (1997).

[4] B. Jiang, J. L. Peng, W. L. Zhong, and L. A. Bursill, *J. Appl. Phys.* **87**, 3462 (2000); W. L. Zhong, B. Jiang, P. L. Zhang, J. M. Ma, H. M. Cheng, and Z. H. Yang, *J. Phys.: Condens Matter* **5**, 2619 (1993); K. Ishikawa, K. Ishikawa, T. Nomura, N. Okada, and K. Takada, *Jpn. J. Appl. Phys.* **35**, 5196 (1996); S. Chattopadhaya, P. Ayyub, V. R. Palkar, and M. Multani, *Phys. Rev. B* **52**, 13177 (1995).

[5] L. M. Eng, H. J. Güntherodt, G. A. Schneider, U. Köpke, and J. Munoz Saldana, *Appl. Phys. Lett.* **74**, 233 (1999).

[6] A. Roelofs, F. Schlaphof, S. Trogisch, U. Böttger, R. Waser, and L. M. Eng, *Appl. Phys. Lett.* 77, 3444 (2000).

[7] K. Franke, J. Besold, W. Haessler, and C. Seegebarth, *Surf. Sci. Lett.* **302**, L283 (1994); T. Hidaka, T. Maruyama, I. Sakai, M. Saitoh, L. A. Wills, R. Hiskes, S. A. Dicarolis, J. Amano, and C. M. Foster, *Integr. Ferroelectr.* **17**, 319 (1997); O. Auciello, A. Gruverman, H. Tokumoto, A. S. Prakash, S. Aggarwal, and R. Ramesh, MRS *Bull.* **23**, 33 (1998); M. Abplanalp, L. M. Eng, and P. Günter, *Appl. Phys. A: Mater. Sci. Process* **66A**, S231 (1998).

[8] A. Gruverman, O. Auciello, and H. Tokumoto, *Ann. Rev. Mater. Sci.* **28**, 101 (1998); A. Gruverman, A. Kholkin, A. Kingon, and H. Tokumoto, *Appl. Phys. Lett.* **78**, 2751 (2001).

[9] U. Hasenkox, C. Mitze, R. Waser, R. Arons, J. Pommer, and G. Güntherrodt, *J. Electroceramics* **3**, 255 (1999).

[10] C. Ganpule, V. Nagarajan, B. Hill, A. Roytburd, E. Williams, R. Ramesh, S. Alpay, A. Roelofs, R. Waser, and L. Eng, *J. Appl. Phys.* **91**, 1477 (2002).

[11] A. Roelofs, T. Schneller, K. Szot, and R. Waser, *Appl. Phys. Lett.* **79**, 3678 (2001).

[12] S. Li, J. A. Eastman, J. Vetrone, C. Foster, R. Newnham, and L. Cross, *J. Appl. Phys.* **36**, 5169 (1997).

19 Piezoelectric Studies at Submicron and Nano Scale

Enrico L. Colla and Igor Stolichnov

Ceramics Laboratory, Materials Institute, Swiss Federal Institute of Technology, 1015 Lausanne, Switzerland.

Abstract

The cycling induced suppression of switchable polarization in ferroelectric thin films capacitors (FeCap) for non-volatile ferroelectric memory (FeRAM) applications remains nowadays a partially unexplained and intriguing phenomena. Due to the very attractive application potentials a large number of research laboratories have attempted, starting from the early 90s, to clarify its origin with a variety of macroscopic characterization techniques. A global analysis of a number of these macroscopic results, enabled one of the authors to predict microscopic features for the onset of fatigue in FeCaps in 1997 [1]. During the same years (early 90s) a powerful analysis technique, to image the piezoelectric activity at the surface of ferroelectric materials at nano scale, was developed by several scientists [2–4]: piezoresponse force microscopy (PFM). This technique soon became one of the most widely used analysis tools for characterizing ferroelectric thin films. In this contribution we will show how PFM has contributed to the description and understanding of the polarization domain behavior in ferroelectrics at sub micron and nano scale, with the help of different significant examples.

19.1 Introduction

19.1.1 Fatigue in FeRAM: macroscopic results invoking nano scale features

The study of ferroelectric thin films during the 90s was strongly motivated by the very promising and partially already realized application of non-volatile ferroelectric memories [5]. The basic element for ferroelectric memories is a small sized ferroelectric capacitor (FeCap), which consists of a ferroelectric material (PZT, SBT, SBNT) and two electrodes (Pt, RuO_2, IrO_2). The technological aspects are therefore very important but the progresses in the last ten years have revealed the existence of features of scientifically high interest too. One of the most intriguing, partially still open questions is related to the fatigue effect, especially for Pt-PZT-Pt based ferroelectric capacitors (FeCap). Many scenarios and models have been proposed [6–12] which more or less can explain the observations, but the suggestion for a realistic mechanism and an ideal solution is still missing.

Figure 19.1 depicts the typical diagram of the evolution of the switching polarization under cycling (N is the number of cycles). The 2 flat lines close to the x-axis are the non-switching charge values ($\pm P_r^{ns}$) for positive and negative voltages. The 2 external lines showing a large variation at $\log(N) = 5 - 6$ represent the switching charge ($\pm P_r^s$). The larger the difference between P_r^s and P_r^{ns}, the easier the discrimination between the state "1" and "0" of a memory cell. The strong decrease of P_r^s after $N = 10^6$ cycles represents therefore a major drawback.

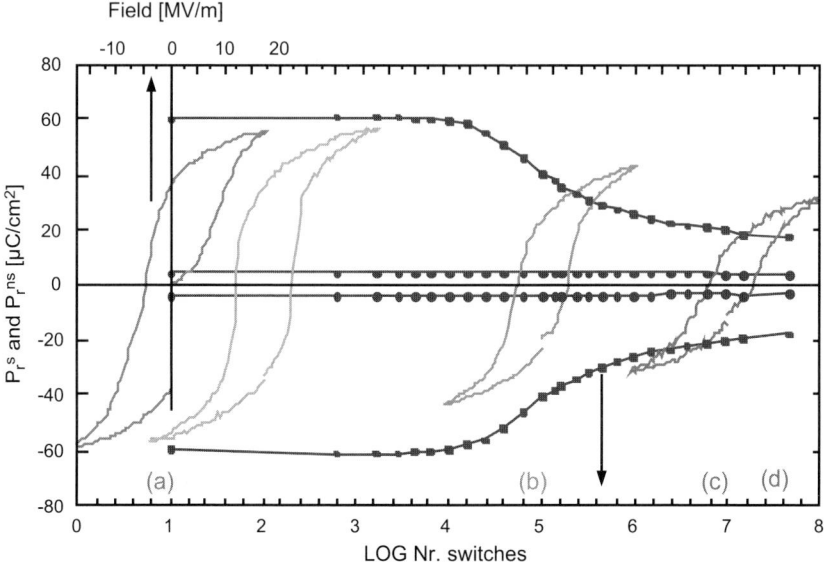

Figure 19.1: Evolution of $P_r^s(N)$, $P_r^{ns}(N)$ and of the hysteresis loops during cycling (N) of the polarization at 50 kHz (The upper x-axis represents the electric field of the first loop).

Figure 19.2, curve (a) and (b) show the fatigue behavior at two different external cycling fields, i.e. 30 ($\approx 4\mathrm{x}E_c$) and 15 MV/m ($\approx 2\mathrm{x}E_c$), respectively. Between $N = 1$ and 10^6 the polarization curves for both fields are nearly identical and after $N = 10^6$ cycles P_r^s is reduced by a factor of 10. Such a behavior is normally interpreted as an indication for "field-independent fatigue". The strong polarization recovery observed at 10^7 cycles for the sample fatigued at 30 MV/m is usually named self-recovery or self-rejuvenation [1–13]. This additional effect is not important for the present discussion.

One can ask the following question: "does the fatigue mechanism show a cumulative character when the sample is cycled by successive fatigue procedures with different fields?" If the statement "field-independent" would be correct, the P_r^s-suppression should be cumulative.

The experimental answer to this question consists of two consecutive fatigue procedures on the same capacitor [1]. To give an experimental answer to this question, two consecutive fatigue procedures on the same capacitor were performed. A FeCap was first fatigued with 150 kV/cm (Figure 19.2 (b)). In the second run, the fatiguing field used was 300 kV/cm. During the latter procedure, the previously suppressed P_r^s was first partially recovered and then suppressed again (Figure 19.2 (c)) according to Figure 19.2 (a), i.e. the fatigue with the

19.1 Introduction

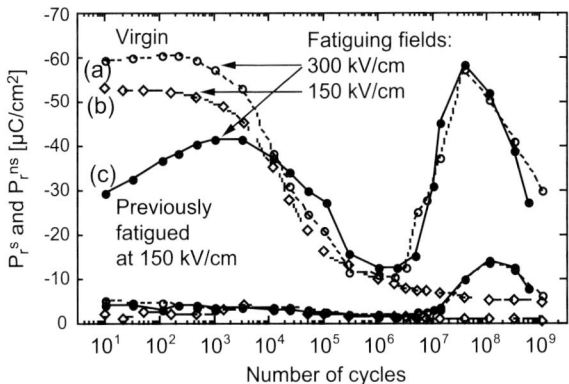

Figure 19.2: Comparison of fatigue measurements at 300 kV/cm of a virgin (a) and a pre-fatigued (c) FECAP. The pre-fatigue procedure, performed at 150 kV/cm and 10^9 cycles (b), has suppressed P_r^s by a factor 10.

higher field of a virgin sample (P_r^s reduction: $\approx$ 1:10). Conclusion: the fatigue effect is not cumulative.

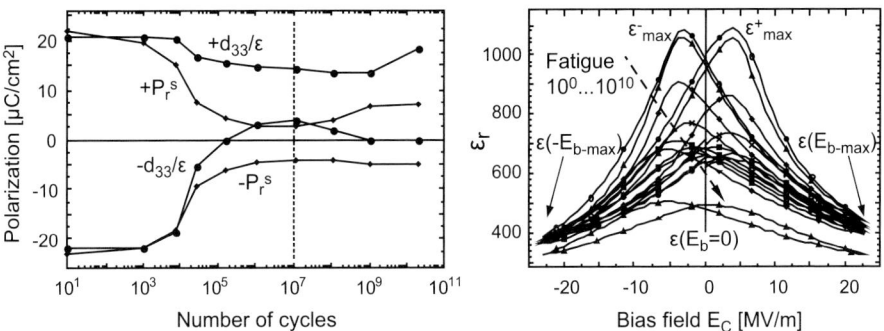

Figure 19.3: Remanent polarization obtained from the $P(E)$-loops and from d_{33}/ε as a function of number of bipolar cycles.

In Figure 19.3 (a) the switching polarization P_r^s obtained by standard ferroelectric measurements and the polarization charge obtained by dividing the normalized piezoelectric coefficient d_{33} by the permittivity ε, are compared. The positive and negative curves are taken after the application of a negative or positive voltage respectively, on the top electrode (TE). In the fatigued state the polarization is not screened or degraded but just locked, predominantly parallel to the field obtained with the positive voltage. The normal switching polarization measurements are not able to differentiate between the two polarizations. The relative % of frozen and switching polarization after 10^7 cycles is 70 % of the polarization frozen "top-to-bottom", 15 % "bottom-to-top" and 15 % still switchable.

Figure 19.2 (b) shows the dielectric permittivity measured during the fatigue procedures [10, 11, 14]. The total measured permittivity in ferroelectric materials is given by the dielectric properties of the lattice and the contribution of the ferroelectric domain walls (DW), i.e. the DW-bending under small AC-fields. The permittivity measured close to the coercive field (about 6.5 MV/m), due to the higher DW-density, must be larger than that measured under stronger bias-field (15 MV/m). In an ideal case, the ferroelectric film at 15 MV/m becomes "single-domain" and only the lattice permittivity should be observed. In these terms the results shown in Figure 19.2 (b) become very instructive. The fatigue independent permittivity measured at 15 MV/m suggests that in the fatigued state, the dielectric properties of the FeCap are fully preserved. On the other hand, since the permittivity measured close to the coercive field drops as the P_r^s values, it can be inferred that fatigue is related to a significant reduction of the number of DW taking part in switching.

The combination of the features mentioned above and others, which can be found in the literature, severely limits the variety of possible microscopic interpretations for fatigue onset. The microscopic fatigue mechanism must have a substantial reversible character, is not cumulative, has the ability to adapt to the used fatiguing field (field self-adjusting), and has no effect on the dielectric properties of the film. According to the present knowledge the inhibition of seeds activity of opposite domains can globally fit the macroscopic features reported. The switching polarization must therefore be suppressed grain-by-grain, giving rise to "ferroelectrically dead areas". The electrical properties of the film remain nearly unaffected by the fatigue process. Figure 19.4 schematizes this interpretation on a sub micron scale [1] and, as it will be shown in the next sections, this feature could be directly observed by means of piezoresponse force microscopy (PFM).

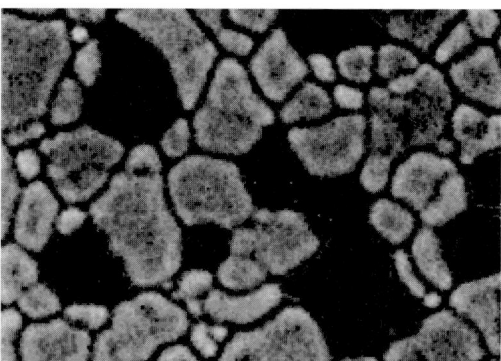

Figure 19.4: Exemplification of the "grain-by-grain" suppression of P_r^s according to a global interpretation of the different macroscopic experimental results. The black areas represent ferroelectric dead portions of the film.

19.1.2 Piezoelectric characterization at nano scale of ferroelectric thin films

There are two main approaches or configurations for the AFM-assisted detection of the local piezoelectric activity (PFM) in ferroelectric thin films for ferroelectric memory applications (FeRAM). The most used one was introduced in the early 90s and uses a conductive AFM-tip as both top electrode and sensor for the induced vibration [2–4]. The second and more recent one [15, 16], uses a normal metallic thin top electrode to apply the electric field and the vibration signal is detected by the AFM-tip above the top electrode. Both approaches present numerous advantages and disadvantages, as widely discussed in the literature, and are quite complementary.

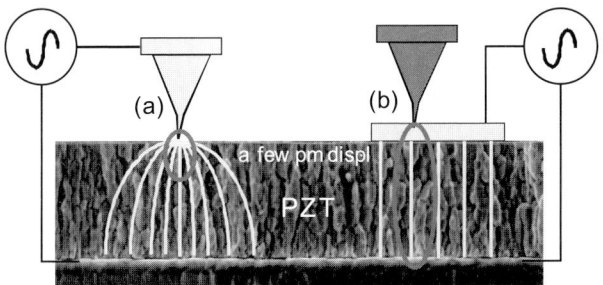

Figure 19.5: (a) AFM tip as TE: Inhomogeneous electric field. Voltage drop at the Tip-Film contact. Possibility of "depth-selective" piezoelectric response (10–50 nm). Induced fatigue probably NOT relevant for real FECAPs. (b) True FECAP: Generation of the relevant fatigue type. The field is homogeneous through the film. The detected signal represents the depth-averaged piezoelectric response.

The conventional PFM approach applies an AC voltage between a conducting cantilever-tip and the electrode on the backside of the sample (Figure 19.5 (a)). The amplitude of the exciting voltage is adjusted below the coercive voltage of the ferroelectric material, while the frequency is chosen below the resonance of the cantilever and far above the upper controlled frequency of the force microscope feedback loop. Because of the inverse piezoelectric effect the applied voltage results in local thickness vibration of the ferroelectric sample due to the out-of-plane component of the polarization. This leads to corresponding oscillations of the cantilever, which can be measured using conventional lock-in techniques. The projection of the polarization vector along the z-direction can therefore be determined. Since our objective is to disclose the origin of fatigue in ferroelectric thin film capacitors (FeCap), this approach suffers of at least two major drawbacks. First, the film cannot be fatigued in a comparable way as in real FeCaps. Second, the exciting field is by far not homogeneous and the detected piezoelectric signal mainly results from the vibration of the very upper layer of the ferroelectric film.

Using a thin top electrode (Figure 19.5 (b)) it is possible to overcome both issues. The external AC voltage applied to the top electrode produces a homogeneous exciting field. The whole thickness of the film contributes to the piezoelectric vibration and it is possible to generate the fatigued state of interest. In general the required amplitude of the AC field to excite the piezoelectric vibration with a top electrode is smaller than without electrode. This aspect combined with the obvious homogeneity of the applied field, allows excluding local switching of the polarization due to excessive electric field.

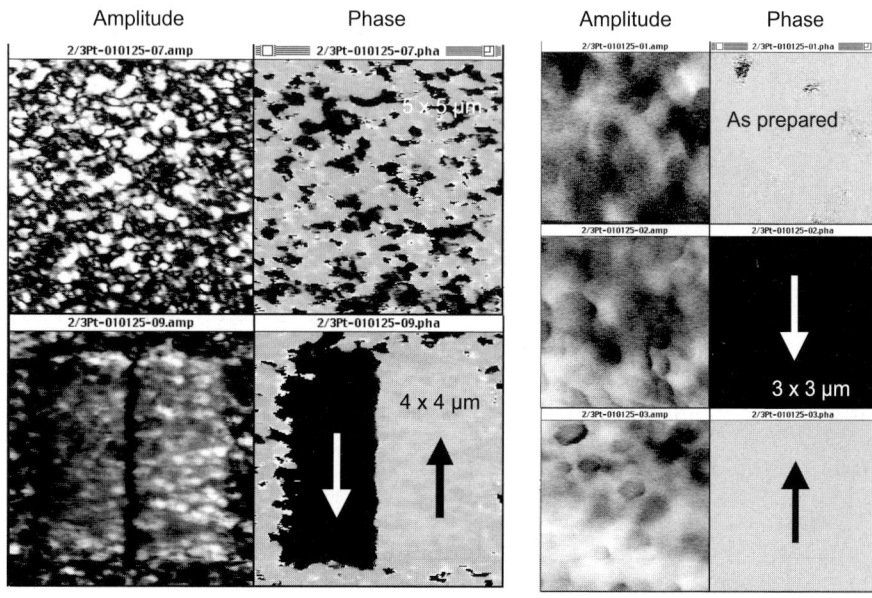

Figure 19.6: Typical nano domain mapping of the polarization using the AFM-tip as top electrode.

Figure 19.6 depicts the typical nano domain mapping of the polarization obtained with the two approaches. On the left side (without top electrode) details of sub micron size can be easily measured. The contrast is very strong and due to the selective response of the upper layer of the ferroelectric film. On the right side (with top electrode) the contrast is less pronounced. The collected piezoelectric vibration at each point is the sum of the ferroelectric film activity through the whole thickness. Unless the film is completely poled, opposite contributions can reduce the detected activity and since the different domains through the film depth are not necessarily aligned, a smearing effect is produced. An additional smearing effect is clearly produced by the presence of the top electrode, which acts as a passive damper. It has to be mentioned that Pt electrodes often consist of sub-micron vertical grains mechanically quite free to vibrate vertically. This peculiarity makes such top electrodes almost transparent for the observation of the piezoelectric activity.

19.2 Investigating cycling induced suppression of switchable polarization in FeCaps

19.2.1 Appearance of frozen polarization nano domains

To investigate the evolution of the frozen polarization nano scale regions during fatigue piezoresponse force microscopy (PFM) is applied in order to excite and determine the vibration phase (orientation) and amplitude of the polarized regions. The selected experimental set-up used for this study consists of carrying out the AFM measurements with and on the top electrode, as explained in the introduction [15]. The AC-field used to excite the vibration was chosen between 1 and 2 MV/m whereas the measured coercive field was $E_c \approx 5$ MV/m. To polarize the FECAP, DC-fields ranging between 17 and 20 MV/m were applied during 100–500 ms. In the fatigue procedures, pulses of magnitude 17 MV/m, width 9 μs and frequencies between 1 and 32 kHz were used. Figure 19.1 shows the fatigue curves $P_r^s(N)$ and $P_r^{ns}(N)$ where N is the number of cycles. The hysteresis loops for the different states (as prepared, virgin-pre-set, initial fatigue and fatigued) are displayed in the same Figure. The first hysteresis loops obtained without pre-set, usually shows initial zero net polarization. It was first verified if the polarization domain configuration obtained by the vibration phase image was compatible with this macroscopic result (Figure 19.7 (a)). Figure 19.7 shows 4 phase images of dimension $10 \times 10 \mu$m. The first image (a) was measured before any electrical treatment and was expected to depict the domain polarization in the "as prepared" state. The dark and bright regions represent areas of opposite phase, i.e. polarization of opposite orientation. By integrating the dark areas one obtains about 52 % of the total surface. This result agrees with the starting position of the first P-hysteresis curve of Figure 19.1 (position (a)).

The homogenous light gray area of Figure 19.7 (b) shows the phase image after the first polarization loop, where the TE was kept at ground and the voltage on the BE was varied according to a triangular function, first at positive then at negative voltages. The measured polarization is therefore oriented "TE to BE" and the signal phase difference to the reference signal was nearly 180°. All the presented images are collected by scanning lines from left to the right, progressing from the bottom to the top of the image. The darker area at the upper part of Figure 19.7 (b) corresponds to a phase change of 180° and was obtained by intentionally reversing the polarization during the image acquisition with the application of a short pulse of +17 MV/m. By this simple procedure it was shown that the polarization could be poled quite homogeneously in both directions. Figure 19.7 (c) depicts the phase image after 10^5 cycles (position (b) on Figure 19.1; ≈ 30 % fatigued) and after the application of $\approx +15$ V on the BE. It is important to point out that the same region, after a poling voltage in the opposite direction (-15 V on BE), was characterized by a completely homogenous phase (not shown here but very similar to the light gray part of Figure 19.7 (b)). Consequently, it is inferred that the dark regions of Figure 19.7 (c) correspond to the still switchable part of the film whereas the bright islands to the frozen ones along the preferred direction. The central part of Figure 19.7 (c) is presented in an enlarged and in 3-dim. mode in Figure 19.8. The phase angle difference between the bright and dark plateaus is nearly 180°. The typical microstructure of the studied films is columnar with sub-μm size grains (100–400 nm). The frozen area must therefore include several grains, but its shape could be influenced by the grain boundaries.

Figure 19.7: Orientations of the polarization for a (a) "as prepared" and (b) poled virgin FECAP. The upper darker part of (b) was obtained by the application of the opposite field. The bright areas in (c) and (d) represent the frozen polarization after 10^5 and 10^7 cycles.

The degree of fatigue obtained by the bright area (Figure 19.9 (c)) is about 23 %, which fits well with the result shown in Figure 19.1 (30 %). Figure 19.7 (d) depicts the situation after 10^7 cycles. The bright regions are the frozen domains and the calculated degree of fatigue is 70 %. As in the 10^5 fatigued state (Figure 19.9 (c)), by poling in the opposite direction a homogenous phase signal is obtained. These results clearly show that the fatigue in Pt-PZT-Pt FeCaps takes place region by region (or grain by grain) and that the orientation of the freezing polarization can have a preferential direction, as it was described in the introduction basing on different macroscopic experimental results [1]. Figure 19.9 (left side) shows the phase images of the FeCap in a highly fatigued state (5×10^7 cycles of 20 MV/m at 32 kHz), taken after the application of opposite poling fields (± 17 MV/m). Surprisingly, at this fatigue stage no preferential orientation seems to exist. In the right picture of Figure 19.9 the difference (black area) between the two previous images is shown. The black area represents the still switching regions of the film. It is interesting to observe the still moving contours of frozen domains embedded in inversely frozen areas. Beside these contours, several regions of significant

19.2 Investigating cycling induced suppression of switchable polarization in FeCaps

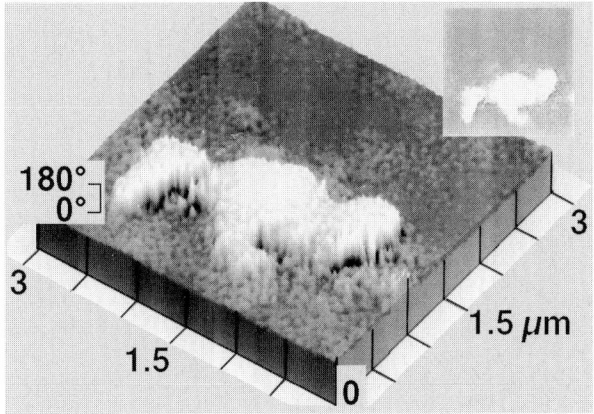

Figure 19.8: Frozen island (bright) in a fatigued sample (10^5) showing preferential direction. The dark region can be fully switched.

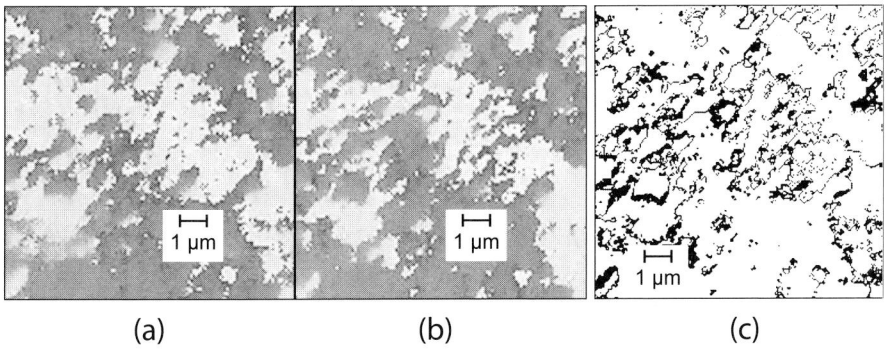

Figure 19.9: (a), (b)) polarization orientations in a fatigued sample (5×10^7) after application of the two opposite poling fields. The frozen regions of opposite orientation are randomly distributed and show size ranging from a few nm to a few μm. (c) Difference between the two patterns ((a) and (b)), representing the remaining switching part of the fatigued FeCap.

size are still contributing to P_s^r. The degree of fatigue estimated using Figure 19.9 was 80–85 %. This value nearly corresponds to that presented in Figure 19.1 (d) and supports the interpretation of the black areas being the remaining switching domains of the fatigued film.

Sub micron investigations have enabled to identify the scale at which a long know macroscopic intriguing effect occurs. The suppression of P_r^s occurs region by region (or grain by grain) where the typical size of the initially frozen regions varies between 100 nm to several μm. The orientation of the frozen nano domains can either have a strong preferential direction (TE to BE in our case) or be equally distributed in both directions. It is not clear which parameters determine the occurrence or not of the preferred orientation configuration. The shape and size of the "surviving" P_r^s-areas in a highly fatigued FECAP in absence of preferred orientation, were also observed. The contours or boundaries of nano domains with opposite direction appear to remain active as well as larger areas with elongated form and size in the sub-μm range.

19.2.2 Nano scale hysteresis loops of fatigued FeCaps

The local phase and amplitude response of the induced piezoelectric vibration of 256 points (30×30 nm) lying on a 6 μm line were measured under varying bias DC-voltage. The 6 μm line was scanned 256 times. At each scan a different bias voltage was applied, starting from $V = 0$ and, step by step, following a triangular shaped voltage cycle (the same used to collect conventional polarization hysteresis loops). Each voltage step $\Delta V = \pm 0.07$ V was applied before measuring the first of the 256 points and the scan duration of the line was 2.5 s. Since the voltage was applied between the top and bottom electrodes (top electrode at ground), the AFM-tip cannot be damaged by high current flows because protected by the top electrode. In general, when using the tip as top electrode, its shape degradation due to melting effects can be a major issue.

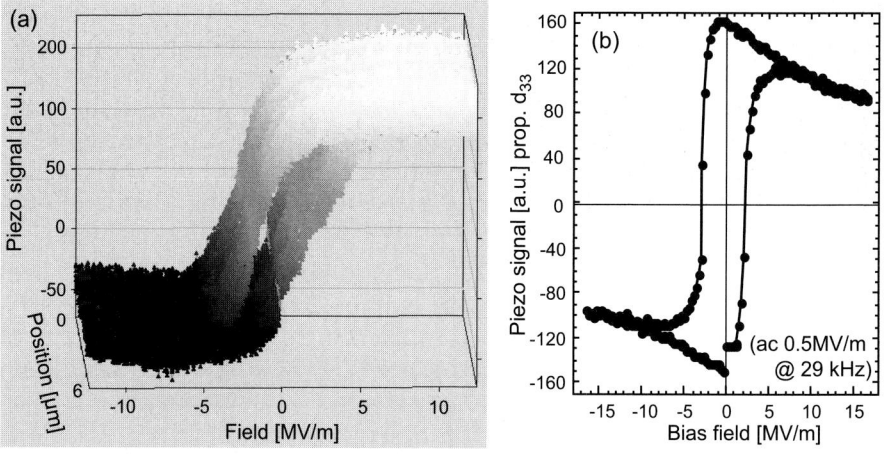

Figure 19.10: (a) Piezoelectric hysteresis loops of 256 points (each approx. 30×30 nm) obtained by combining phase and amplitude of the induced piezoelectric vibration. (b) Average of the 256 loops [17].

By combining the amplitude with the phase information, the piezoelectric hysteresis loops of each point can be calculated and is depicted in 3-dim in Figure 19.10(a). Figure 19.10(b) shows the averaged piezoelectric loop of the 256 points. Since the piezoelectric response (amplitude) is in first approximation proportional to the effective d_{33} this averaged loop can be compared with that obtained with a double beam interferometer and therefore used to

19.2 Investigating cycling induced suppression of switchable polarization in FeCaps

calibrate the response of the AFM tip in order to extract quantitative data concerning the local effective d_{33}.

The local piezoelectric loops were used to analyze the fatigued state. Figure 19.11 shows the phase maps $f(x,y)$ of a 6×4 μm region within a fatigued FeCap (10^8 switching cycles) after negative (left) and positive (right) poling. The central part of Figure 19.11 represents the evolution of the phase signal $f(x,E)$ of an horizontal line (0 < x < 6 μm) lying in the lower part of the 6×4 μm region (see the horizontal arrows) by varying the external bias field E. The regions, which are still switchable, correspond to the bright areas of the right 6×4 μm picture. The dark regions of this picture represent therefore the only fatigued ones. This shows the existence of a preferential direction of the frozen polarization (N-oriented).

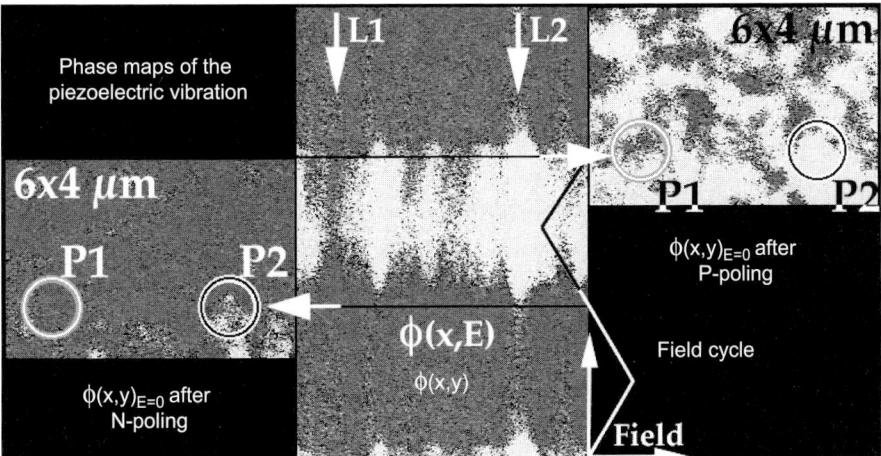

Figure 19.11: Phase maps $f(x,y)$ of a 6×4 μm region of a fatigued FeCap (10^8 cycles) after negative (left) and positive (right) poling and evolution map of the piezoelectric phase signal $f(x,E)$ (central picture) under varying (triangular shape) electric field E of the horizontal line indicated by the horizontal arrows. P1, P2, L1 and L2 are discussed in detail later.

The following features of the central picture should be emphasized:

1. the fatigued state is highly inhomogeneous (phase and amplitude)

2. at higher electric fields almost all the points show a phase change

3. the degree of fatigue at the observed scale (30×30 nm) appears to vary smoothly

Feature 1 is not surprising and confirms the results shown in the previous section on "region by region" onset of fatigue. Feature 2 and 3 are the most intriguing ones and for a better understanding we first consider the amplitude maps of the poled states (Figure 19.12).

The discussion of the amplitude maps will be limited to the most basic observations. The first concerns the larger amplitude (brighter) in the N-poled state (except for the bottom right

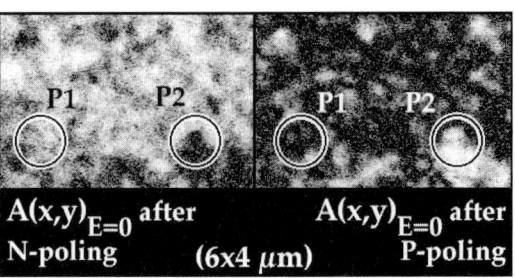

Figure 19.12: Amplitude maps $A(x, y)$ of the piezoelectric vibration corresponding to the respective phase maps of Figure 19.11.

corner area). This confirms the existence of a preferential direction for the freezing regions, which can give, rise to a mechanical competition when the still "alive" regions are switched into the opposite direction (P-poled). This competition globally inhibits the piezoelectric vibration. The second observation concerns the lack of strongly active regions for both poling states. This implies that the frozen regions must have a much "thinner texture" than the size and shape of the dark regions shown in the right picture of Figure 19.11. Still switchable regions should coexist. In addition, most of the still switching large regions appear very dark in the N-poled state (Figure 19.12 left) and very bright in the P-state, indicating that their polarization is also partially locked. This can be well seen in the bottom right corner of both pictures of Figure 19.12. All these features indicate that the onset of fatigue occurs "region by region" as expected but that a dead region might still contain small switching regions whereas still active regions might contain already fatigued very small regions. This can explain the observed change in phase for higher bias fields and the apparently smooth variation of the degree of fatigue.

The piezoelectric hysteresis loops of two selected points, P1 and P2 (Figure 19.11 and Figure 19.12), were analyzed more in detail and are shown in Figure 19.13 (the lines L1 and L2 in Figure 19.11 represent their phase evolution under varying electric field). For P1 the polarization is fully frozen whereas in P2 a significant part of the polarization is still switching. However, P1 appears to consist of a majority of N-frozen regions coexisting with P-frozen ones, because the measured amplitude is much smaller then the maximal one (as a result of opposite piezoelectric vibrations). P2 appears also to be composed of a minority of P-frozen regions coexisting with still switching regions due to its strong asymmetry (shifted upwards).

The results of this section show that the dead micro areas may consist of an unbalanced coexistence of opposite frozen nano domains and might still contain small volumes with switching polarization. Still active areas also might contain already fatigued very small regions. This can explain the observed change in phase for higher bias fields, the apparently smooth variation of the local degree of fatigue and the asymmetric behavior of the amplitude.

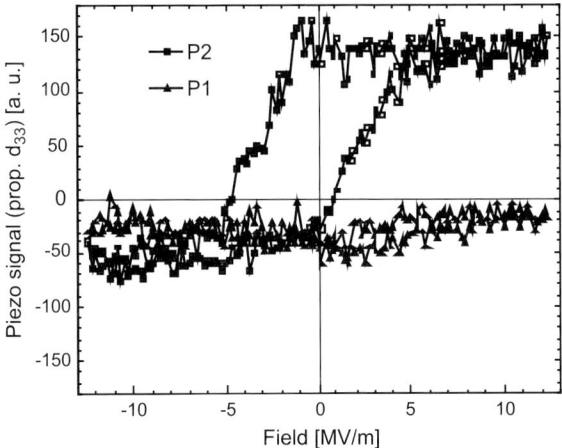

Figure 19.13: Piezoelectric loops proportional to $d_{33}(E)$ for two 30×30 nm regions (P1 and P2 in Figure 19.11) of a fatigued FeCap (10^8 cycles).

19.3 Size effect on the polarization patterns in μ-sized ferroelectric film capacitors

19.3.1 Downscaling of ferroelectric capacitors

Effects of ferroelectric capacitor size on the spontaneous polarization are of high practical importance for memory applications. Ferroelectric film capacitors (FeCaps) for high-density memories have to retain acceptable switching properties as the lateral size of the capacitor scales down to the sub micron range. However, a decrease of the FeCap size below some critical value provokes a substantial deterioration of the switching characteristics [18] and formation of the preferential polarization state [19]. These effects have been attributed to the lateral damage of the FeCap provoked by etching [18], electric field non-uniformity at the edges and accumulation of defects at the lateral surface [19]. In this section we study the effect of variation of the ferroelectric capacitor size on the polarization distribution using PFM. It is demonstrated that the nature of size effects observed when reducing FeCap size can be rather complicated and its understanding may require analysis of the strain-induced switching in combination with the depolarization phenomena.

For this study PZT films with (111) and (100) orientation grown on Pt (111) bottom electrodes by chemical solution deposition were used. In the case of (100) orientation a 10 nm lead titanate buffer layer was used in order to enhance (100) nucleation. The films with Zr/Ti ratio of 40/60 contained dopants including Ca, Sr and La [20]. The thickness of (111) and (100) oriented films was 135 nm and 150 nm, respectively. IrO$_x$ top electrodes with different sizes down to 0.5×0.5 μm^2 have been patterned by reactive ion etching. The electroded PZT film was etched approximately to a half of its original thickness.

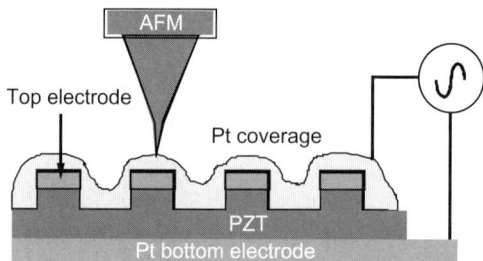

Figure 19.14: Experimental setup for study of the polarization distribution in FeCaps with size down to sub micron range. AFM cantilever is used only for piezoresponse mapping, whereas AC and DC voltage is applied through the independent probe.

The nano scale map of the polarization distribution was obtained by PFM using the method with the top electrode [11]. It was therefore possible to visualize the distribution of polarization averaged across the film thickness over the area of the FeCap. The technical problem of application of the signal to very small FeCaps (sizes down to $0.5 \times 0.5 \, \mu m^2$) arranged in arrays through the independent probe was resolved by deposition of Pt-dots with diameter 0.6 mm and thickness of 30 nm over the structured PZT film. The electrical probe was in contact with this additional top Pt overlay as depicted in the schema of the measurement (Figure 19.14). The sputtered non-annealed Pt dots must have a rather poor contact with the etched PZT film outside the FeCap areas but a good contact with the FeCap top electrodes. Thus, the applied AC-voltage will induce measurable piezoelectric vibration only within the FeCap areas as shown in Figure 19.15.

19.3.2 Size induced polarization instability

Figure 19.15 and Figure 19.16 show the maps of phase and amplitude of piezoelectric response for the FeCaps of $2 \times 3 \, \mu m^2$ and $0.5 \times 0.5 \, \mu m^2$-size, respectively. The distribution of piezoresponse phase and amplitude across the FeCap area characterizes the polarization orientation and its vertical homogeneity. Each picture contains four pairs of images representing the phase (top) and amplitude (bottom) of the piezoresponse measured in four consecutive runs performed immediately one after the other on the same FeCap as specified below. In the first measurement (Figure 19.15 (a) and Figure 19.16 (a)) the small AC-voltage was applied with a bias of +3 V to the bottom electrode. The DC-voltage, which is about 5 times larger than the coercive voltage, is expected to orient the polarization bottom-to-top. Then the DC voltage was switched off and the amplitude and phase images have been taken again (run II, Figure 19.15 (b) and Figure 19.16 (b)). In the run III the same measurement was repeated, while the DC bias of -3 V applied to the bottom electrode (Figure 19.15 (c) and Figure 19.16 (c)). Run IV, where the piezoresponse images are taken again after the DC voltage of -3 V has been removed, concludes the series of the four measurements (Figure 19.15 (d) and Figure 19.16 (d)).

19.3 Size effect on the polarization patterns in μ-sized ferroelectric film capacitors

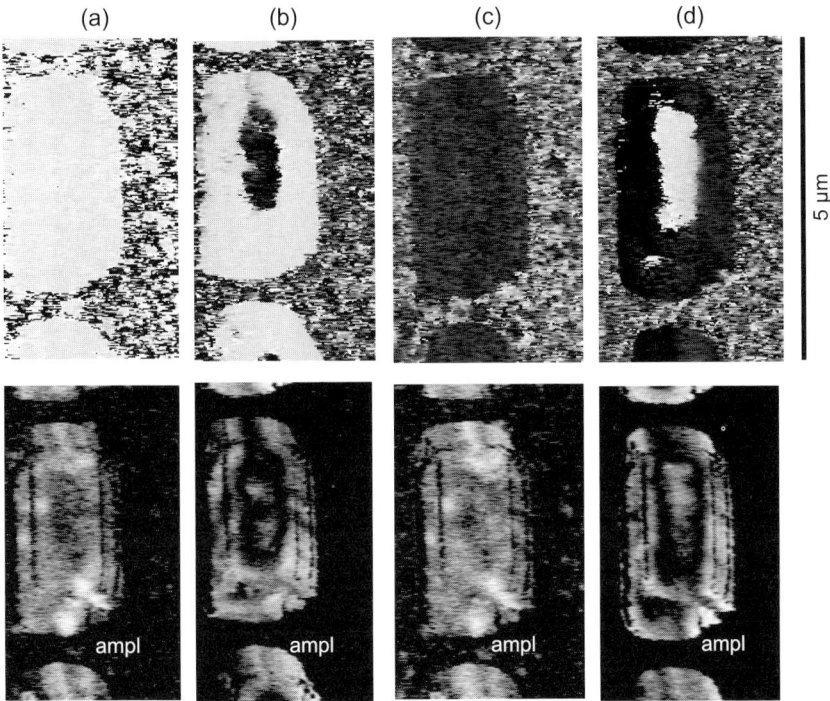

Figure 19.15: Phase (upper row) and amplitude (lower row) of piezoelectric response for the $2\times 3\,\mu m^2$-FeCap. (a) DC voltage of +3 V is applied during the measurements, (b) the same measurement repeated after DC voltage has been switched off, (c) the measurement taken while the DC voltage of –3 V applied, (d) measurement taken after –3 V DC voltage has been switched off.

The most remarkable features of the polarization pattern in the $2\times 3\,\mu m^2$-size FeCaps is the inversion of the polarization orientation in the center of both positively (Figure 19.15 (b)) and negatively (Figure 19.15 (d)) poled capacitors at the absence of the DC-bias voltage. With the presence of the bias voltage the polarization is forced to align with the external electric field and the phase of piezoelectric vibration is homogeneous which is seen as uniform "white" (Figure 19.15 (a)) or "black" (Figure 19.15 (c)) phase images.

However, the completely poled state appears not to be stable without the supporting DC voltage. Once the poling DC voltage switches off, the vertical component of the polarization in the center of the FeCap changes its sign. This effect is observed for both polarities where the phase images are characterized by a central area with the complementary color (Figure 19.15 (b) and (d)). Comparison of the maps of local piezoelectric response for FeCaps with size $2\times 3\,\mu m^2$ (Figure 19.15) and $0.5\times 0.5\,\mu m^2$ (Figure 19.16) reveals a pronounced size dependence of the polarization patterns. Specifically, $0.5\times 0.5\,\mu m^2$ poled FeCaps exhibit no sign of polarization inversion in the center. The set of images in Figure 19.16 shows nearly uni-

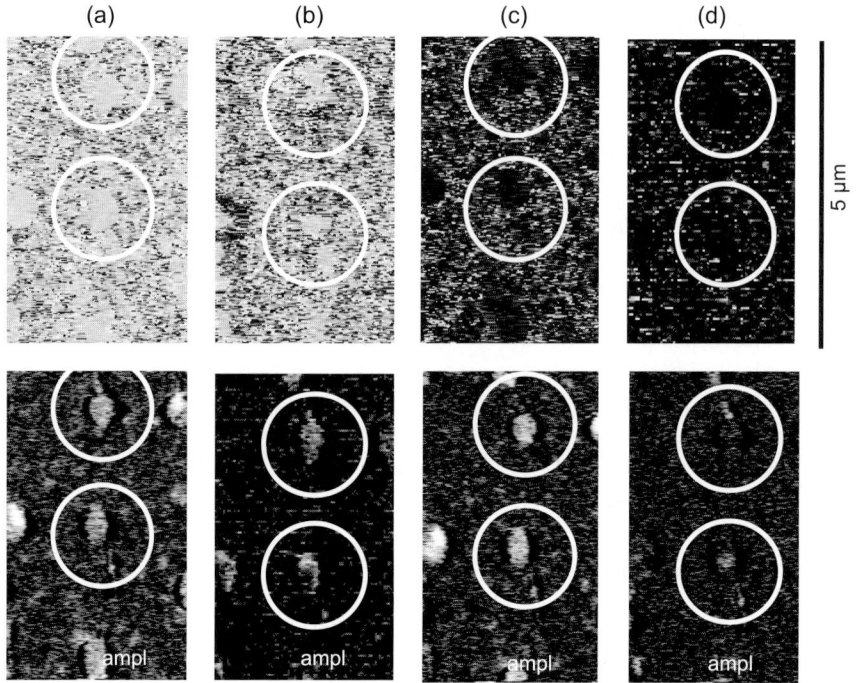

Figure 19.16: Phase (upper row) and amplitude (lower row) of piezoelectric response for the $0.5 \times 0.5\,\mu m^2$ FeCap. (a) DC voltage of +3 V is applied during the measurements, (b) the same measurement repeated after DC voltage has been switched off, (c) the measurement taken while the DC voltage of −3 V applied, (d) measurement taken after −3 V DC voltage has been switched off.

form polarization distributions both under poling voltage and without any DC voltage applied. Since in Figure 19.16 the contrast between the FeCap and the surrounding non-electroded area is not as high as in Figure 19.15, the areas containing FeCaps in Figure 19.16 have been marked with white circles.

To study the effect of PZT film orientation on the polarization patterns we collected amplitude and phase piezoresponse maps for (100)-oriented $2 \times 3\,\mu m^2$ PZT film capacitors (Figure 19.17) under the same conditions that were used for (111)-oriented FeCaps. Images in Figure 19.17 show phase (left) and amplitude (right) piezoresponse of the capacitor prepoled with DC voltage of +4 V (Figure 19.17 (a)) or −4 V (Figure 19.17 (b)) applied to the bottom electrode. No DC-bias was applied to the FeCap during the measurements. Comparing polarization distribution across the (111)-oriented and (100)-oriented capacitors one concludes that the polarization pattern depends on the film orientation. In particular, inversion of the vertical component of polarization observed for (111) FeCaps cannot be traced in the images for (100) FeCaps. In the latter case the color of the phase image is uniform except for several spots, which can be attributed to the technical deficiency of the measurement.

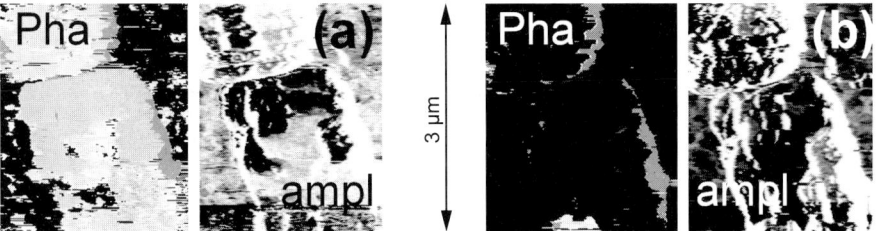

Figure 19.17: Phase (left) and amplitude (right) of piezoelectric response for the (100)-oriented PZT FeCap with size $2\times 3\,\mu m^2$. (a) FeCap prepoled with voltage of +4 V applied to the bottom electrode, (b) FeCap prepoled with voltage of -4 V applied to the bottom electrode.

19.4 Direct observation of inversely-polarized frozen nanodomains in fatigued FeCaps

19.4.1 Removable electrodes

The electrical switching of a uniformly poled ferroelectric material consists of two stages: the nucleation of incipient nanodomains of the new polarization direction, and the growth of these nanodomains to full size polarization domains. The activation energy for the nucleation of nanodomains is orders of magnitude larger than that for their growth [21]. Due to inhomogeneities, defects, and other irregularities, poled ferroelectric materials contain usually residuals of oppositely polarized domains, which strongly enhance the switching process. For this reason, the coercive fields of the switching process typically measured in ferroelectric materials are much lower than the thermodynamic phase transition predicted by theory.

In thin ferroelectric films the nuclei of oppositely polarized domains, which determine triggering of the polarization reversal, are located next to irregularities at the electrode-ferroelectric interface. In conventionally prepared capacitors they are likely to occur close to the interface with the top electrode, since the quality of this interface is inferior to that of the interface with the bottom electrode. A series of experiments have led to a recent conclusion that polarization fatigue, namely the suppression of switching polarization induced by external electric field cycling, is a result of the blocking of oppositely polarized nanodomains by trapped charges [10, 11] rather than pinning of domain walls through the ferroelectric bulk [8] (interface scenario [22, 23] vs. bulk scenario, respectively). In other words, the suppression of the switching polarization is associated with the inhibition of opposite nanodomains at the ferroelectric film interface near the top electrode. This concept is known as the "interface scenario" of the polarization fatigue [13, 15]. This scenario implies that in the fatigued state, regions containing oppositely poled domains are located within a thin layer just beneath the top electrode. The existence of these regions, although supported by different indirect experimental results [10] has not been directly observed so far, and is the focus of this chapter.

AFM-assisted detection of the local piezoelectric activity with a conductive AFM-tip is obviously the ideal tool to directly verify the existence and the expected properties of such a

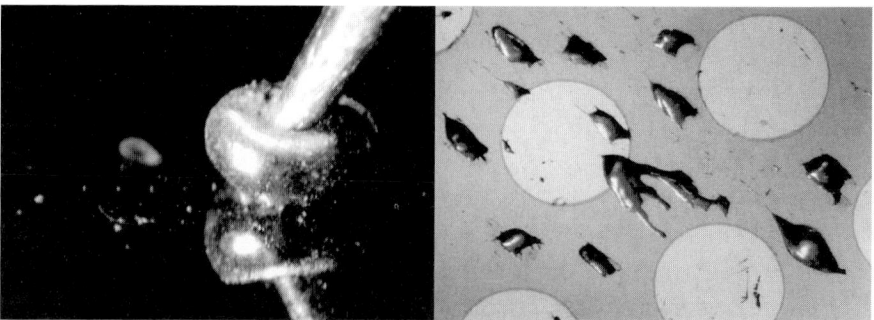

Figure 19.18: Two metals (liquid at room temperature) were used as top electrode. Hg (Mercury) provides bad wetting on the PZT surface (figure left) and therefore insufficient electric contact. Ga (Gallium) shows quite good wetting properties on the PZT surface (drops deposited at about 60°C) enabling reasonable P-loops measurements and fatigue onset.

layer [2–4]. However, the preparation of the real fatigued state requires the presence of a top electrode, which clearly impedes the direct access to the film surface after fatigue.

In order to overcome this problem, a recently developed new approach was used, based on removable metallic liquid top electrodes, i.e. Gallium electrodes. Drops of size smaller than 0.5 mm were deposited at 60°C (Figure 19.18). At this conditions Ga appears to sufficiently wet the surface of the ferroelectrics, resulting in acceptable electric contact (Figure 19.18 (a)). The obtained polarization hysteresis loops (Figure 19.19 (b)) are asymmetric (different bottom and top electrodes) but the collected remanent polarization is the same as that measured with deposited Pt top electrodes. The use of Ga as top electrode enabled therefore to merge the following experimental advantages:

1. to cycle the polarization with a homogenous electric field and provoke a suppression of the switching polarization, comparable to that normally obtained in true ferroelectric capacitors

2. to have a direct access with a conductive AFM-tip to the PZT-film surface of the fatigued areas by electrode removal and to dig the polarization states of the upper layers of the film.

19.4.2 Inversely-polarized nanodomains

A conventional $<111>$ oriented $Pb(Zr_{0.45}Ti_{0.55})O_3$ (PZT) ferroelectric film, deposited on $Pt/TiO_2/SiO_2/Si$ substrate by sol-gel spin casting and Pt-sputter top electrodes, was used in this study [24]. The PZT film thickness was 175 nm. The experimental set-up used for the AFM assisted piezoresponse imaging study of the PZT film consisted of a Park Scientific Instrument "Autoprobe CP" AFM combined with a lock-in amplifier to excite and collect the piezoelectric vibration response amplitude and phase.

19.4 Direct observation of inversely-polarized frozen nanodomains in fatigued FeCaps

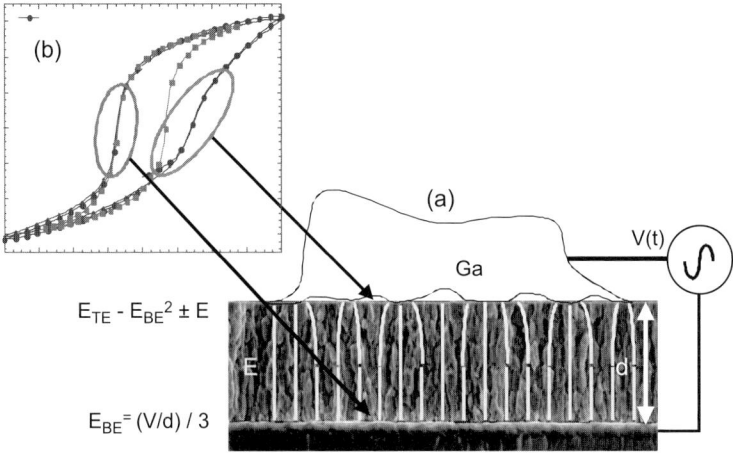

Figure 19.19: Bottom interface: sees a homogeneous field reduced by factor 3. Top interface: the contact is not homogeneous and the field also not. The nucleation is probably influenced.

Figure 19.20 shows the piezoresponse images (phases on the left side, amplitudes on the right side) of a fatigued Pt-PZT-Pt structure and subsequent application of a positive (upper images) and a negative (lower images) poling voltage. A small AC-voltage is applied to the Pt-top-electrode and the associated homogeneous electric field through the PZT-film induces the piezoelectric vibration, which is detected by the AFM-tip through the Pt-top-electrode.

The frozen areas are best revealed in Figure 19.20 (c), which depicts the phase response after the application of a negative voltage. The dark regions represent the still switchable domains, the bright regions represent the frozen domains. Since the preferred orientation of the frozen polarization corresponds to the positive poled state, no contrast is observed in Figure 19.20 (a) (the black border lines were added according to contrast of Figure 19.20 (c)). The phase of the signal, which represents the polarization orientation, is determined by the phase of vibration signals averaged through the whole film depth. With this experimental configuration (deposited Pt top electrode) it is therefore impossible to measure the polarization state as a function of the depth and therefore, it is impossible to observe the presence of inhibited nanodomains with opposite polarization at the top surface of the ferroelectric layer. Next, we perform the AFM measurements on the same ferroelectric film without Pt-sputtered top electrode and using a conductive AFM-tip as top electrode. Figure 19.21 shows the piezoelectric response of an $8 \times 4\,\mu$m area containing a fatigued larger area (left side) and a small virgin region for reference purposes (rough triangle on the right side). The fatigued state was reached after 3×10^4 cycles with a Ga drop as top electrode, which was subsequently removed giving direct access to the film surface. The poling was performed by applying 10 V DC with the AFM-tip. The piezoelectric signal was obtained by exciting the vibration with a small AC-voltage (1.5 V_p) on the AFM-tip. The small triangular virgin area behaves as expected. The polarization can be reversed and the piezoelectric signal is quite strong and homogeneous. The fatigued region shows the expected frozen areas. The polarization cannot be switched

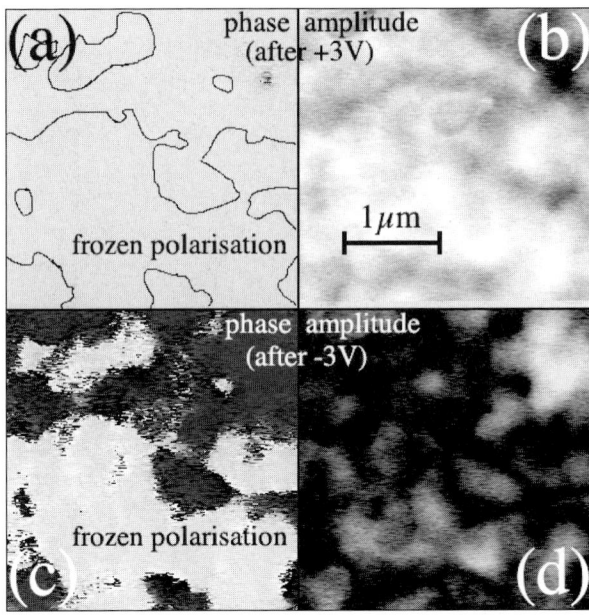

Figure 19.20: Piezoelectric vibration maps of phase ((a), (c)) and amplitude ((b), (d)) ($3 \times 3 \, \mu m^2$) of a fatigued Pt-PZT-Pt structure after positive ((a), (b)) and negative ((c), (d)) poling. Bright and dark phase areas correspond to bottom-to-top and top-to-bottom polarization orientations, respectively.

everywhere and a preferential orientation clearly exists. However, compared to the results shown in Figure 19.20, the preferential direction of the frozen regions is inversed (the uniform dark color, Figure 19.21 (c), means top-to-bottom polarization, while the uniform bright color, Figure 19.20 (a), means bottom to top polarization).

The small AC-field applied by the conductive tip is expected to mainly excite the very upper layer of the PZT-film [25], where are located the predicted frozen nanodomains with opposite polarization state (Figure 19.22, dark triangles). The equivalent frozen regions, if measured using the AFM configuration with Pt-top electrode (Figure 19.20), would show the piezoelectric response averaged over the whole film depth (Figure 19.22, right), impeding the direct observation of the blocked domains. According to Figure 19.22, an increasing AC-voltage is expected to induce more and more detectable piezoelectric vibrations from the layers below the inhibited seeds. Figure 19.24 shows the evolution of the piezoelectric response for increasing AC-voltages applied with the conductive AFM-tip, on a fatigued PZT-thin film area. After poling bottom-to-top with −10 V DC on the AFM-tip, the $4 \times 4 \, \mu m$ maps were taken at AC-voltages from 1.5 to 5 V peak and finally back to 1.5 V_p (Figure 19.23).

The maps on the left side show the phase, i.e. the polarization orientation. Those on the right side show the corresponding amplitude and the used AC-voltages is indicated. Of particular interest is the evolution of the marked dark area in Figure 19.24 top-left. This area should

19.4 Direct observation of inversely-polarized frozen nanodomains in fatigued FeCaps

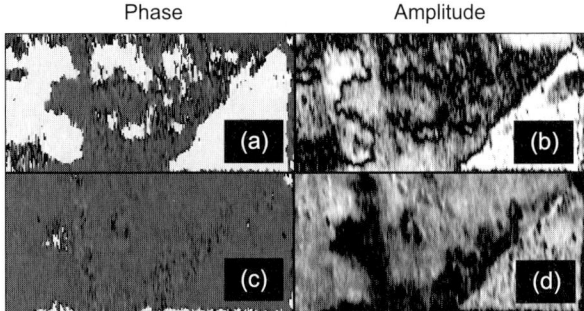

Figure 19.21: Piezoelectric vibration ((a), (c)) phase- and ((b), (d)) amplitude-maps ($8 \times 4\ \mu m^2$) of a PZT-film area. The triangular contrast on the right side of each map (under the letters) represent a still virgin area, used as reference. The rest of the surface was fatigued with 3×10^4 cycles using a Ga drop as top electrode (removed afterwards). (a), (b) and (c), (d) were poled with the AFM-tip under -10 V and +10 V, respectively (polarization-orientation as in Figure 19.20).

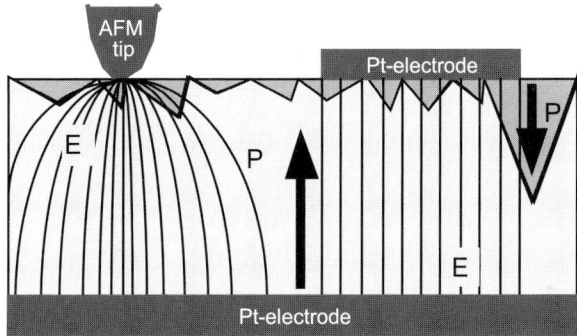

Figure 19.22: Proposed polarization configuration in a fatigued PZT thin film. The upper dark triangles represent the polarization frozen nanodomains. The averaged polarization has the bottom-to-top direction. The AC-electric field applied by a conductive AFM-tip selectively excites the upper layer of the film (left side).

represent an ensemble of inhibited nanodomains oriented top-to-bottom. Between 1.5 and 2.5 V_p there are no substantial differences apart the expected general increase of piezoelectric signal amplitude and the consequent better definition of the phase image (better signal-to-noise ratio). At 3.5 V_p the size of the dark area starts shrinking (Figure 19.24, 3.5 V_{AC}, right). At this point the growing contribution from the lower part of the film, which has opposite polarization, becomes observable. By further increasing the AC-voltage to 5 V_p the net response of that area mainly switches to the opposite direction (Figure 19.24 bottom-left). The corresponding signal amplitude is decreased (Figure 19.24 bottom-right), probably as a consequence of the competing displacements of the opposite oriented layers. The last measurement

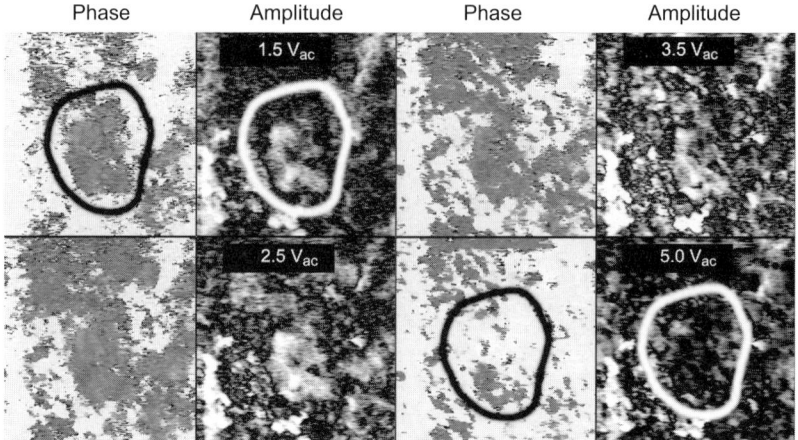

Figure 19.23: Phase (left side) and amplitude (right side) maps (4×4 μm) of the piezoelectric response at 1.5 Vac after the AC-voltage increase sequence.

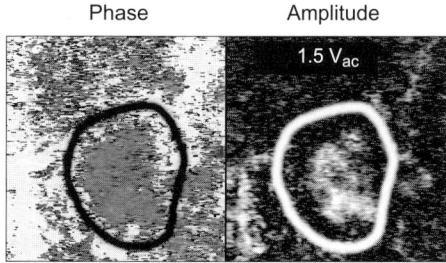

Figure 19.24: Phase (left side) and amplitude (right side) maps (4×4 μm) of the piezoelectric response of a fatigued area after removal of the top electrode (Ga drop) for increasing AC voltage on the conductive AFM-tip.

was again performed at 1.5 V_p to verify whether the polarization states were influenced or not by the 5 V_p AC-field. The phase and amplitude maps (Figure 19.23) were practically identical to the initial ones (Figure 19.24, top) indicating that the polarization states were not influenced. In conclusion, the existence of frozen nanodomains at the upper surface of a fatigued ferroelectric capacitor could be confirmed and directly observed. The predicted configuration of the polarization in a fatigued film could also be established. In addition, it is shown that PZT-thin films with Ga-drops as top electrode can be fatigued as standard Pt-PZT-Pt capacitors (relevant fatigue for FeRAM applications) and their subsequent removal allows the direct access to the intact polarization states close to the film surface by means of AFM-assisted piezoelectric activity detection.

Bibliography

[1] E. L. Colla, A. K. Tagantsev, D. V. Taylor, and A. L. Kholkin, *J. Korean Phys. Soc.* **32**, 1353 (1998).

[2] A. Gruverman, O. Auciello, J. Hatano, and H. Tokumoto, *Ferroelectrics* **184**, 3191 (1996).

[3] T. Hidaka, T. Maruyama, M. Saitoh, N. Mikoshiba, M. Shimizu, T. Shiosaki, L. A. Wills, R. Hiskes, S. A. Dicarolis, and J. Amano, *Appl. Phys. Lett.* **68**, 2358 (1996).

[4] T. Hidaka, T. Maruyama, I. Sakai, M. Saitoh, L. A. Wills, R. Hiskes, S. A. Dicarolis, J. Amano, and C. M. Fosters, *Integrated Ferroelectrics* **17**, 319 (1997).

[5] O. Auciello, J. F. Scott, and R. Ramesh, *Physics Today* **51**, 22 (1998).

[6] T. Mihara, H. Watanabe, and C. A. p. d. Araujo, *Jnp. J. Appl. Phys.* **33**, 3996 (1994).

[7] P. K. Larsen, G. J. M. Dormans, D. J. Taylor, and P. J. v. Veldhoven, *J. Appl. Phys.* **76**, 2405 (1994).

[8] W. L. Warren, D. Dimos, B. A. Tuttle, G. E. Pike, R. W. Schwartz, P. J. Clew, and D. C. McIntyre, *J. Appl. Phys.* **77**, 2623 (1995).

[9] J. J. Lee, C. L. Thio, and S. B. Desu, *J. Appl. Phys.* **78**, 5073 (1995).

[10] E. Colla, A. L. Kholkin, D. Taylor, A. K. Tagantsev, K. Brooks, and N. Setter, *Microelectronic Engineering* **29**, 145 (1995).

[11] E. L. Colla, A. K. Tagantsev, D. V. Taylor, and N. Setter, *Appl. Phys. Lett.* **72**, 2478 (1998).

[12] A. Gruverman, O. Auciello, and H. Tokumoto, *Appl. Phys. Lett.* **69**, 3191 (1996).

[13] A. K. Tagantsev, I. Stolichnov, E. L. Colla, and N. Setter, *J. Appl. Phys.* **90**, 1387 (2001).

[14] E. L. Colla, A. K. Tagantsev, D. V. Taylor, and A. L. Kholkin, *Integrated Ferroelectrics* **18**, 19 (1997).

[15] E. L. Colla, S. Hong, D. V. Taylor, A. K. Tagantsev, K. No, and N. Setter, *Appl. Phys. Lett.* **72**, 2763 (1998).

[16] S. Hong, J. Woo, H. Shin, J. U. Jeon, Y. E. Pak, E. L. Colla, N. Setter, E. Kim, and K. No, *J. Appl. Phys.* **89**, 1377 (2001).

[17] E. L. Colla, A. Raake, D. V. Taylor, and N. Setter, *Integrated Ferroelectrics* **27**, 1215 (1999).

[18] T. Sakoda et al., *Jpn. J. Appl. Phys.* **40**, 2911 (2001).

[19] M. Alexe, C. Harnagea, D. Hesse, and U. Gösele, *Appl. Phys. Lett.* **79**, 242 (2001).

[20] L. Kammerdiner, T. Davenport, and D. Hadnagy, Patent US5969935 (1999).

[21] A. K. Tagantsev, C. Pawlaczyk, K. Brooks, and N. Setter, *Integrated Ferroelectrics* **4**, 1 (1994).

[22] I. K. Yoo and S. B. Desu, *Materials Science &Engineering B (Solid State Materials for Advanced Technology)* B **13**, 319 (1992).

[23] H. N. Al-Shareef, K. R. Bellur, O. Auciello, and A. I. Kingon, *Integrated Ferroelectrics* **5**, 185 (1994).

[24] I. Stolichnov, A. Tagantsev, S. Gentil, S. Hiboux, P. Muralt, N. Setter, J. Cross, and M. Tsukada, MRS *Symp. Proc.* **596**, 387 (2000).

[25] C. Durkan, M. E. Welland, D. P. Chu, and P. Migliorato, *Phys. Rev.* B **60**, 16198 (1999).

Authors

Beige, Horst: beige@physik.uni-halle.de

Böttger, Ulrich: boettger@iwe.rwth-aachen.de

Buchal, Christoph: c.buchal@fz-juelich.de

Cho, Yasuo: cho@riec.tohoku.ac.jp

Colla, Enrico: enrico.colla@epfl.ch

Damjanovic, Dragan: dragan.damjanovic@epfl.ch

Eng, Lukas: eng@iapp.de

Hoffmann, Michael J.: Michael.Hoffmann@ikm.uni-karlsruhe.de

Imlau, Mirco: Mirco.Imlau@uni-osnabrueck.de

Kleemann, Wolfgang: kleemann@uni-duisburg.de

Klein, Norbert: n.klein@fz-juelich.de

Prume, Klaus: prume@aixacct.com

Roelofs, Andreas: Andreas.Roelofs@seagate.com

Rosenman, Gil: gilr@eng.tau.ac.il

Saito, Keisuke: keisuke.saito@bruker-axs.jp

Schmitz, Thorsten: schmitz@aixacct.com

Stephenson, Brian: stephenson@anl.gov

Trolier-McKinstry, Susan: STMcKinstry@psu.edu

Whatmore, Roger W.: R.W.WHATMORE@cranfield.ac.uk

Index

Accelerating voltages 200
Acoustic phonons 22
Al_2O_3 107
Amorphous photo-resist layer 217
Anisotropic
    crystals 77
    properties 18
Antiferrodistortive 153
Application specific silicon integrated circuit (ASIC) 221
Atomic force microscopy (AFM) 189, 332
    High voltage AFM 190
    imaging 309
    low and high voltage 195
    low voltage technique 193
Axis of revolution 82

$Ba(Mg_{1/3}Ta_{2/3})O_3$ 107
$Ba_xSr_{1-x}TiO_3$ 107
Ba-Nd-Ti-O 107
Ba-Zn-Ta-O 107
Barium titanate 17
$BaTiO_3$ 18, 26
Berlincourt meters 46
Biaxial indicatrix 83
Birefringence 84
Bismuth-layer-structured ferroelectric (BLSF) 122
Boltzmann's constant 171
Born approximation 96
Bose statistics 104
Bound surface charge density 243
Bragg
    angle 126, 179
    condition 149
    peaks 155
Brillouin
    scattering 292
    zone 105
Broad-band test devices 109
Built-in polarization 246
Burns temperature 278
Butterfly loop 35, 146

Ca-Ti-Nd-Al 107
Calcite rhomb 79
Cantilever 211
Capacitance-voltage curve 33, 73
Capillary forces 242
$CaTiO_3$ 107
Chaotic behavior 263
Charge amplifiers 254
Chess board 201
Clausius-Mossotti
    equation 15
    relation 106
Clockwise response-field 257
Clusters 200
Coercive
    field 17
    voltage 59
Cole-Davidson-type exponent 297
Collimator 124
Comparison of noise and signal 225
Complex
    impedances 108
    reflection and transmission coefficients 108
Controlling chaos 269
Converse piezoelectric effect 252
Cooper pairs 101
Coplanar
    ridge structure waveguide 93
    transmission lines 114
Core-Shell structures 28

Coulomb force 243
Coupled electro-mechanical 257
Coupling
    coefficient 43, 170
    ports 111
    strength 111
Courtney 114
Critical incident angle 126
Cross-sectional TEM 132
Crystal
    bulk 180
    truncation rod (CTR) 154
Crystalline system 11
Curie-von-Schweidler behavior 36
Curie-Weiss law 17, 19, 282
Current step method 58
Czochralski technique 172

De-aging procedure 69
Debye
    -Scherrer ring 125
    relaxation 103
Depolarization
    energy 214
    field 190
Depth
    -selective piezoelectric response 359
    profile 127
Dielectric
    constant 14, 92
    function 16
    losses 16
    nonlinear series-resonance circuit 263
    permittivity 358
    polarization 13
    properties 227
    surface layer 248
    susceptibility 14, 166
Diffractometer 119
Diffuse phase transitions 27
Dip-pen nanolithography (DPN) 211
Domain
    boundary 347
    broadening 212
    coalescence 215
    elongation 206
    energy 203, 214
    forward growth 202
    in $BaTiO_3$ single crystal 306
    inverted $\sim$ 321
    reorientation 144
    shape invariant 206
    structure 31
    surface energy 214
        density 207
    switching 149
    two-dimensional structure 197
Domain wall
    in PZT 310
    motion 33, 209
    polarization 15
Doping 27
Double
    beam interferometry 48
    refraction 78
Downscaling of ferroelectric capacitors 367
Duffing equation 264
Dynamic
    hysteresis measurement 59
    switching 249

Effective
    coefficients 47
    dielectric polarization P_z 241
    electro-optic coefficient 177
    microwave index 93
    trap density 177
Eigenmode 112
Elastic compliance 43
Electrical
    breakdown 87
        channels 191
    characterization of ferroelectrics 53
    conductivity 211
    resistivity 231
    stray field energy 30
Electromechanical
    hysteresis 251
    properties 138
Electron
    -hole competition factor 179
    beam (EB) lithography 198
    injection 246
    quasi-drop 202
Electronic
    harmonic oscillator 84
    polarization 15
Electrooptic

coefficient 85, 172
effect 83
media 164
needle 90
tensor 85
Ellipsoid equation 86
Ellipsometry 244
Elliptically polarized 84
Energy minimum 215
Excitation signal 54
External screening process 216
Extrinsic contribution 40, 138, 149

Fast pulse measurements 340
Fatigue 355
excitation signal 67
measurement 66
FeRAM 62
chain ∼ 338
Ferroelectric
domain 30
breakdown 189
limit 352
materials 24, 27
phase transition 154
polarization 17
switching 35
transition 155
Fiber-texture 125
Filling factor 112
Finite element method (FEM) 337
First order phase transition 20
Flipchip 116
Fluorite-SBTN 131
Focused ion beam (FIB) 330
Form immobile electron 217
Free charge injection 208
Free energy 19, 149, 212
Frequency dependence 16
Frozen polarization nano domains 361

Gain factor 177
Gaussian beam 89
Gibb's free energy 141
Ginzburg-Landau theory 18
Glan-Thomson prism 172
Grain
diameter 350
size effects 26

Grating vector 170
Grazing incidence X-ray diffraction (GIXRD)
121, 122, 132, 134
Griffiths temperature 278

H. A. Lorentz 83
Half-ellipsoidal domain 213
Half-wave symmetry 256
Hamiltonian 279
Harmonic oscillators 103
Helmholtz double layer 246
Heteroepitaxial 154
HF etching 200
High-resolution synchrotron
X-ray diffraction 137, 145
High-resolutionsynchrotron 149
High-temperature
XRD 143
superconductor (HTS) 101
Higher order nonlinear dielectric microscopy
312
Hooke 94
Huyghens 78

Image charges 204
Impedance 100
Imprint 68
In situ X-ray studies 151, 153, 154
Inclination 81
Indirect electron beam exposure 194
Infrared light 97
Initial state of polarization 338
Integrated optics 89
Interaction energy 204
Interaxial angle 128
Interference patterns 88
Intrinsic
contribution 40, 138
deformation 149
strain 146, 147
unipolar strain 147
Inversely-polarized nanodomains 372
Ionic polarization 15
Ising model 275
Isotropic substance 82

Johnson noise 225

L-α-alanin doped TGS-crystal 270

LaAlO$_3$ 107
Landau-Ginzburg-Devonshire theory 18, 157, 192, 233, 275
Laser 93
Lateral domain size 213
Lattice
    deformation 149
    disorder 277
    distortion 141, 143, 149
    potential 33
    vibrations period 203
Lead titanate nano-grain 348
Leakage measurement 65
Light
    amplification 171
    damping 171
    propagation 77
LiNbO$_3$ 189
Linear stripe 157
Linearly polarized 84
LiTaO$_3$ (lithium tantalate) 319
    congruent ∼ (CLT) 320
    stoichiometric ∼ (SLT) 320
Local
    dielectric constant 248
    electric field 14
    perturbations 181
Lock-in techniques 253
Lorentz 94
    polarization 125
Loss tangent 73, 99
Low temperature co-fired ceramics (LTCC) 108, 115
    process 108
Lyddane-Sachs-Teller relation 23

Mach-Zehnder 88
Macroscopic switching 255
Maxwell's equations 95, 100
Measurement
    methods 53
    of $e_{31,f}$ 50
    of physical parameters 227
    set-up 56
Mechanical stresses 149
Meissner effect 101
Metalorganic chemical vapor deposition (MOCVD) 151
Microphonic noise 227

Micropolar regions 28
Microstrip line 114
Microwave losses 103
Microwaves 91, 99
Mobile nanoelectrodes 193
Monochromator 119
Monoclinic phase 142–145
Monolithic microwave integrated circuits (MMICs) 114
Monte Carlo simulation 198
Morphotropic composition 138
Morphotropic phase boundary (MPB) 24, 121, 138
    -PZT 129, 134
Multibounce 119, 120
Multidiffractions 119
Multiple tip arrays 210
Multipole filter 110
Mutual compensation 208

Nano-scale 359
    180° structure 307
    ferroelectric domain 308
    hysteresis loops 364
Nanodomain
    formation in LiTaO$_3$ 320
    mapping 360
    reversal 191
    structure, rewritable 194
    tailoring 189
Negative piezoelectric phase 258
Noise-equivalent power NEP 225
Non-contact interaction 243
Non-destructive 241
Non-invasive 241
Nonlinear
    dielectric
        coefficients 313
        imaging 304
    dynamics 264
    optics 93
    resonant system 264
    wave equation 95
Normal-incidence laser 152
Nucleation activation energy 203

Off-band rejection 113
Ohm's law 100
On-off-amplitude modulator 88

Index

One-dimensional domain configuration 197
Optic
    axis 80
    phonons 22
    sign 79
Optical
    characterization 77
    field 94
    methods 45
    modulator electrode 92
    pathlength 87
    rectification 98
    retarder 88
    second harmonic 201
    wave 91
    waveguides 89
Optocommunication applications 92
Orientation polarization 15

Parallel domain writing 217
Parasitic holograms 169
$PbTiO_3$-$PbZrO_3$ (PZT) 24
Penetration depth 199
Periodic refractive index 196
Phase
    -reversal electrodes 93
    delay 87
    diagram for PZT 24
    difference 90
    identification 127
    matching 98
    mismatch 90
    modulators 87
    portrait of a harmonic oscillator 265
    portraits 266
    transition 18, 23, 166
    velocity 91
Phonon spectrum 22
Photo-induced light scattering 163
Photonic
    devices 191
    sensor 253
Photorefraction 164
Photovoltaic effect 168
Piezo-imaging 309
Piezoelectric
    coefficient 11, 45, 47
    constants 39
    displacement 249
    effect 39, 226, 242, 344
    measurements 46
    property determination 231
    resonance 43
Piezoelectricity 11, 39
Piezoresponse force microscopy (PFM) 195, 343, 359
    High-resolution (PFM) 292
Pinning center 255
Planar
    resonators 114
    transmission line 109
    tuneable HTS 108
Plastic deformation 208
Pneumatic loading methods 49
Pockels effect 83
Point defects 103
Polar
    nanoregions 277
    structure 180
Polarization
    configuration in a fatigued PZT thin film 375
    density 94
    irreversible ∼ 32
    order parameter 280
    reversal 146, 193
    reversible ∼ 32
    switching 253
Polycapillaries 119, 125
Polyetheretherketone (PEEK) 129
Positive saturation 59
Prepolarization pulse 60
Pseudo-Voigt function 139
Pulse measurement 61
Pulsed lasers 97
Pyroelectric
    ceramics 221
    ceramics thin films 234
    coefficient 226, 229
    detectors 222
    devices 221
    effect 221
    materials 232
    response 222
    thin films 221
PZT-SKN 145

Quadratic electro-optic effect 280

Quartz reference sensor 254
Quasi-static
    hysteresis loop 63
    measurements 45

Random-field Ising model 279
Real-time studies 152
Reference
    capacitor 56
    resistance 57
Refractive index 81, 170, 171, 280
Relative permittivity 171
Relaxed remanent polarization 58
Relaxor ferroelectrics 28, 166, 275
Remanent polarization 17, 58
Removable electrodes 371
Resolution function 122
Resonator 110
Retention measurement 71
Rochelle salt 270
Rocking curve 134
    scan 126

Sapphire 105
Satellite intensities 156, 158
    odd-order positions 156
Sawyer Tower method 56
SBTN 119, 122, 131, 134
Scanning
    nonlinear dielectric microscope (SNDM) 303
    probe microscopy (SPM) 50, 210
Scattered optical field 96
Screening 190
Second
    -harmonic
        generation 96
        radiation 97
    order
        nonlinear optics 96
        phase transition 19
Secondary pyroelectric coefficient 226
Seed scattering 183
Self
    -polarization 244
    -polarized polarization mechanism 31
    -pumped phase conjugation 164
Shunt method 57
Single cell capacitors 329

Size
    effect 351
    induced polarization instability 368
Small signal measurements 73
SNDM
    domain engineering system 320
    imaging 309
Soft mode concept 22
Space charge polarization 15
Specific detectivity D^* 225
Split-post resonator 114
Spontaneous polarization 12, 172
$SrBi_2Ta_2O_9$ 25
$SrTiO_3$ 107
    thin film varactors 116
Static hysteresis measurement 63
Step shaped voltage waveform 65
Strain 39
    contribution 147
    irreversible $\sim$ 34
    reversible $\sim$ 34
Stray capacitance 337
Stress 39
String-like domains 202
Stripe
    domain formation 158
    domains 155
    period 157
Stripline 110
Strontium-barium-niobate 163
Sub-unit-cell accuracy 152, 154
Substitution 27
Superlattices 192
Surface
    impedance 100
    phase diagram 153
Switching
    current 35
    polarization 355
    time 35
Symmetry classes 86
Synchrotron x-ray scattering 151

Tapping 242
$TE_{01\delta}$ 113
Temperature coefficient 106
TGS-crystal 267, 270
Thermal
    capacity H 232

conductance G_T 232
diffusivity κ 224
properties 231
Thermally stimulated currents (TSC) 228
Three-dimensional measurement technique 316
TiO_2 107
Tip apex capacitance 206
Topographical images 346
Trap density 171
Treatment signal 58
Tuneability 99
Tungsten-bronze structure 165

Ultrahigh memory storage 194
Uniaxial
    crystals 80
    indicatrix 80
    relaxors 166
Unloaded quality factor 112

Vibration direction 81

Virtual ground
    measurements 329
    method 57
Vogel-Fulcher
    (*V-F*) equation 288
    temperature 288
Volume holographic memories 164

Wave
    fronts 78
    vector 169
Wavelet 79
Whispering gallery 113
    modes 112

X-ray diffraction (XRD) 137, 138, 141
    reciprocal space mapping (XRD-RSM) 121
XPS 335

Zr-Sn-Ti-O 107